Electric Machinery
and Transformers

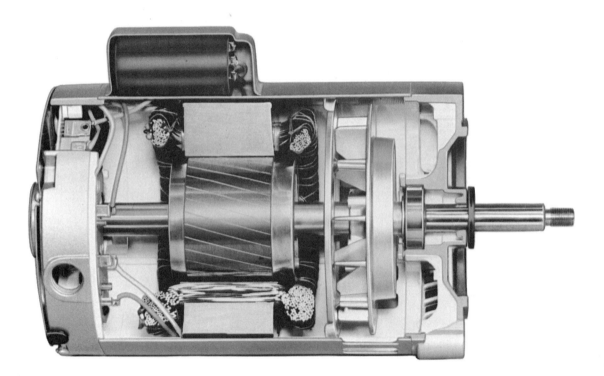

Capacitor Motor (Courtesy of Franklin Electric)

BHAG S. GURU

HÜSEYIN R. HIZIRÖGLU

Department of Electrical and Computer Engineering
GMI Engineering & Management Institute
Flint, Michigan

Harcourt Brace Jovanovich, Publishers

Technology Publications

San Diego　　New York　　Chicago　　Austin　　Washington, D.C.
London　　Sydney　　Toronto

To my parents and my family—
  Bhag S. Guru

To my parents—
  Hüseyin R. Hiziroglu

ISBN 0-15-520945-0

Library of Congress Number 87-81096

Printed in the United States of America

Contents

Preface

This book presents the fundamentals of basic electrical machines—transformers, direct-current generators and motors, synchronous motors and generators, three-phase and single-phase induction machines—and some special purpose machines. It is intended for a one-semester course on electrical machines. However, since such topics as electromechanical energy conversion, winding techniques, dynamics, and various electronic control schemes are included, the book is also suitable for a two-semester course.

Throughout the book, our basic philosophy is to analyze a machine on a physical basis in order to provide a clear understanding of and intuitive feeling for how it operates under steady- as well as transient-states. Then the mathematics is used to express the quantitative relationships. We assume the reader has no previous acquaintance with the fundamental operations of these machines but does possess a background in mathematics that includes the study of linear differential equations and the analysis of electric circuits. Each machine is represented by its equivalent circuit and is then analyzed using basic circuit laws and theorems. In addition to the equivalent circuit approach, wherever we felt it would be more instructive and helpful to the student's understanding, we have discussed the principle of operation and the analysis of the electric machines on the basis of electromagnetic field theory.

To illustrate how theory may be applied to obtain quantitative results, numerous examples have been worked out in the text in detail. Example problems were chosen to enhance the student's understanding of a particular section of the text. Considerable care has been exercised in the development of problems at the end of each chapter. These problems constitute an important part of the text and should form an integral part of the study. Students are required to use intuitive reasoning to solve the problems.

In Chapter 1, the fundamental principles of dc and ac circuits as well as electromagnetic field theory and its application to magnetic circuits are reviewed. Chapter 2 points out the importance of electromechanical energy conversion principles. The operating principles of single-phase, three-phase and autotransformers are clearly stated in Chapter 3. In addition, the types of losses that can occur in transformers and the measures to be taken to keep these losses to a minimum for optimal operating conditions are explained. Direct-current generators and their operating characteristics are introduced in Chapter 4. Chapter 5 deals with different types of dc motors, their operating characteristics, and conventional methods of speed control. Synchronous generators are discussed in Chapter 6; the differences between cylindrical and salient-pole rotors are highlighted. This chapter also explains the parallel operation of these generators. Logically following the synchronous generators are the synchronous motors in Chapter 7. The typical operating characteristics, *V*-curves, and their operations as synchrounous condensers to improve the overall power factor are emphasized. Chapter 8 sheds light on polyphase induction motors. Although the discussion is limited to three-phase motors, the equations are so general that they easily can be extended to multiphase motors. Single-phase motors are examined in Chapter 9. Methods to determine the performance of single-phase motors with both windings are included. Some insights into the workings of shaded-pole and universal motors are also given. Permanent-magnet motors such as steppers and brushless dc motors are covered in Chapter 10. Also included in this chapter are linear induction and hysteresis motors. Chapter 11 explores the dynamics of electrical machines using Laplace transforms as well as numerical solutions to state equations. The last chapter, Chapter 12, prepares the student to investigate new horizons in the control of electrical machines using solid-state devices.

Throughout the book we have emphasized the use of the SI unit system. However, in some examples, we also have employed the English Unit System since it is still widely used in the United States.

The winding techniques of electrical machines are often omitted in today's courses due to time limitations. Nevertheless, to highlight the different winding techniques for dc and ac electric machines, we have added Appendices A and C, respectively. The development of the revolving field is discussed in Appendix B. The winding distribution factor is covered in Appendix D and Appendix E aids in the conversion of units from one system to another. Since we have extensively utilized Laplace transform techniques in Chapter 11, Appendix F contains a table of Laplace transforms for common functions.

Since, in most books, boldface characters are used extensively for vector and phasor quantities, the instructor has to adopt his or her own notations while teaching the course. In some cases, students may be confused as to which notation should be followed. The problem is compounded when the symbols are not used consistently. In order to eliminate this confusion, we have adopted a consistent notation system in this text. Capital letters are used exclusively for dc quantities and for the rms values of ac variables, whereas lowercase script letters are employed for the instantaneous values of time-dependent variables. For representations of a vector, a phasor, and a complex quantity we have incorporated an arrow, a tilde, and a caret on top of the letters, respectively.

In the phasor diagrams throughout this book we have omitted the use of j operator since all the diagrams are drawn in complex plane with real and imaginary quantities along horizontal and vertical axes, respectively.

We have attempted to explain the fundamental behavior and the operating principles of electric machines in an elucidative manner. We have achieved our goal if and only if the reader is completely satisfied with our presentation. Any comments or suggestions in this regard are not only highly welcomed but will also form the basis for future revisions.

We want to thank the following reviewers whose comments and suggestions were invaluable to us in refining the text: Warren R. Hill, The University of Southern Colorado; P. Erik Liimatta, Anne Arundel Community College; Roger P. Webb, Georgia Institute of Technology; Elias G. Strangas, The University of Missouri—Rolla, and George Gella, The Ohio State University. We are indebted to our colleagues Joseph L. Costello, Albert O. Simeon, and Mohammad Torfeh who either read parts of the book or taught from earlier versions of it. Our students' comments also contributed significantly to the development of the book. We are grateful to those companies that have been very cooperative in providing us with the photographs: Bodine Electric Company, Franklin Electric, Lab-Volt Systems, Universal Electric, Inc., Century Electric, Inc., Clifton Precision, Electric Apparatus Company, ESAS Elektrik San, Turkey, Hampden Engineering Corporation, RMR Corporation, and POWEREX, Inc.

We are grateful to Susan Mease, Philip Menzies and Ben Wentzell of Harcourt Brace Jovanovich, Inc., and Laura Cleveland of WordCrafters, Inc., for their keen work during the developmental stages of the book.

Bhag S. Guru
Hüseyin R. Hiziroğlu

CHAPTER 1

Basic Principles of Electric Circuits and Electromagnetics

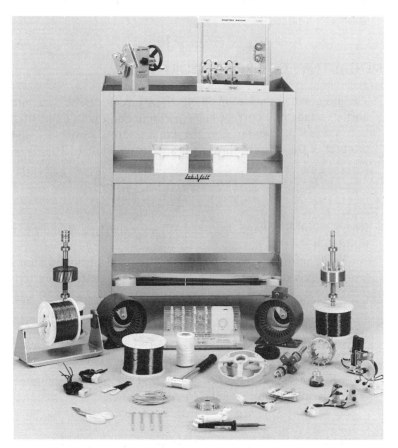

Parts and Tools Used to Assemble a Rotating
Machine (Courtesy of Lab-Volt)

Certain basic principles govern the behavior of all electrical machines. Knowledge of these principles is necessary to understand the basic operations of all types of electrical machines. This chapter is therefore devoted to the study of these principles.

To understand the basic operation of electrical machines, we must understand how to analyze alternating-current (ac) and direct-current (dc) electrical circuits, electromagnetic fields, and mechanics. It is, however, difficult to assume that a student understands all three areas at the time he or she is introduced to electrical machines. Most often, a student has already had encounters with electrical circuit theory and mechanics and that is why most books on electrical machines treat the different topics from a simple circuits point of view with a minimum use of electromagnetic fields. On the other hand, electrical machines can also be taught by following the electromagnetic fields approach. It is difficult to assess which approach results in better understanding of electrical machines. Therefore, in this book every effort is made to develop the theory from both points of view.

1.1   FUNDAMENTAL LAWS OF CIRCUIT THEORY

The unknown quantities in an electrical circuit can be determined by using Kirchhoff's voltage and current laws in conjunction with Ohm's law.

Ohm's law states that the voltage drop across a resistor is equal to the product of the current through it and its resistance. That is,

$$v = iR \tag{1.1}$$

where v is the voltage drop across the resistance R and i is the current flowing through it.

Kirchhoff's voltage law states that the algebraic sum of all the voltages around any closed path in a circuit is equal to zero. In other words,

$$\sum_{m=1}^{n} v_m = 0 \tag{1.2}$$

Kirchhoff's current law states that the algebraic sum of all the currents at any node in a circuit is equal to zero. That is,

$$\sum_{m=1}^{n} i_m = 0 \tag{1.3}$$

■ EXAMPLE 1.1

Determine the currents i_1 and i_2 in the circuit shown in Fig. 1.1 using Kirchhoff's laws.

Solution

By Kirchhoff's current law (KCL) at node B, we have

$$i_3 - i_1 + i_2 = 0$$

or
$$i_3 = i_1 - i_2 \qquad (1.4)$$

The application of Kirchhoff's voltage law (KVL) to the closed-loop $ABEF$ yields

$$-100 + 5i_1 + 10i_3 = 0$$

Substituting for i_3 from Eq. (1.4) we obtain

$$15i_1 - 10i_2 = 100 \qquad (1.5)$$

Another application of KVL to the closed-loop $BCDE$ yields

$$10i_2 - 10i_3 = 0$$

or
$$20i_2 - 10i_1 = 0 \qquad (1.6)$$

Solving Eqs. (1.5) and (1.6), we get

$$i_1 = 10 \text{ amperes (A)}$$

and
$$i_2 = 5 \text{ A}$$

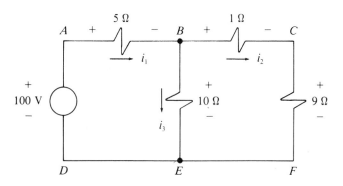

FIGURE 1.1
Circuit for
Example 1.1

Although all electrical circuits can be analyzed by applying Kirchhoff's voltage and current laws, the resulting equations become cumbersome as the complexity of the electrical circuit increases. Therefore, we need other mathematical techniques to solve complex but practical problems. Among many techniques proposed, we will often use what is commonly known as *Thevenin's theorem*. Thevenin's theorem reduces a complicated circuit as viewed from the load side into an independent voltage source, V_T, in series with an equivalent resistance, R_T, where V_T is the open-circuit voltage between the two nodes and R_T is the ratio of open-circuit voltage to short-circuit current. If the electrical circuit consists of independent sources only, the equivalent resistance can also be obtained by looking in at the terminals, with the voltage sources replaced by short circuits and the current sources replaced by open circuits.

The open-circuit voltage is obtained by removing the load impedance and analyzing the rest of the circuit. The short-circuit current is the current passing through the nodes when the load is replaced by a short circuit. Let us demonstrate the usefulness of Thevenin's theorem with the following example.

■ EXAMPLE 1.2

For the electrical circuit shown in Fig. 1.2, determine the voltage across and current through the 2-ohm (Ω) resistor.

Solution

Assume that a and b are the nodes where the load resistance of 2 Ω is connected. To apply Thevenin's theorem at terminals a and b, remove the load resistor from the circuit and determine the open-circuit voltage V_{ab}. From the application of Kirchhoff's voltage law, the open-circuit voltage, for the circuit shown in Fig. 1.3, is

$$V_{ab} = \frac{12}{4 + 4} \, 4 = 6 \text{ volts (V)}$$

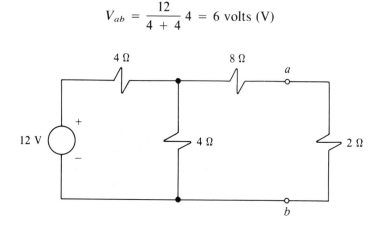

FIGURE 1.2
Circuit for
Example 1.2

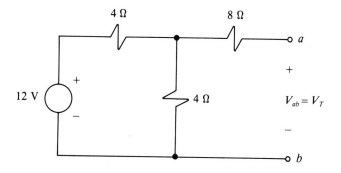

FIGURE 1.3
Circuit in Figure
1.2 with
Terminals *a* and
b Open-
Circuited

Since we have an independent voltage source, the Thevenin's equivalent resistance can be obtained by replacing the voltage source by a short circuit as shown in Fig. 1.4. It is evident from the circuit that

$$R_T = 8 + \frac{4 \times 4}{8} = 10 \ \Omega$$

The Thevenin's equivalent circuit is given in Fig. 1.5. The current through the

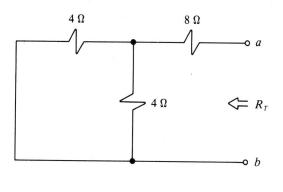

FIGURE 1.4
Circuit in Figure
1.3 with
Independent
Voltage Source
Short-Circuited

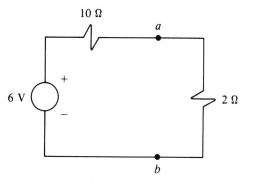

FIGURE 1.5
Thevenin's
Equivalent
Circuit of the
Circuit Shown in
Figure 1.2

load resistance is

$$I_L = \frac{6}{10 + 2} = 0.5 \text{ A}$$

and the corresponding load voltage is

$$V_L = 0.5 \times 2 = 1 \text{ V}$$

1.2 ALTERNATING-CURRENT (AC) CIRCUITS

Electrical energy can be transmitted over long distances only at high voltage levels due to economic and physical restrictions on the current-carrying conductors. On the other hand, power generators cannot be constructed above certain voltage levels, again because of economic constraints. Consequently, electrical energy is generated at low voltage levels and stepped up to high-voltage transmission levels by means of power transformers. At the consumer end of the transmission line, the voltage is stepped down to low levels for safe utilization. As we shall study in Chapter 3, transformers can operate only with time-varying voltages. Therefore, it becomes necessary to generate, transmit, and utilize electrical energy with alternating currents due to economic reasons. However, for the last several years with the development of high-voltage solid-state devices, direct-current (dc) transmission at high voltages has been attracting attention. But, regardless of the type of transmission voltage, electrical energy is produced and utilized in most applications in the form of alternating currents. For these reasons, in this section we will examine the basic concepts of sinusoidal ac circuit analysis.

A sinusoidal voltage waveform can be mathematically represented as

$$v(t) = V_m \cos(\omega t + \phi) \tag{1.7}$$

where V_m is the maximum value or the magnitude of the sinusoidal waveform, ω is the angular frequency in radians/second (rad/s), ϕ is the phase position of the waveform in radians with respect to a given reference frame, and $v(t)$ is the instantaneous value of the waveform at any instant of time, as illustrated in Fig. 1.6.

The angular frequency, ω, can be expressed as

$$\omega = 2\pi f \tag{1.8}$$

where f is the frequency in hertz (Hz). The time required to complete one cycle of the waveform is known as the time period and is given as

$$T = \frac{1}{f} \tag{1.9}$$

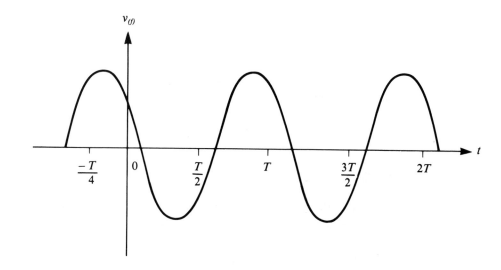

FIGURE 1.6
Voltage
Waveform
Given by Eq.
1.7

The average and root-mean-square (rms) values of a periodic waveform are defined as

$$V_{av} = \frac{1}{T} \int_0^T v(t)\, dt \tag{1.10}$$

and

$$V_{rms} = \sqrt{\frac{1}{T} \int_0^T [v(t)]^2\, dt} \tag{1.11}$$

respectively. Using these definitions, the average value of a sinusoidal function is zero, while its rms value is

$$V_{rms} = \frac{V_m}{\sqrt{2}} \tag{1.12}$$

1.2.1   Sinusoidal Steady-state Response of Electric Circuits

The total response of a circuit under the influence of sources consists of transient as well as steady-state response. Such a response can be obtained by solving the integrodifferential equations obtained from the circuit. However, if we are interested in the steady-state response of a circuit with sinusoidal sources only, we can transform the integrodifferential equations into a set of algebraic equations using phasors.

A sinusoidal waveform

$$v(t) = V_m \cos(\omega t + \phi) \tag{1.13}$$

can be expressed in phasor form as

$$\tilde{V} = Ve^{j\phi} = V\underline{/\phi} \qquad (1.14)$$

where V is the rms value of the voltage waveform. In this book, we will express all phasor quantities in terms of their rms values.

The following example illustrates the use of phasors in determining the steady-state response of a circuit.

■ EXAMPLE 1.3

Determine the expression for the instantaneous voltage drop across the inductor in the series circuit shown in Fig. 1.7. Also draw the complete phasor diagram.

Solution

The voltage source, $v(t) = 14.14 \cos(2t)$, can be expressed in phasor form as $\tilde{V} = 10\underline{/0°}$.

The impedances of the individual elements at an angular frequency of 2 rad/s are

$$\hat{Z}_R = 1 \ \Omega$$

$$\hat{Z}_L = j\omega L = j(2)(2) = j4 \ \Omega$$

$$\hat{Z}_c = -\frac{j}{\omega c} = -\frac{j}{2 \times 0.5} = -j1 \ \Omega$$

The circuit can now be redrawn in phasor form as shown in Fig. 1.8. Since

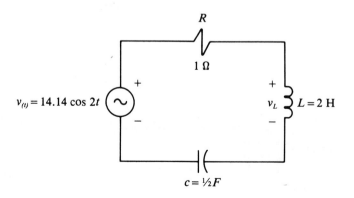

FIGURE 1.7
Circuit for
Example 1.3

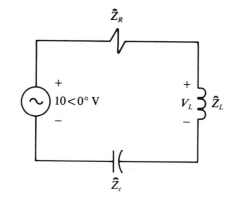

FIGURE 1.8
Phasor Domain
Equivalent
Circuit of the
Circuit Shown in
Figure 1.7

all the impedances are in series, the total circuit impedance is

$$\hat{Z} = \hat{Z}_R + \hat{Z}_L + \hat{Z}_c$$
$$= 1 + j4 - j1 = 1 + j3 = 3.1622\underline{/71.56°}$$

The current in the circuit is given by

$$\tilde{I} = \frac{\tilde{V}}{\hat{Z}} = \frac{10\underline{/0°}}{1 + j3} = 3.1622\underline{/-71.56°} \text{ A}$$

or, in the time domain,

$$i(t) = \sqrt{2} \times 3.1622 \cos(2t - 71.56°) \text{ A}$$

The voltage drop across the inductive impedance is

$$\tilde{V}_L = \tilde{I}\hat{Z}_L = (3.1622\underline{/-71.56°})(j4) = 12.65\underline{/18.44°} \text{ V}$$

or, in the time domain,

$$v_L(t) = 17.89 \cos(2t + 18.44°) \text{ V}$$

In this circuit the voltage across the inductor leads the applied voltage by an angle of 18.44° or 0.322 rad. Figure 1.9 shows the corresponding phasor diagram.

1.2.2 Power Concept in an AC Circuit

Let us consider the voltage and current waveforms in a load as $v(t) = \sqrt{2} \text{ V}$ $\cos\omega t$ and $i(t) = \sqrt{2} I \cos(\omega t - \phi)$, respectively. The instantaneous power ab-

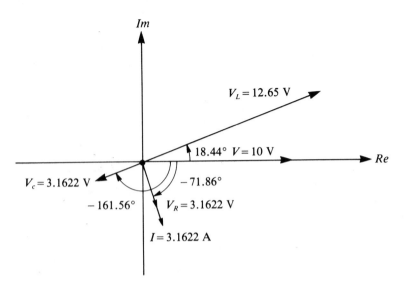

FIGURE 1.9
Phasor Diagram
of Example 1.3

sorbed by the load is

$$p(t) = v(t)i(t) = 2VI \cos \omega t \cos(\omega t - \phi) \tag{1.15}$$

The average value is

$$P_{av} = \frac{1}{T} \int_0^T p(t) \, dt = \frac{\omega}{2\pi} \int_0^{2\pi/\omega} 2VI \cos \omega t \cos(\omega t - \phi) \, dt \tag{1.16}$$
$$= VI \cos \phi$$

where V and I are the rms values of the voltage and current and ϕ is the phase angle between them.

In terms of phasors, we can define the complex power, $\hat{S}$, as

$$\hat{S} = \tilde{V}\tilde{I}^* \tag{1.17}$$

where $\tilde{V}$ is the voltage phasor and $\tilde{I}^*$ is the complex conjugate of the current phasor. The complex power can also be expressed as

$$\hat{S} = P + jQ \tag{1.18}$$

where P is the average power in watts (W) and Q is the reactive power in VAR. If $\tilde{V} = V\underline{/0°}$ and $\tilde{I} = I\underline{/-\phi°}$, the complex power is

$$\hat{S} = (V\underline{/0°})(I\underline{/\phi°}) = VI \cos \phi + jVI \sin \phi \tag{1.19}$$

From Eqs. (1.18) and (1.19), it is evident that

$$P = VI \cos \phi \qquad (1.20)$$

and

$$Q = VI \sin \phi \qquad (1.21)$$

The term $\cos \phi$ in Eq. (1.20) is known as the power factor of the system. The magnitude of the complex power, $\hat{S}$, is known as the apparent power and is expressed in VA. A representation of P, Q, and S in a complex plane is known as the power diagram, as depicted in Fig. 1.10.

■ EXAMPLE 1.4

An inductive load draws a current of 10 A at a power factor of 0.7 when connected to a 100-V, 60-Hz source. Calculate (a) the apparent power, active power, and reactive power drawn by the load, and (b) the shunt capacitance that must be connected to the load so that the overalll power factor across the load terminals will be unity.

Solution

(a) For the circuit shown in Fig. 1.11, the applied voltage is $V = 100\underline{/0°}$. Since the power factor is 0.7, the phase angle between the voltage and the current in the circuit is $\phi = \cos^{-1}(0.7) = 45.57°$. Since the load is inductive, the current must lag the reference applied voltage. That is, $\bar{I} = 10\underline{/-45.57°}$. The complex

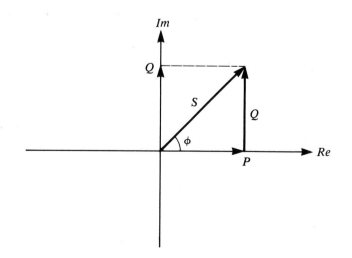

FIGURE 1.10
Power Diagram
or Power
Triangle

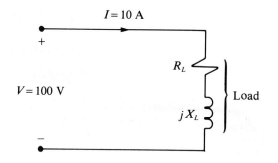

FIGURE 1.11
Circuit for
Example 1.4

power is

$$\hat{S} = \tilde{V}\tilde{I}^* = (100\underline{/0°})(10\underline{/45.57°})$$

$$= 1000\underline{/45.57°} \text{ VA}$$

$$= 1000 \cos(45.57°) + j1000 \sin(45.57°)$$

$$= 700 + j714 \text{ VA}$$

Thus,

Apparent power: $S = 1000$ VA

Active power: $P = 700$ W

Reactive power: $Q = 714$ VAR

(b) To get a unity power factor, the total reactive power must be zero. This can be achieved by injecting reactive power into the load by the shunt capacitor. The complex power expression is

$$\hat{S} = 700 + j714 + 100I_c\underline{/-90°} = 700 + j(714 - 100I_c)$$

where $I_c\underline{/90°}$ is the current through the capacitor parallel to the load, as shown in Fig. 1.12.

Setting the reactive power equal to zero, we obtain

$$I_c = 7.14 \text{ A}$$

Since $V = \dfrac{1}{\omega C} I_c$, the required value of the capacitance, C, is

$$C = \frac{I_c}{\omega V} = \frac{7.14}{2\pi \times 60 \times 100} = 189\mu F \quad \text{(microfarads)}$$

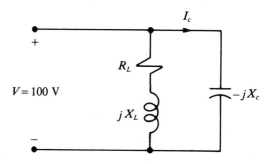

FIGURE 1.12
Compensated
Power Factor
Circuit for
Example 1.4

1.3   THREE-PHASE CIRCUITS

A balanced three-phase system consists of three sinusoidal ac voltages and/or currents with equal amplitudes and 120° phase shifts with respect to one another, as shown in Fig. 1.13 with positive phase sequence of *ABC*. In a three-phase system the circuit elements can be either Y- or Δ-connected. A balanced Y-connected load supplied by a balanced three-phase source is illustrated in Fig. 1.14. The phase voltage across the load can be expressed as

$$\hat{V}_{AN} = V_p \underline{/0°} \tag{1.22a}$$

$$\tilde{V}_{BN} = V_p \underline{/-120°} \tag{1.22b}$$

$$\tilde{V}_{CN} = V_p \underline{/120°} \tag{1.22c}$$

where phase *a* is taken as a reference with V_p as the rms value of voltage for each of the phases. If $\hat{Z} = Z\underline{/\theta°}$ is the load impedance per phase, the phase currents are

$$\tilde{I}_{AN} = \frac{\tilde{V}_{AN}}{\hat{Z}} = \frac{V_p}{Z} \underline{/-\theta°} = I_p \underline{/-\theta°} \tag{1.23a}$$

$$\tilde{I}_{BN} = \frac{\tilde{V}_{BN}}{\hat{Z}} = \frac{V_p}{Z} \underline{/-\theta° - 120°} = I_p \underline{/-\theta° - 120°} \tag{1.23b}$$

$$\tilde{I}_{CN} = \frac{\tilde{V}_{CN}}{\hat{Z}} = \frac{V_p}{Z} \underline{/-\theta° + 120°} = I_p \underline{/-\theta° + 120°} \tag{1.23c}$$

where
$$I_p = \frac{V_p}{Z} \tag{1.24}$$

is the magnitude of the current in each of the phases.

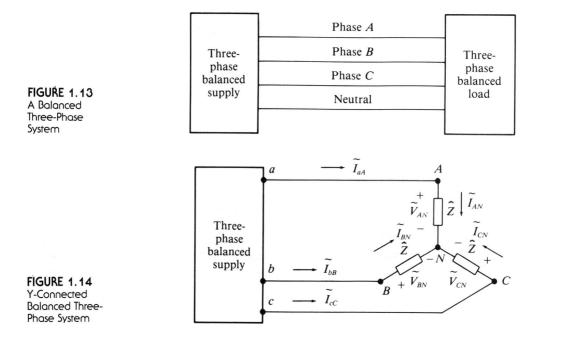

FIGURE 1.13
A Balanced
Three-Phase
System

FIGURE 1.14
Y-Connected
Balanced Three-
Phase System

The line voltage, V_{AB}, from line A to line B can be obtained as

$$\tilde{V}_{AB} = \tilde{V}_{AN} - \tilde{V}_{BN} \tag{1.25}$$

With the help of Eq. (1.22), Eq. (1.25) can be written in simplified form as

$$\tilde{V}_{AB} = \sqrt{3} \, V_p \underline{/30°} = V_l \underline{/30°} \tag{1.26a}$$

Similarly,
$$\tilde{V}_{BC} = \sqrt{3} \, V_p \underline{/-90°} = V_l \underline{/-90°} \tag{1.26b}$$

and
$$\tilde{V}_{CA} = \sqrt{3} \, V_p \underline{/150°} = V_l \underline{/150°} \tag{1.26c}$$

where
$$V_l = \sqrt{3} \, V_p \tag{1.27}$$

is the magnitude of the line voltage between any two lines of the three-phase system.

In a Y-connected circuit the phase current is the same as the line current. That is,

$$I_l = I_p \tag{1.28}$$

In a Δ-connected system, as shown in Fig. 1.15, the magnitudes of the phase and line voltages are the same (i.e., $V_l = V_p$). For a positive phase sequence, we can

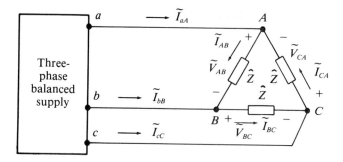

FIGURE 1.15
Δ-Connected
Balanced Three-
Phase System

express the line voltages as

$$\tilde{V}_{AB} = V_p\underline{/0°} \tag{1.29a}$$

$$\tilde{V}_{BC} = V_p\underline{/-120°} \tag{1.29b}$$

$$\tilde{V}_{CA} = V_p\underline{/120°} \tag{1.29c}$$

Once again, if $\hat{Z} = Z\underline{/\theta°}$ is the impedance of each of the three phases, the phase currents are

$$\tilde{I}_{AB} = \frac{\tilde{V}_{AB}}{\hat{Z}} = \frac{V_p}{Z}\underline{/-\theta°} = I_p\underline{/-\theta°} \tag{1.30a}$$

$$\tilde{I}_{BC} = \frac{\tilde{V}_{BC}}{\hat{Z}} = \frac{V_p}{Z}\underline{/-\theta° - 120°} = I_p\underline{/-\theta° - 120°} \tag{1.30b}$$

$$\tilde{I}_{CA} = \frac{\tilde{V}_{CA}}{\hat{Z}} = \frac{V_p}{Z}\underline{/-\theta° + 120°} = I_p\underline{/-\theta° + 120°} \tag{1.30c}$$

The line currents $\tilde{I}_{aA}$, $\tilde{I}_{bB}$, and $\tilde{I}_{cC}$ can be obtained by applying KCL at nodes A, B, and C, respectively. At node A,

$$\tilde{I}_{aA} = \tilde{I}_{AB} - \tilde{I}_{CA}$$

$$= \sqrt{3}\,I_p\underline{/-\theta° - 30°} = I_l\underline{/-\theta° - 30°} \tag{1.31a}$$

Likewise,

$$\tilde{I}_{bB} = \sqrt{3}\,I_p\underline{/-\theta° - 150°} = I_l\underline{/-\theta° - 150°} \tag{1.31b}$$

$$\tilde{I}_{cC} = \sqrt{3}\,I_p\underline{/-\theta° + 90°} = I_l\underline{/-\theta° + 90°} \tag{1.31c}$$

where

$$I_l = \sqrt{3}\,I_p$$

The total complex power in a balanced three-phase system can be expressed as

$$\hat{S} = 3\tilde{V}_p\tilde{I}_p^*$$ (1.32a)

$$= \sqrt{3}\ \tilde{V}_l\tilde{I}_l^*$$ (1.32b)

in terms of phase or line quantities of the system.

Example 1.5 illustrates the use of the per phase system in analyzing three-phase circuits.

■ EXAMPLE 1.5

A balanced three-phase, 480-V (line), 60-Hz supply feeds a Y-connected load with an impedance of $4 + j3\ \Omega$ per phase. Determine (a) the line and the phase currents, and (b) the total active power, reactive power, and apparent power absorbed by the load.

Solution

(a) Since a balanced load is fed from a balanced three-phase power supply, as shown in Fig. 1.16, we can analyze the circuit on a per phase basis as given in Fig. 1.17 where V_p is the per phase voltage and is taken as a reference for phase

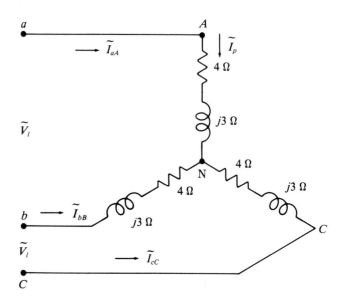

FIGURE 1.16
Three-Phase
Circuit Diagram
for Example 1.5

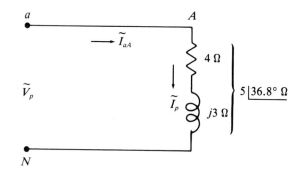

FIGURE 1.17
Per Phase Circuit
Representation
of the System
Given in Figure
1.16

A. That is,

$$\tilde{V}_p = \frac{480}{\sqrt{3}} \underline{/0^\circ} = 277.13 \underline{/0^\circ} \text{ V}$$

Phase impedance: $\hat{Z} = 4 + j3 = 5 \underline{/36.8^\circ}\ \Omega$

Phase current: $\tilde{I}_p = \dfrac{\tilde{V}_p}{\hat{Z}} = \dfrac{277.13 \underline{/0^\circ}}{5 \underline{/36.8^\circ}} = 55.42 \underline{/-36.8^\circ}$ A

    Since the phase current is the same as the line current in a Y-connected circuit, the current through the line *A* is

$$\tilde{I}_{aA} = 55.42 \underline{/-36.8^\circ} \text{ A}$$

For a positive phase sequence, the currents through lines *B* and *C* are

$$\tilde{I}_{bB} = 55.42 \underline{/-156.8^\circ} \text{ A}$$

and

$$\tilde{I}_{cC} = 55.42 \underline{/83.2^\circ} \text{ A}$$

respectively.

(b) Total complex power: $\hat{S} = 3\tilde{V}_p\tilde{I}_p^*$

$$= 3 \times 277.13 \underline{/0^\circ} \times 55.42 \underline{/36.8^\circ}$$

$$= 36,897.95 + j27,603.19$$

Apparent power: $S = 46,080.3$ VA   *magnitude only*

Active power: $P = 36,897.95$ W

Reactive power: $Q = 27,603.19$ VAR

1.4 ELECTROMAGNETIC FIELD THEORY

Most electric machines are analyzed using equivalent-circuit models. The circuit concept, even though it is simple to use and easy to explain, is only approximate. In many cases, the equivalent-circuit parameters are assumed to be constant even when the machine is operating under different loads. The effect of magnetic saturation is taken care of only approximately. For exact analysis of these machines, we must resort to the use of electromagnetic field equations. The purpose of this section is to review the basic concepts of field theory.

The fundamental theory of electromagnetic fields is based on Maxwell's equations, which govern the relationship for such fields. Maxwell's equations are generalizations of laws based on experiments. Our aim is not to trace the history of these experiments but to present them in the form that is most useful from the application point of view.

The four Maxwell's equations in the differential or the point form are

$$\nabla \times \hat{E} = \frac{\partial \vec{B}}{\partial t} \tag{1.33}$$

$$\nabla \times \hat{H} = \vec{J} + \frac{\partial \vec{D}}{\partial t} \tag{1.34}$$

$$\nabla \cdot \vec{B} = 0 \tag{1.35}$$

$$\nabla \cdot \vec{D} = \rho \tag{1.36}$$

where $\hat{E}$ = electric-field intensity in volts per meter (V/m)
  $\hat{H}$ = magnetic-field intensity in amperes per meter (A/m)
  $\vec{B}$ = magnetic-flux density in teslas or Webers per square meter (T or Wb/m^2)
  $\vec{D}$ = electric-flux density in coulombs per square meter (C/m^2)
  $\vec{J}$ = electric (volume) current density in amperes per square meter (A/m^2)
  ρ = electric (volume) charge density in coulombs per cubic meter (C/m^3)

The preceding equations are tied together by the law of conservation of charge, which is also known as the equation of continuity. That is,

$$\nabla \cdot \vec{J} = -\frac{\partial \rho}{\partial t} \tag{1.37}$$

To explain all electromagnetic phenomena, we must add the **Lorentz force** equation, which is given by

$$\vec{F} = q(\hat{E} + \vec{v} \times \vec{B}) \tag{1.38}$$

where $\vec{F}$ is the force in newtons (N).

The magnetic-flux density can be expressed in terms of the magnetic-field intensity as

$$\vec{B} = \mu_0 \mu_r \vec{H} \tag{1.39}$$

where μ_r is the relative permeability of the medium and μ_0 is the permeability of free space. In our treatment of the subject, we will only consider a homogeneous, isotropic, and linear medium. As a consequence, permeability will be considered as a scalar quantity.

With the help of Stoke's theorem, Eq. (1.33) can be expressed in integral form as

$$\oint_c \vec{E} \cdot \vec{dl} = - \int_s \frac{\partial \vec{B}}{\partial t} \cdot \vec{ds} \tag{1.40}$$

where s is the total surface bounded by contour c, as shown in Fig. 1.18.

The line integral on the left side gives us the total induced voltage, e, in a closed loop. Thus,

$$e = - \int_s \frac{\partial \vec{B}}{\partial t} \cdot \vec{ds} \tag{1.41}$$

This equation gives the induced electromotive force (emf) in a stationary closed loop due to a time rate of change of magnetic flux density, $\vec{B}$. For this reason, the induced voltage is referred to as the transformer emf, and Eq. (1.41) is known as the *transformer equation*.

If the loop is also moving with a velocity $\vec{v}$ as shown in Fig. 1.19, there will be an additional induced emf in it, which is given by

$$e_m = \oint_c (\vec{v} \times \vec{B}) \cdot \vec{dl} \tag{1.42}$$

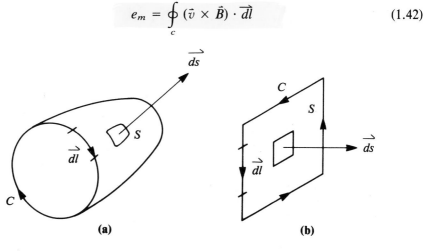

FIGURE 1.18
Appropriate
Direction of $\vec{dl}$
Associated with
$\vec{ds}$

(a)   (b)

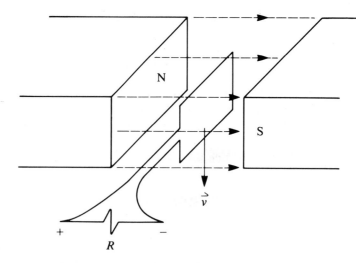

FIGURE 1.19
Loop Moving
Downward in a
Constant
Magnetic Field

and is known as the *motional emf* or the *speed voltage*. In a special case where $\vec{v}$, $\vec{B}$, and the conductor of length $\bar{l}$ are mutually perpendicular, $\vec{v}$ is the same for all parts of the conductor, and $\vec{B}$ is uniformly distributed, the motional emf becomes

$$e_m = vBl \tag{1.43}$$

Since this voltage is induced by the motion of the conductor in a magnetic field, it is also known as the voltage induced due to flux-cutting action.

For a closed loop moving in a magnetic field, the induced voltage must be equal to the sum of the transformer emf and the motional emf. That is,

$$e = -\int_s \frac{\partial \vec{B}}{dt} \cdot \vec{ds} + \oint_c (\vec{v} \times \vec{B}) \cdot \vec{dl} \tag{1.44}$$

Equation (1.44) can be written in simplified form as

$$e = -\frac{\partial \phi}{\partial t} \cdot N \tag{1.45}$$

$$\phi = \int_s \vec{B} \cdot \vec{ds} \tag{1.46}$$

where

is the total flux passing through the area of the loop. Equation (1.45) is known as *Faraday's law of induction*. This equation yields the induced emf in a closed loop having only one turn. If the number of turns in the loop is N_c, the induced voltage will be N_c times as much.

■ EXAMPLE 1.6

A coil having 1000 turns is immersed in a magnetic field varying uniformly from 100 milliwebers (mWb) to 20 mWb in 5 seconds(s). Determine the induced voltage in the coil.

Solution

Change in flux: $\Delta\phi$ = final flux − initial flux

$$= 20 - 100 = -80 \text{ mWb}$$

Change in time: $\Delta t = 5 \text{ s}$

Induced emf:

$$e = -N_c \frac{d\phi}{dt} \cong -N_c \frac{\Delta\phi}{\Delta t}$$

$$= -1000 \times \frac{-80 \times 10^{-3}}{5} = 16 \text{ V}$$

■ EXAMPLE 1.7

A square loop with sides 10 cm long is immersed in a peak magnetic-field intensity of 100 A/m varying sinusoidally at a frequency of 50 megahertz (MHz) as shown in Fig. 1.20. The plane of the loop is perpendicular to the direction of magnetic field. A voltmeter is connected in series with one side of the loop. What is the reading on the voltmeter?

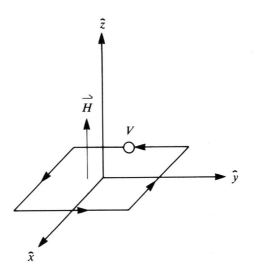

FIGURE 1.20
Figure for
Example 1.7

Solution

Since the loop is stationary, the voltage induced will be the transformer emf. Let us assume that the magnetic-field intensity is directed along the $\hat{z}$ axis and the loop is in the xy plane. The magnetic-field intensity can be expressed as

$$\bar{H} = 100 \sin \omega t \; \hat{z} \; \text{A/m}$$

where $\omega = 2\pi f$ and $f = 50$ MHz. Assuming the permeability of the medium is the same as that of free space, the magnetic-flux density can be expressed as

$$\bar{B} = \mu_0 H = 100\mu_0 \sin \omega t \; \hat{z}$$

The transformer emf is

$$e = - \int_s \frac{\partial \bar{B}}{\partial t} \cdot \overrightarrow{ds}$$

where $\overrightarrow{ds} = dx \; dy \; \hat{z}$. Thus,

$$e = -100 \; \mu_0 \int_{x=-0.05}^{0.05} \int_{y=-0.05}^{0.05} \frac{\partial}{\partial t} (\sin \omega t) \, dx \, dy$$

$$= -100\mu_0\omega \cos \omega t \int_{x=-0.05}^{0.05} dx \int_{y=-0.05}^{0.05} dy$$

$$= -100 \; \mu_0\omega \cos \omega t \; (0.1)(0.1)$$

$$= -100 \times 4\pi \times 10^{-7} \times 2 \times \pi \times 50 \times 10^6 \times 0.1 \times 0.1 \cos \omega t$$

$$= -39.48 \cos \omega t$$

The voltmeter will only read the rms value of the induced voltage. Therefore, the reading on the voltmeter would be

$$E = \frac{39.48}{\sqrt{2}} = 29.92 \; \text{V}$$

■ **EXAMPLE 1.8**

A square loop with each side equal to a meters is rotating with an angular velocity of ω rad/s in a magnetic-flux density that varies as $\bar{B} = B_m \sin \omega t \; \hat{z}$. The axis of the loop is at a right angle to the magnetic field. Determine the induced voltage using (a) the motional and transformer emf concept and (b) Faraday's law of induction.

Solution

Let us assume that the axis of the loop is along the y axis, as shown in Fig. 1.21. Consider the loop, at any time t, making an angle θ with the x axis. Let us first determine the motional emf.

$$e_m = \oint_c (\vec{v} \times \vec{B}) \cdot \overrightarrow{dl}$$

For the top conductor C, it is evident that

$$\vec{v} \times \vec{B} = vB \sin \theta \, \hat{y}$$

where $v = (a/2)\omega$ and $\theta = \omega t$. The only component of $\overrightarrow{dl}$ that results in nonzero induced voltage must be along $\hat{y}$ direction. Thus, the motional emf induced in conductor C is

$$e_{mc} = \int_{-a/2}^{a/2} vB \sin \omega t \, dy$$

But $B = B_m \sin \omega t$. Thus,

$$e_{mc} = vB_m \sin^2 \omega t \int_{-a/2}^{a/2} dy = avB_m \sin^2 \omega t = \frac{a^2}{2} B_m \omega \sin^2 \omega t$$

Similarly, we can prove that the voltage induced in conductor D would be

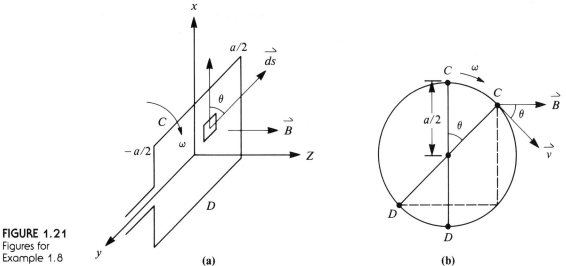

FIGURE 1.21
Figures for
Example 1.8

(a)

(b)

equal to that of conductor C. Thus, the total motional emf becomes

$$e_m = a^2 B_m \omega \sin^2 \omega t$$

The transformer emf is

$$e_t = - \int_s \frac{\partial \vec{B}}{\partial t} \cdot \vec{ds}$$

Since

$$\vec{B} = B_m \sin \omega t \, \hat{z}$$

$$\frac{\partial \vec{B}}{\partial t} = B_m \omega \cos \omega t \, \hat{z}$$

Thus,

$$e_t = - B_m \omega \cos \omega t \int_s \hat{z} \cdot \vec{ds}$$

From Fig. 1.21,

$$\int_s \hat{z} \cdot \vec{ds} = a^2 \cos \omega t$$

Therefore,

$$e_t = - a^2 B_m \omega \cos^2 \omega t$$

The total voltage induced in the coil is

$$e = e_m + e_t = - a^2 B_m (\cos^2 \omega t - \sin^2 \omega t)$$

$$= - a^2 B_m \omega \cos 2\omega t$$

Thus, the induced voltage in the coil pulsates with twice the angular frequency.

(b) *Faraday's law:* The induced emf is

$$e = - \frac{d\phi}{dt}$$

where

$$\phi = \int_s \vec{B} \cdot \vec{ds} = a^2 B_m \sin \omega t \cos \omega t$$

$$= \frac{1}{2} a^2 B_m \sin 2\omega t$$

and

$$\frac{d\phi}{dt} = a^2 B_m \omega \cos 2\omega t$$

Thus,

$$e = - a^2 B_m \omega \cos 2\omega t$$

The induced voltage obtained using Faraday's law is the same as was calculated using motional and transformer emfs. As is evident, the application of Faraday's law involved less computations.

With the help of Stoke's theorem, Eq. (1.34) can be expressed in integral form as

$$\oint_c \hat{H} \cdot \vec{dl} = \int_s \hat{J} \cdot \vec{ds} + \int_s \frac{\partial \vec{D}}{\partial t} \cdot \vec{ds} \tag{1.47}$$

The total current on the right side of Eq. (1.47) is the sum of the conduction current, $\int_s \hat{J} \cdot \vec{ds}$, and the displacement current, $\int_s (\partial \vec{D}/\partial t) \cdot \vec{ds}$. If we limit our discussion to dc or low frequencies, such as power frequencies of 50 or 60 Hz, the displacement current is of no significance for conductors as compared to the conduction current and therefore can be dropped. As a consequence, Eq. (1.47) can be rewritten as

$$\oint_c \hat{H} \cdot \vec{dl} = \int_s \hat{J} \cdot \vec{ds} \tag{1.48}$$

which is known as *Ampere's law*. Ampere's law states that the line integral of magnetic-field intensity around a closed path is equal to the total current enclosed. Example 1.9 illustrates the application of Ampere's law.

■ EXAMPLE 1.9

A cylindrical conductor carries a uniformly distributed direct current of 10 A through its cross-sectional area of 10 cm². (a) Determine the magnetic field intensity at a point $0.25R$ from the axis of the conductor, where R is the radius of the conductor. (b) Calculate the magnetic field intensity on the surface of the conductor. (c) What is the magnetic field intensity 10 m away from the axis of the conductor?

Solution

Let us assume that the conductor is long and the current flow is in the z direction, as shown in Fig. 1.22a. The radius of the conductor is

$$R = \sqrt{\frac{10}{\pi}} = 1.784 \text{ cm}$$

(a) The magnetic-field intensity is required at a point $r = 0.25 \times 1.784 = 0.446$ cm away from the axis of the conductor. To determine the total current enclosed,

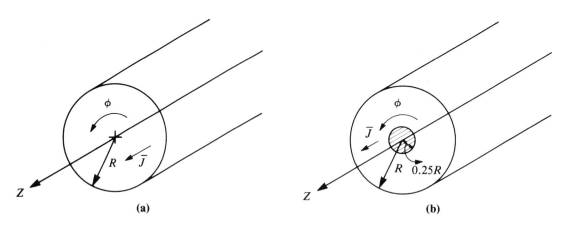

FIGURE 1.22
(a) Representation of the Conductor in Cylindrical Coordinates for Example 1.9; (b) Region where the Magnetic-Field Intensity Will Be Calculated for Example 1.9

let us first calculate the current density, $\vec{J}$, in the conductor as

$$\vec{J} = \frac{I}{A}\hat{z} = \frac{10}{10}\hat{z} = \hat{z}\,1\ \text{A/cm}$$

The total current enclosed by the circle of radius 0.446 cm, as shown by the hatched area in Fig. 1.22b, can be calculated as

$$\int_s \vec{J} \cdot \vec{ds} = I_T = 1.0 \times \pi \times 0.446^2 = 0.625\ \text{A}$$

The application of Ampere's law, Eq. (1.48), yields

$$\int_0^{2\pi} \vec{H} \cdot r d\phi\ \hat{\phi} = 0.625$$

Thus, $\qquad H_\phi = \dfrac{I_T}{2\pi r} = \dfrac{0.625}{2\pi \times 0.446 \times 10^{-2}} = 22.3\ \text{A/m}$

(b) The radius of the closed path is the same as that of the conductor. The total current enclosed is 10 A. Thus,

$$H_\phi = \frac{I_T}{2\pi r} = \frac{10}{2\pi \times 1.784 \times 10^{-2}} = 89.21\ \text{A/m}$$

(c) Since the closed contour is at a radius of 10 m from the axis of the conductor, the total current enclosed is the same as the current in the conductor. Therefore,

$$H_\phi = \frac{I_T}{2\pi r} = \frac{10}{2\pi \times 10} = 0.159 \text{ A/m}$$

1.5 MAGNETIC MATERIALS AND THEIR PROPERTIES

All magnetic materials fall into three main categories, diamagnetic, paramagnetic, and ferromagnetic, on the basis of their magnetization characteristics. The magnetization characteristic is a plot of magnetic-flux density versus the magnetic-field intensity. Such a plot is also referred to as the *B–H* curve for the magnetic material.

The magnetization characteristics for vacuum, paramagnetic, and diamagnetic materials are simply straight lines because of their constant permeabilities. The permeability of diamagnetic material is lower than the permeability of vacuum, while that of the paramagnetic material is higher. However, these differences are so small that for all practical purposes the permeability of diamagnetic and paramagnetic materials can be approximated as equal to the permeability of vacuum. The *B–H* curves for these materials are shown in Fig. 1.23. The relative

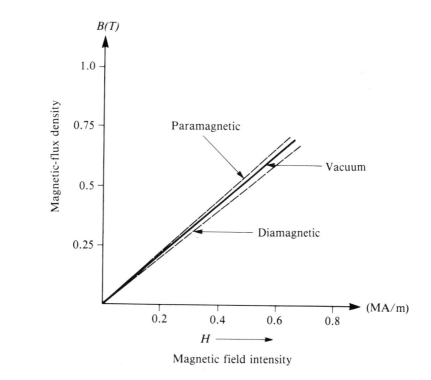

FIGURE 1.23
Magnetic
Characteristics of
Diamagnetic
and
Paramagnetic
Materials

TABLE 1.1
Typical Values
of Relative
Permeability

Material	Relative Permeability	Type
Silver	0.999 980	Diamagnetic
Copper	0.999 991	Diamagnetic
Water	0.999 991	Diamagnetic
Vacuum	1.0	Nonmagnetic
Air	1.000 038	Paramagnetic
Aluminum	1.000 023	Paramagnetic
Palladium	1.000 8	Paramagnetic

permeabilities of some of the materials falling into these categories are given in Table 1.1.

The magnetization characteristics for ferromagnetic materials are quite different than those just discussed. For an originally unmagnetized ferromagnetic material, such as iron, the *B–H* curve is shown in Fig. 1.24. This curve is usually referred to as the initial magnetization characteristic. It must be noted that an increase in *H* causes *B* to rise slowly at first, then rapidly, then very slowly again, and finally to flatten off. The permeability of the material at any point on the curve is the ratio of *B* and *H* at that point. The general behavior of the permeability as a function of the magnetic-field intensity is shown in Fig. 1.25.

If we take a ferromagnetic material in the shape of a ring and wind two coils around it as shown in Fig. 1.26, the application of time-varying current in one

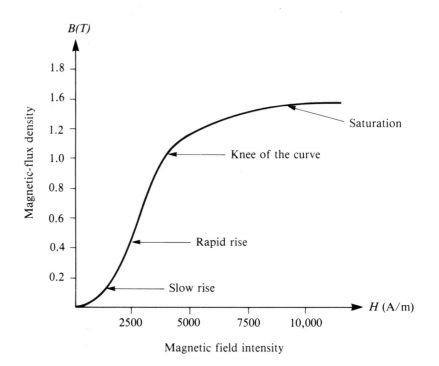

FIGURE 1.24
Typical
Magnetization
Curve

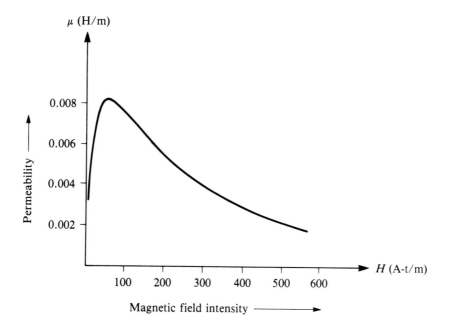

FIGURE 1.25
Permeability as
a Function of
Magnetic-Field
Intensity for
Silicon Steel

coil will induce a voltage in the other coil. The current-carrying coil establishes the magnetic field intensity, $\vec{H}$, which is taken as an independent variable. The voltage induced in the other coil helps to determine the changes in the flux and thereby in the magnetic-flux density, $\vec{B}$, inside the ring. With this arrangement we can determine the magnetization characteristic of a magnetic material.

Let us slowly increase H by increasing the current in the coil and plot the corresponding increase in B. As pointed out earlier, the B–H plot will follow the initial magnetization curve. When the B–H curve starts leveling off, we consider that the flux density has reached its maximum value, B_m, and the magnetic material is fully saturated. The corresponding value of the magnetic-field intensity is H_m.

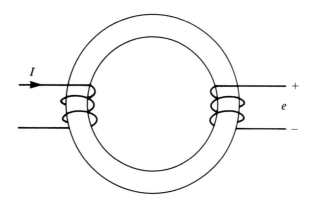

FIGURE 1.26
An Arrange-
ment to
Determine the
B–H Curve of a
Ferromagnetic
Material

If we now slowly decrease H by decreasing the current in the coil, we find that the curve does not retrace itself but follows a different path, as shown in Fig. 1.27. Also, B does not decrease as rapidly as it was increasing. This irreversibility is termed *hysteresis,* which simply means that B lags H. When the current in the coil is zero, that is H is equal to zero, there still exists some magnetic-flux density in the material. This is called the *remanence* or the *residual-flux density* and is denoted by B_r in Fig. 1.27.

To reduce the flux density in the magnetic material to zero, we have to reverse the direction of the current in the coil. The value of H that brings B to zero is known as *coercive force, H_c.* Any further increase in H in the reverse direction will cause the magnetic material to be magnetized with the opposite polarity. If we continue to increase H in that direction, the magnetic-flux density will also increase rapidly at first and then flatten off as saturation approaches. The magnitude of the maximum flux density attained in either direction will be the same.

As the current or the H field is brought to zero, the magnetic-flux density in the magnetic material will again be equal to its residual magnetism. Again, we have to reverse the direction of the current in the coil in order to bring the magnetic-flux density in the magnetic material to zero. The current in the coil is now in the same direction as it was at the outset. Any further increase in the current will start magnetizing the specimen with the original polarity.

We have gone through a complete cycle of magnetization. The loop so traced

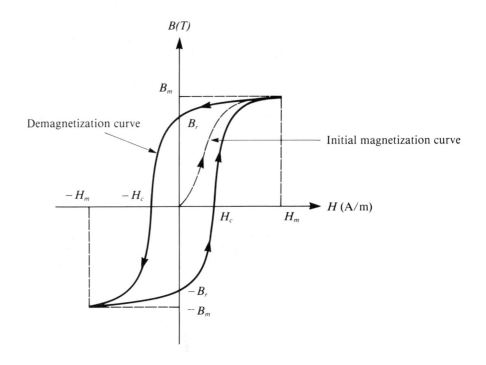

FIGURE 1.27
Hysteresis Loop
for a
Ferromagnetic
Material

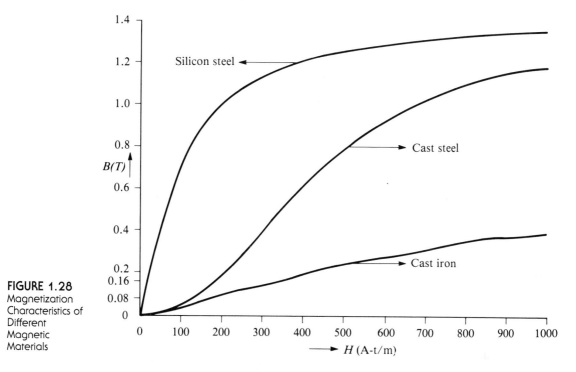

FIGURE 1.28
Magnetization
Characteristics of
Different
Magnetic
Materials

by the magnetization curve is referred to as the *hysteresis loop*. Typical magnetization curves for cast iron, cast steel, and annealed silicon steel are given in Fig. 1.28.

The shape of the hysteresis loop depends on the type of magnetic material. On the basis of the hysteresis loop, magnetic materials can be classified as hard or soft. Hard magnetic materials exhibit high remanence and large coercive force. Such materials are used for permanent magnets. Magnetically soft materials possess very low remanence and low coercive force and are used in the construction of electric machines.

1.6 MAGNETIC CIRCUITS

Whenever the magnetic flux is confined to a well-defined path consisting of group of materials, magnetic or nonmagnetic in nature, a magnetic circuit is formed. Figure 1.29 shows a basic magnetic circuit, which consists of magnetic material and a small air gap. A coil of N turns carrying current I is wound around it. The application of Ampere's law along the dashed path shown in the figure yields

$$\oint_c \vec{H} \cdot \vec{dl} = NI \tag{1.49}$$

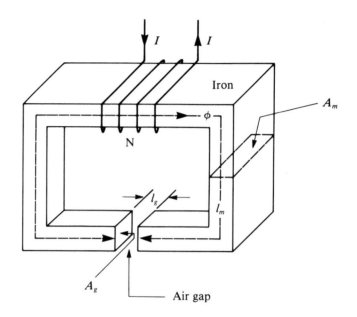

FIGURE 1.29
Series Magnetic
Circuit Consisting
of Iron and Air

The magnetic-field intensity $\vec{H}$ can be conveniently expressed in terms of the magnetic flux at any point along the path using $\vec{B} = \mu\vec{H}$ and $\phi = BA$, where A is the cross-sectional area of the circuit at the point being considered, as

$$\oint_c \frac{\phi\, dl}{\mu A} = NI \tag{1.50}$$

Since we are dealing with a series magnetic circuit, we expect the magnetic flux ϕ to be essentially constant at any point in the circuit. Therefore, Eq. (1.50) can be written as

$$\phi \oint_c \frac{dl}{\mu A} = NI \tag{1.51}$$

Equation (1.51) enables us to determine the magnetic flux in terms of the parameters of the magnetic circuit. It appears to be similar to the equation for a series electric circuit (i.e., $V = IR$). By analogy, we can define the magnetomotive force (mmf), $\mathscr{F}$, as

$$\mathscr{F} = NI \tag{1.52}$$

and the reluctance, $\mathscr{R}$, as

$$\mathscr{R} = \oint_c \frac{dl}{\mu A} \tag{1.53}$$

In terms of the preceding definitions, Eq. (1.51) can be expressed as

$$\mathscr{F} = \phi\mathscr{R} \tag{1.54}$$

For Fig. 1.29, Eq. (1.53) can be approximated as

$$\begin{aligned}\mathscr{R} &= \frac{l_m}{\mu_m A_m} + \frac{l_g}{\mu_g A_g} \\ &= \mathscr{R}_m + \mathscr{R}_g\end{aligned} \tag{1.55}$$

where l_m and l_g are the mean lengths, A_m and A_g are the cross-sectional areas, and μ_m and μ_g are the permeabilities of the magnetic material and air gap, respectively. The equivalent circuit representation of the magnetic circuit of Fig. 1.29 is given in Fig. 1.30.

The reciprocal of the reluctance is called the *permeance* of the magnetic circuit. In terms of the magnetic circuit parameters, the permeance is then given by

$$\mathscr{P} = \frac{\mu A}{l} \tag{1.56}$$

1.6.1   Self- and Mutual Inductances

Let us consider a simple magnetic circuit in the form of a ring, as shown in Fig. 1.31, surrounded by n coils. Let ϕ_{ij} be the flux linking the ith coil and produced by the jth coil. The total flux linking the ith coil is then

$$\phi_i = \sum_{j=1}^{n} \phi_{ij} \tag{1.57}$$

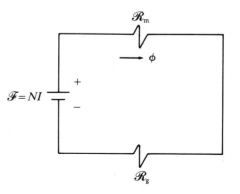

FIGURE 1.30
Equivalent-
Circuit
Representation
of the Magnetic
Circuit Shown in
Figure 1.29

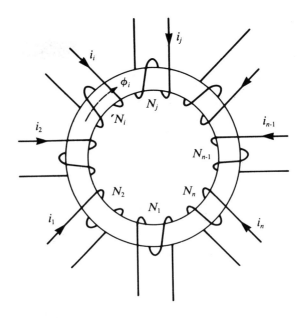

FIGURE 1.31
Magnetic Circuit
Formed by n
Coils

The emf induced in the ith coil is

$$e_i = -N_i \frac{d\phi_i}{dt} = -\sum_{j=1}^{n} N_i \frac{d\phi_{ij}}{dt} \tag{1.58}$$

Since the only changes in ϕ_{ij} are those that result from changes in the currents, Eq. (1.58) can be rewritten as

$$e_i = -\sum_{j=1}^{n} N_i \frac{di\phi_j}{di_j} \cdot \frac{di_j}{dt} \tag{1.59}$$

When $i \neq j$, we can define

$$M_{ij} = N_i \frac{d\phi_{ij}}{di_j} \tag{1.60}$$

as the *mutual inductance* between coils i and j. It must be noted that only coil j is carrying current i_j to produce the flux that is linking coil i. It can be shown that the mutual inductance M_{ji} between the coils j and i is the same as M_{ij}.

On the other hand, when $i = j$, Eq. (1.60) can be written as

$$M_{ij} = N_i \frac{d\phi_{ii}}{di_i} \tag{1.61}$$

where M_{ii} is known as the *self-inductance* of the coil and can be written simply as L_i. That is,

$$L_i = N_i \frac{d\phi_{ii}}{di_i} \qquad (1.62)$$

Let us cosider a magnetic circuit with two coils as shown in Fig. 1.32. A current i_1 in coil 1 produces a magnetic flux

$$\phi_{11} = \frac{N_1 i_1}{\mathcal{R}_1} \qquad (1.63)$$

From Eq. (1.62), the self-inductance of coil 1 is

$$L_1 = \frac{N_1^2}{\mathcal{R}_1} \qquad (1.64)$$

where $\mathcal{R}_1$ is the reluctance of the magnetic path. Similarly, when coil 2 is carrying current i_2, the self-inductance of coil 2 can be expressed as

$$L_2 = \frac{N_2^2}{\mathcal{R}_2} \qquad (1.65)$$

A portion of the flux ϕ_{22} produced by the current i_2 in coil 2 links coil 1. In other words, we can write

$$\phi_{12} = k_2 \phi_{22} \qquad (1.66)$$

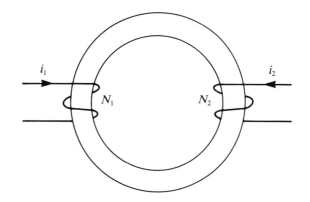

FIGURE 1.32
Magnetic Circuit
with Two Coils

where $k_2 \leqslant 1$. The mutual inductance between coils 1 and 2 can be written as

$$M_{12} = N_1 \frac{d\phi_{12}}{di_2} \tag{1.67}$$

But

$$\phi_{12} = k_2 \frac{N_2 i_2}{\mathcal{R}_2} \tag{1.68}$$

Thus,

$$M_{12} = k_2 \frac{N_1 N_2}{\mathcal{R}_2} = k_2 L_2 \frac{N_1}{N_2} \tag{1.69}$$

Similarly, we can write

$$M_{21} = k_1 L_1 \frac{N_2}{N_1} \tag{1.70}$$

Since $M_{12} = M_{21} = M$, from Eqs. (1.69) and (1.70) we obtain

$$\begin{aligned} M &= \sqrt{k_1 k_2 L_1 L_2} \\ &= k\sqrt{L_1 L_2} \end{aligned} \tag{1.71}$$

where $k = \sqrt{k_1 k_2}$ is known as the *coupling coefficient*.

■ EXAMPLE 1.10

A uniformly distributed toroidal coil of 400 turns is wound on an iron ring of square cross section. The inner radius is 10 cm and the outer radius is 12 cm. The relative permeability of the magnetic material is 1500. Determine the inductance of the coil.

Solution

The geometry of the problem is illustrated in Fig. 1.33. From Ampere's law,

$$\oint_c \vec{H} \cdot \vec{dl} = \int_s \vec{J} \cdot \vec{ds}$$

At any radius r such that $a < r < b$, where a and b are the inner and outer radii of the toroid, we find

$$\oint_c \vec{H} \cdot \vec{dl} = 2\pi r H_\phi$$

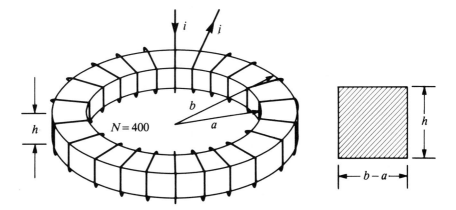

FIGURE 1.33
Geometry for
Example 1.10

If i is the current in the coil, the total current enclosed by the closed path will be

$$\int_s \vec{J} \cdot \vec{ds} = Ni$$

Therefore,

$$H_\phi = \frac{Ni}{2\pi r}$$

The magnetic-flux density is

$$B_\phi = \mu H_\phi = \frac{\mu Ni}{2\pi r}$$

The total flux in the square ring is

$$\phi = \int_s \vec{B} \cdot \vec{ds} = \frac{\mu Ni}{2\pi} \int_{r=a}^{b} \frac{1}{r}\, dr \int_{z=0}^{h} dz = \frac{\mu Ni}{2\pi} h \ln\left(\frac{b}{a}\right)$$

where h is the height of the square ring. That is, $h = b - a$. Thus, the exact expression for the inductance is

$$L = N\frac{d\phi}{di} = \frac{\mu N^2}{2\pi} h \ln\left(\frac{b}{a}\right)$$

Substituting the values, we get

$$L = \frac{4\pi \times 10^{-7} \times 1500 \times 400^2 \times 2 \times 10^{-2}}{2\pi} \ln\left(\frac{12}{10}\right) = 0.175 \text{ henrys (H)}$$

1.6.2 Analysis of Magnetic Circuits

We can make use of Eq. (1.54) in solving linear magnetic circuits. Since the permeability of the magnetic circuit is known, we can represent the magnetic circuit by an equivalent circuit using the reluctance concept. For the nonlinear magnetic circuits, the total mmf requirements can be easily determined by making use of field equations and the *B–H* characteristics for the magnetic materials. On the other hand, if the total mmf is known, there is no easy way to calculate the flux in any part of a nonlinear magnetic circuit. Such a problem can only be solved using iterative technique, as described next.

At the outset, we can assume that a large percentage of the mmf is needed for the nonmagnetic regions, and with that assumption we can proceed to calculate the flux density and the flux in that region. From the continuity of the flux lines, we will then know the flux in the magnetic region and can determine the mmf requirements to sustain that flux. Now we can add all the required mmfs and see if they are less than, equal to, or more than what is given. On that basis, we can make a second educated guess and solve again until the calculated mmf requirements are very close to the applied mmf.

■ EXAMPLE 1.11

We wish to establish an air-gap flux density of 1.0 T in the magnetic circuit as shown in Fig. 1.29. If $A_g = A_m = 40$ cm^2, $l_g = 0.5$ mm, $l_m = 1.2$ m, $N = 100$ turns, and $\mu_r = 2500$, determine the current in the coil using (a) the reluctance concept, and (b) field equations.

Solution

(a) Since the relative permeability is given as a constant, we can use the reluctance concept. The magnetic flux in the air gap is

$$\phi_g = B_g A_g = 1.0 \times 40 \times 10^{-4} = 0.004 \text{ Wb}$$

The magnetic flux in the magnetic region is the same as that in the air gap for a series magnetic circuit. That is,

$$\phi_m = \phi_g = 0.004 \text{ Wb}$$

The reluctance of the air gap is

$$\mathcal{R}_g = \frac{l_g}{\mu_g A_g} = \frac{0.5 \times 10^{-3}}{4\pi \times 10^{-7} \times 40 \times 10^{-4}} = 99,472 \text{ A/Wb}$$

The reluctance of the magnetic region is

$$\mathcal{R}_m = \frac{l_m}{\mu_m A_m} = \frac{1.2}{4\pi \times 10^{-7} \times 2500 \times 40 \times 10^{-4}} = 95{,}493 \text{ A/Wb}$$

The required mmf is

$$\mathcal{F} = NI = \phi(\mathcal{R}_m + \mathcal{R}_g)$$

$$= 0.004(99{,}472 + 95{,}493) = 779.86 \text{ ampere-turns (A-t)}$$

Therefore, the current in the coil is

$$I = \frac{\mathcal{F}}{N} = \frac{779.86}{100} = 7.799 \text{ A}$$

(b) Using field equations, the magnetic field intensity in the air gap is

$$H_g = \frac{B_g}{\mu_g} = \frac{1.0}{4\pi \times 10^{-7}} = 795{,}775 \text{ A-t/m}$$

The mmf requirements for the air gap region are

$$\mathcal{F}_g = H_g l_g = 795{,}775 \times 0.0005 = 379.89 \text{ A-t}$$

Since the area of the magnetic region is the same as that of the air, the flux density in the magnetic region is $B_m = 1.0$ T. Therefore, the magnetic-field intensity in the magnetic region is

$$H_m = \frac{B_m}{\mu_m} = \frac{1.0}{4\pi \times 10^{-7} \times 2500} = 318.31 \text{ A-t/m}$$

The mmf requirements for the magnetic region are

$$\mathcal{F}_m = H_m l_m = 318.31 \times 1.2 = 381.97 \text{ A-t}$$

Thus, the total mmf requirement is

$$\mathcal{F} = \mathcal{F}_g + \mathcal{F}_m = 397.89 + 381.97 = 779.86 \text{ A-t}$$

which is the same as obtained in part (a).

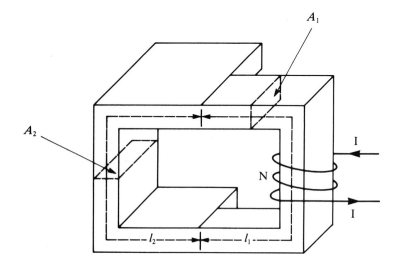

FIGURE 1.34
Magnetic Circuit
for Examples
1.12 and 1.13

■ EXAMPLE 1.12

A cast-steel core, shown in the Fig. 1.34, has the following pertinent dimensions: $A_1 = 40$ cm^2, $A_2 = 80$ cm^2, $l_1 = 0.5$ m, and $l_2 = 1.2$ m. Find the mmf required to produce a flux of 0.0032 Wb in the core.

Solution

Since there is only one flux path, the total flux in both regions must be the same. The flux densities are

$$B_1 = \frac{\phi}{A_1} = \frac{0.0032}{40 \times 10^{-4}} = 0.8 \text{ T}$$

$$B_2 = \frac{\phi}{A_2} = \frac{0.0032}{80 \times 10^{-4}} = 0.4 \text{ T}$$

Magnetic intensities from the *B–H* curve (Fig. 1.28) for the cast-steel core are

$$H_1 = 510 \text{ A-t/m} \quad \text{and} \quad H_2 = 300 \text{ A-t/m}$$

The total mmf requirement is

$$NI = H_1 l_1 + H_2 l_2 = 510 \times 0.5 + 300 \times 1.2 = 615 \text{ A-t}$$

Sometimes, it pays to make use of a tabulation scheme like the following table, which demonstrates the clarity and ease with which the solutions can be obtained.

Part No.	Flux (Wb)	Area (m²)	B (T)	H (A-t/m)	l (m)	A-t
1	0.0032	40 × 10⁻⁴	0.8	510	0.5	255
2	0.0032	80 × 10⁻⁴	0.4	300	1.2	360
					Total mmf =	615

In Example 1.12 it was rather easy to determine the total mmf needed to establish the required flux in the simple series circuit. The complexity of the problem becomes quite obvious when we are required to determine the flux and the flux densities for a given mmf as outlined next.

■ EXAMPLE 1.13

A magnetic circuit, similar to that shown in Fig. 1.34, is made of silicon steel with the following dimensions: $l_1 = 40$ cm, $l_2 = 75$ cm, $A_1 = 4$ cm², and $A_2 = 10$ cm². A 400-turn coil carrying a current of 0.3 A is wound around the magnetic path. What are the flux densities in the two regions?

Solution

Since there is only one magnetic path, the flux in both the regions must be the same. Total applied mmf is $400 \times 0.3 = 120$ A-t. Let us assume that one-half the applied mmf is needed for region 1. With this assumption, we can now calculate the magnetic-field intensity, H_1; the magnetic-flux density, B_1, from the $B–H$ curve for silicon steel; and the total flux in region 1. Since the flux in region 2 is the same as that in region 1, the flux density, magnetic-field intensity, and total mmf requirements for region 2 can then be determined. The sum of the two mmfs is then compared with the applied mmf. In case these mmfs are not very close to each other, another educated guess is in order and the process is repeated again.

First Try

Part No.	Flux (Wb)	Area (m²)	B (T)	H (A-t/m)	l (m)	A-t
1	0.00036	4 × 10⁻⁴	0.9	150	0.4	60
2	0.00036	10 × 10⁻⁴	0.36	50	0.75	37.5
					Total mmf =	97.5

The calculated mmf is smaller than the applied mmf, the difference being 22.5 A-t. That gives us a clue to make the second guess, because most of the 22.5 A-t must be consumed in region 1 due to its relatively high flux density. In that case, let us assume that the total mmf requirement for region 1 is 80 A-t.

Second Iteration

Part No.	Flux (Wb)	Area (m^2)	B (T)	H (A-t/m)	l (m)	A-t
1	0.0004	4×10^{-4}	1.0	200	0.4	80
2	0.0004	10×10^{-4}	0.4	55	0.75	41.3
					Total mmf =	121.3

Since the calculated mmf is close to the applied mmf, there is no need of further iterations.

■ EXAMPLE 1.14

Figure 1.35 shows a parallel magnetic circuit made of silicon steel. All the pertinent dimensions are given in centimeters. A coil having 400 turns is wound around the center leg. What must be the current in the coil in order to establish a flux density in the air-gap region of 0.8 T? Assume uniform flux distribution and neglect leakage and fringing.

Solution

The air-gap area is $10 \times 4 = 40$ cm². The magnetic flux in the air-gap region CD is $B_g A_g = 3.2$ mWb. This is also the flux in the regions ABC and DEF. Thus we can calculate the total mmf requirements for the two magnetic regions and the air gap. The region $AHGF$, being parallel, must have the same mmf requirements. This helps in the calculation of the flux in this region. The flux in the center leg is made of these two fluxes. The mmf drop for the center leg can now be determined. The total mmf requirement for the complete circuit is the sum of the mmf requirements for the center leg and one of the other two legs. Let us now proceed to determine the current in the coil.

The mean-length calculations are

Air-gap CD = 0.1 cm (given)
Path DEF = $(120 - 10)/2 + (80 - 0.5 - 10)/2 = 89.95$ cm
Path ABC = same as path DEF = 89.95 cm
Path $AHGF$ = $2(120 - 10)/2 + (80 - 10) = 180$ cm
Path AF = $80 - 10 = 70$ cm

Region	Flux (mWb)	A (cm²)	B (T)	H (A-t/m)	l (cm)	A-t	$\mathcal{R}$
CD	3.2	40	0.8	636,620	0.1	636.6	198,938
ABC	3.2	40	0.8	120	89.75	107.9	33,719
DEF	3.2	40	0.8	120	89.75	107.9	33,719
		mmf requirements for the above paths are =				852.4	
AHGF	9.6	80	1.2	473	180.0	852.4	88,792
AF	12.8	112	1.14	300	70.0	210.0	16,406
		mmf requirements for the complete circuit =				1062.4	

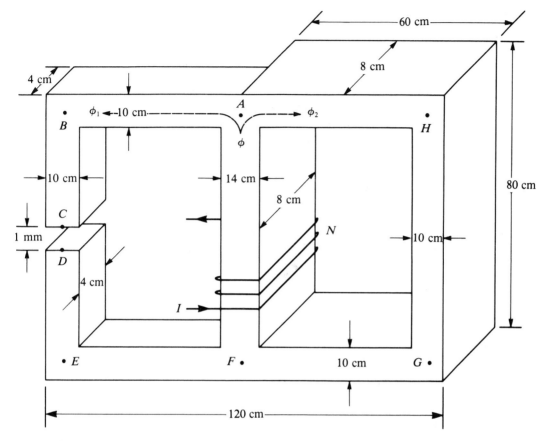

FIGURE 1.35
Magnetic Circuit for Example 1.14

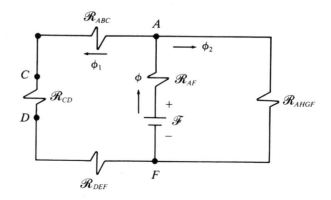

FIGURE 1.36
Equivalent-
Circuit
Representation
of the Magnetic
Circuit of Figure
1.35

The current in the coil is 1062.4/400 = 2.656. The exact equivalent circuit is shown in Fig. 1.36.

■ EXAMPLE 1.15

The cross section of a 2-pole dc machine with its pertinent dimensions in millimeters is illustrated in Fig. 1.37. Analyze the magnetic circuit and determine the required mmf per pole to establish a flux of 5.37 mWb in the air gap. The poles and the rotor are punched from silicon steel sheet, while the yoke is of cast steel. Use the *B–H* curves given in Fig. 1.28.

Solution

Each pole of a dc machine has the same number of turns, N, with the common current, I. The mmf produced by each winding is $\mathscr{F} = NI$. This mmf sets up the required flux on a per pole basis. Knowing the flux in the air-gap region, we can determine the flux densities in all sections of the machine. It is obvious from Fig.

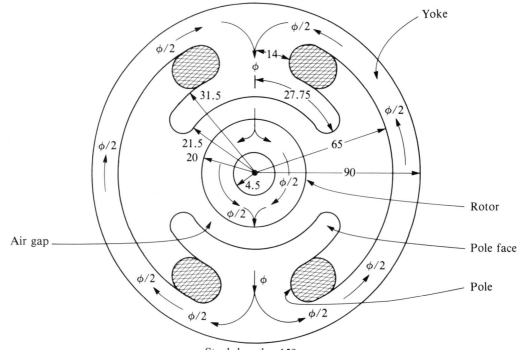

FIGURE 1.37
Magnetic Circuit of a dc Machine for Example 1.15

1.37 that the flux provided by each pole is the same in the pole, pole face, and air-gap region. However, it divides equally when it flows through the rotor and yoke. The mmf drop in either half of the yoke or the rotor must be the same.

Let $\mathcal{R}_p$, $\mathcal{R}_f$, $\mathcal{R}_g$, $\mathcal{R}_r$, and $\mathcal{R}_y$ be the reluctances of the pole, pole face, air gap, rotor, and yoke section, respectively. In terms of the reluctances, the magnetic circuit can be completely represented by an equivalent circuit, as shown in Fig. 1.38. From the equivalent circuit it is apparent that

$$2\mathcal{F} = \phi(2\mathcal{R}_p + 2\mathcal{R}_f + 2\mathcal{R}_g) + \frac{\phi}{2}(\mathcal{R}_r + \mathcal{R}_y) \qquad (1.72)$$

or $$\mathcal{F} = (\mathcal{R}_p + \mathcal{R}_f + \mathcal{R}_g + 0.25\mathcal{R}_r + 0.25\mathcal{R}_y)\phi \qquad (1.73)$$

where the flux, ϕ, is known and the reluctance of the magnetic circuit can be determined from its physical dimensions and known permeability. However, we are given the *B–H* curves of the magnetic materials. Therefore, we can use the field approach directly to calculate the mmf drops in each section of the magnetic circuit. The reluctance can then be obtained by dividing the mmf drop by the flux.

To carry out the magnetic-circuit calculations, we will make use of the tab-

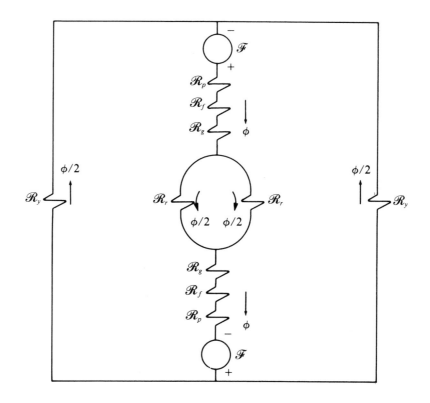

FIGURE 1.38
Equivalent-
Circuit
Representation
of the Magnetic
Circuit of Figure
1.37

ulation scheme explained earlier. The mean lengths of the magnetic sections are:

Pole: $l_p = 65 - 31.5 = 33.5$ mm

Pole face: $l_f = 31.5 - 21.5 = 10.0$ mm

Air gap: $l_g = 21.5 - 20.0 = 1.5$ mm

Rotor: $l_r = \dfrac{\pi}{2}(20 + 4.5) + 20 - 4.5 = 53.98$ mm

Yoke: $l_y = \dfrac{\pi}{2}(65 + 90) + 90 - 65 = 268.47$ mm

The cross-sectional areas are:

Pole: $A_p = 2 \times 14 \times 150 = 4200$ mm^2

Pole face: $A_f = 2 \times 27.75 \times 150 = 8325$ mm^2

Air gap: $A_g = 2 \times 27.75 \times 150 = 8325$ mm^2

Rotor: $A_r = (20 - 4.5) \times 150 = 2325$ mm^2

Yoke: $A_y = (90 - 65) \times 150 = 3750$ mm^2

Note that the cross-sectional area of the air gap is approximated to be equal to the area of the pole face. This assumption is not far-fetched and will become evident during our discussion of dc machines.

The mmf requirements are tabulated next. From the table, the total mmf requirements are 1722.6 for two poles. Thus, the mmf per pole must be 861.3 A-t. Note that most of the mmf is expended in the air-gap region.

Part Name	ϕ (mWb)	A (mm^2)	B (T)	H (A-t/m)	l (mm)	mmf (A-t)	$\mathcal{R}$
Pole	5.37	4200	1.28	600	33.5	20.1	3,743
Pole face	5.37	8325	0.645	85	10.0	0.85	158
Air gap	5.37	8325	0.645	513,275	1.5	769.91	143,373
Rotor	2.685	2325	1.15	325	53.48	17.38	6,473
Air gap	5.37	0325	0.645	513,275	1.5	769.91	143,372
Pole face	5.37	8325	0.645	85	10.0	0.85	158
Pole	5.37	4200	1.28	600	33.5	20.1	3,743
Yoke	2.685	3750	0.716	460	268.47	123.5	45,995
		Total mmf requirements for 2 poles			=	1722.6	
		mmf requirements per pole			=	861.3	

1.7 MAGNETIC LOSSES

When a magnetic material is subjected to a time-varying flux, there is some energy loss in the material in the form of magnetic losses. Magnetic losses are also known as the iron or core losses. Core losses consist of eddy-current and hysteresis losses.

The *eddy-current loss* in the magnetic material is caused by the flow of currents, known as the eddy currents, that are the result of induced voltage in the magnetic material by the time-varying magnetic flux, as shown in Fig. 1.39. These currents not only result in the loss of energy in the magnetic material but also exert a demagnetization effect on the core. The demagnetization action increases as the center of the cross section of the magnetic material is approached. The overall effect of demagnetization is the crowding of the magnetic flux toward the outer surface of the magnetic material. This results in a nonuniform distribution of the magnetic flux in the core. The effect of eddy currents, and thereby eddy-current loss, can be reduced to a great extent by breaking the eddy-current path with thin strips of magnetic material (laminations), as shown in Fig. 1.40. There is a layer of insulation which creates a high resistance path from one lamination to the other. The thinner the laminations, the smaller will be the flow of eddy currents. Although it is impossible to totally eliminate the eddy-current losses, these losses can effectively be reduced to an acceptable level.

Eddy-current loss has been found to depend on lamination thickness, t; the frequency, f; maximum flux density, B_m; and the volume of the magnetic material, V. Eddy-current loss can be expressed as

$$P_e = k_e f^2 t^2 B_m^2 V = k_e f^2 t^2 B_m^2 \frac{M}{\delta} \qquad (1.74)$$

where k_e is a constant depending on the units used in the equation and a function of the resistivity of the material, M is the mass of the core, and δ is the density of the magnetic material.

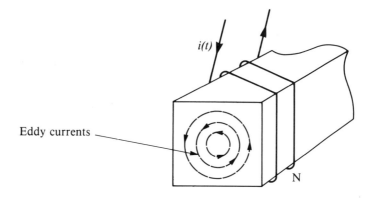

FIGURE 1.39
Eddy Currents in
a Magnetic
Material

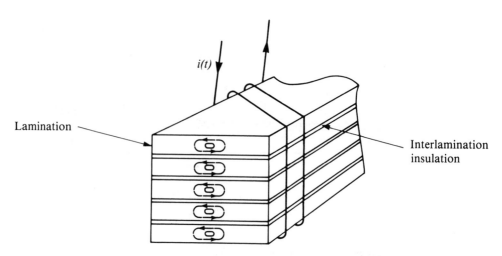

Lamination

Interlamination
insulation

FIGURE 1.40
Laminated
Magnetic Core
to Reduce the
Effects of Eddy
Currents

$i(t)$

Hysteresis loss results from continuous reversals in the alignment of the magnetic particles in the magnetic material. The energy required to cause these reversals is the hysteresis loss. Experimental data have shown that hysteresis loss depends on the volume of the magnetic material, V; maximum flux density, B_m; and the frequency, f. It can be expressed as

$$P_h = k_h f B_m^n V = k_h f B_m^n \frac{M}{\delta} \tag{1.75}$$

where k_h is a constant and depends on the units used in the equation and the material, and n, which is known as the Steinmetz exponent, has been found to vary from 1.5 to 2.5.

The core loss of a magnetic material is then the sum of hysteresis loss and eddy-current loss or

$$
\begin{aligned}
P_c &= V(k_h f B_m^n + k_e t^2 f^2 B_m^2) \\
&= \frac{M}{\delta} (k_h f B_m^n + k_e t^2 f^2 B_m^2)
\end{aligned}
\tag{1.76}
$$

which, for given values of M, δ, and t, may be written

$$P_c = K_h f B_m^n + K_e f^2 B_m^2 \tag{1.77}$$

If the core loss is to be broken down into its constituent parts P_h and P_e, the coefficients K_h and K_e and the exponent n must be found. We need three independent equations to evaluate three independent quantities. To do so, the core losses are measured at two frequencies with two flux densities, as demonstrated by Example 1.16.

■ EXAMPLE 1.16

The following data were obtained on a thin sheet of silicon steel. What are the hysteresis and eddy-current losses?

Frequency (Hz)	Flux Density (T)	Core Loss (W/kg)
25	1.1	0.4
25	1.5	0.8
60	1.1	1.2

Solution
Substituting in Eq. (1.77), we obtain

$$30.25K_e + 1.1^n K_h = 0.016 \tag{1.78}$$

$$56.25K_e + 1.5^n K_h = 0.032 \tag{1.79}$$

$$72.6K_e + 1.1^n K_h = 0.02 \tag{1.80}$$

Subtracting Eq. (1.78) from Eq. (1.80), we get

$$K_e = 94.451 \times 10^{-6}$$

Eliminating K_e in Eqs. (1.78) and (1.79) and solving for n, we have

$$n = 2.296$$

From Eq. (1.79),

$$K_h = 0.0105$$

The computed eddy-current and hysteresis losses are tabulated as follows:

f (Hz)	B (T)	P_h (W/kg)	P_e (W/kg)	P_c (W/kg)
25	1.1	0.327	0.072	0.398
25	1.5	0.666	0.133	0.799
60	1.1	0.784	0.411	1.195

These calculations show that the eddy-current loss is smaller than the hysteresis. Usually, the hysteresis loss is 50% to 70% of the total core loss.

1.8   PERMANENT MAGNETS

The rapid development of permanent-magnet materials and their commercial availability has made them very attractive in various applications of electromechanical energy-conversion devices. A material is termed a permanent-magnetic material if it retains some flux density, known as the residual flux density B_r, after it has been magnetized. To restore the material to its virgin state, it must be demagnetized by applying the field intensity in the opposite direction. The value of the field intensity required to reduce the residual flux density to zero is known as the *coercive force*. Therefore, for proper utilization of a permanent magnet, we must concentrate our attention on its demagnetization characteristics, that is, the magnetic behavior of a material depicted in the second quadrant of the hysteresis loop. Figure 1.41 shows comparisons of the demagnetization characteristics of Alnico, ceramic, and Samarium based rare-earth cobalt magnets. The figure shows that there is a dramatic difference between the properties of magnetic materials belonging to different groups of alloys. In addition, the permanent magnet designer makes use of another curve called the $B-H$ or energy-product curve. The demagnetization and the energy-product curves are plotted on the same graph as shown in Fig. 1.42. We can make maximum use of the

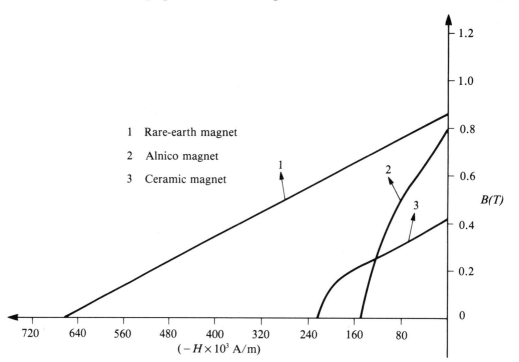

FIGURE 1.41
Demagnetization Characteristics of Alnico, Ceramic, and Rare-Earth Magnets

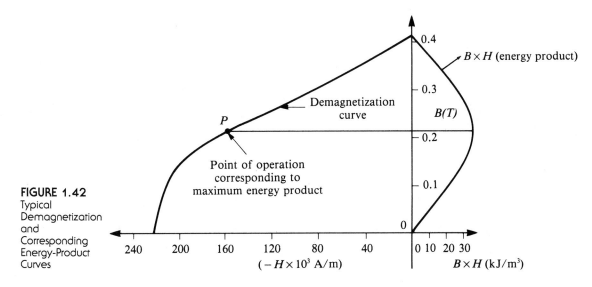

FIGURE 1.42
Typical
Demagnetization
and
Corresponding
Energy-Product
Curves

magnetic material by operating the magnet at its maximum energy-product level, which is shown by the horizontal dashed line in Fig. 1.42.

Figure 1.43 shows a series magnetic circuit containing a permanent magnet (shown shaded), magnetic material such as steel, and an air gap. The application of Ampere's law to the magnetic circuit yields

$$H_m l_m + H_s l_s + H_g l_g + H_s l_s = 0 \qquad (1.81)$$

where H_m, H_g, and H_s are the magnetic-field intensities in the permanent magnet, air gap, and steel, respectively. In a series magnetic circuit, the magnetic flux is

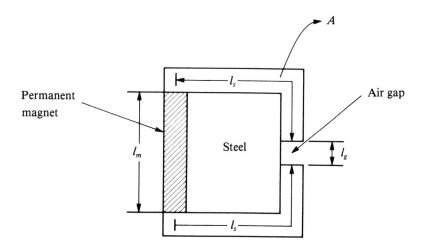

FIGURE 1.43
Series Magnetic
Circuit

essentially the same. Therefore,

$$B_m A_m = B_g A_g = B_s A_s \qquad (1.82)$$

where B_m, B_s, and B_g are the flux densities in the magnet, steel, and air gap, respectively. A_m, A_s, and A_g are the cross-sectional areas, as shown in the figure. Since $B = \mu H$, we can write

$$H_g = \frac{B_m A_m}{\mu_0 A_g}$$

and

$$H_s = \frac{B_m A_m}{\mu_s A_s}$$

Equation (1.81) can be written as

$$H_m = -\frac{1}{l_m}\left(\frac{l_g}{\mu_0}\frac{A_m}{A_g} + \frac{2l}{\mu_s}\frac{A_m}{A_s}\right) B_m \qquad (1.83)$$

This relationship is known as the *operating line,* and its intersection with the demagnetization curve yields the *operating point.* If the permeability of the magnetic material used to form the magnetic path is sufficiently high, Eq. (1.83) may then be approximated as

$$H_m = -\frac{1}{\mu_0}\frac{l_g}{l_m}\frac{A_m}{A_g} B_m \qquad (1.84)$$

For a special case when $A_m = A_g$, Eq. (1.84) can be written as

$$B_m = \frac{\mu_0 l_m}{l_g} H_m \qquad (1.85)$$

where H_m is the magnitude of the coercive force corresponding to B_m. But $B_m = B_g$ from Eq. (1.82). Thus,

$$B_g = \frac{\mu_0 l_m}{l_g} H_m$$

or

$$B_g^2 = \mu_0 \frac{l_m A_m}{l_g A_g} H_m B_m$$

or

$$B_g H_g V_g = B_m H_m V_m \qquad (1.86)$$

where V_g and V_m are the volumes of the air gap and magnet, respectively. Equation (1.86) emphasizes the fact that the energy available in the air gap is maximum when the point of operation corresponds to the maximum energy product of the magnet.

■ EXAMPLE 1.17

The dimensions for the magnetic circuit given in Fig. 1.43 are as follows: $l_g = 1$ cm, $A_g = A_m = A_s = 10$ cm^2, $l_s = 50$ cm, and $\mu_r = 3000$ for steel. What is the minimum length of the magnet to establish the maximum energy in the air gap? Neglect all leakages and fringing effects.

Solution

To establish the maximum energy in the air gap, the point of operation must correspond to the maximum energy product of the magnet. Substituting the values in Eq. (1.83), we obtain

$$H_m = -\frac{8223}{l_m} B_m$$

From the demagnetization and energy-product curves as plotted in Fig. 1.44, B_m

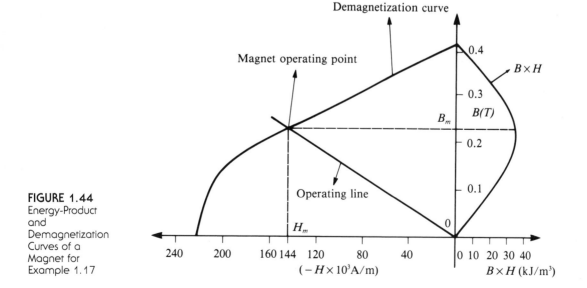

FIGURE 1.44
Energy-Product and Demagnetization Curves of a Magnet for Example 1.17

= 0.23 T and $H_m = -144$ kA/m. Thus, the minimum length of the magnet is

$$l_m = -\frac{8223}{H_m} B_m$$

$$= \frac{8223 \times 0.23}{144 \times 10^3} = 0.0131 \text{ m} \quad \text{or} \quad 1.31 \text{ cm}$$

■ EXAMPLE 1.18

Calculate the induced voltage in a single-turn coil rotating at 100 rad/s in a permanent-magnet system, as shown in Fig. 1.45. Both rotor and yoke are made of steel with a relative permeability of 3000. The demagnetization and energy-product curves for the magnet are given in Fig. 1.47.

Solution

We will use the subscripts m, g, a, and s for the magnet, air gap, steel rotor on which the coil is mounted, and the steel yoke that provides the return path for the magnetic flux. From Ampere's law,

$$H_m l_m + H_g l_g + H_a l_a + H_g l_g + H_m l_m + H_s L_s = 0$$

or
$$2H_m l_m = -(2H_g l_g + H_a l_a + H_s l_s) \tag{1.87}$$

First we calculate the point of operation of the magnet. As is evident from Fig. 1.46, there are two magnetic paths for the flux. From symmetry, the flux in each point would be half of the total flux. That is, if ϕ is the flux per path, the total

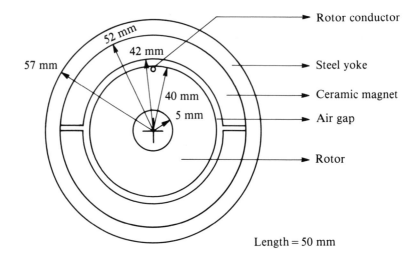

FIGURE 1.45
Magnetic
Geometry for
Example 1.18

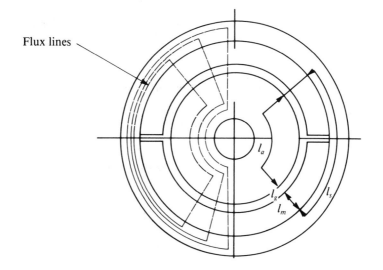

FIGURE 1.46
Flux Distribution
and Mean
Lengths of the
Magnetic Path

flux available from the magnet will be 2ϕ. Focusing our attention on one magnetic path, we have

$$\phi = B_m A_m = B_s A_s = B_g A_g = B_a A_a$$

where the areas are defined for one path only.

Equation (1.87) can be expressed, in terms of B_m, as

$$2H_m l_m = -\left(2\frac{l_g}{\mu_0}\frac{A_m}{A_g} + \frac{l_a}{\mu_a}\frac{A_m}{A_a} + \frac{l_s}{\mu_s}\frac{A_m}{A_s}\right)B_m \qquad (1.88)$$

The mean lengths are:

Magnet:　　　　　$l_m = 52 - 42 = 10$ mm

Air gap:　　　　　$l_g = 2.0$ mm

Rotor (Fig. 1.46):　$l_a = 17.5 + 17.5 + (2\pi \times 22.5/4) = 70.34$ mm

Yoke (Fig. 1.46):　$l_s = 2.5 + 2.5 + (2\pi \times 54.5/4) = 90.61$ mm

The cross-sectional areas are:

Magnet:　$A_m = (52 + 42) \times \dfrac{2\pi}{4} \times 50 = 3691.37$ mm^2

Rotor:　　$A_a = 35 \times 50 = 1750$ mm^2

Air gap: $A_g = (42 + 40) \times \dfrac{2\pi}{4} \times 50 = 3220.13 \text{ mm}^2$

Yoke: $A_s = 5 \times 50 = 250 \text{ mm}^2$

Substituting the values of mean lengths and the cross-sectional areas in Eq. (1.88), we get

$$H_m = -202{,}158.27 B_m$$

From the demagnetization curve of the magnet (Fig. 1.47), the operating point is at

$$B_m = 0.337 \text{ T}$$

Thus, the flux per path is

$$\phi = 0.337 \times 3691.37 \times 10^{-6} = 0.001244 \text{ Wb}$$

Hence, the flux provided by each magnet is

$$\phi_T = 2\phi = 0.002488 \text{ Wb}$$

As the coil is rotating with a uniform angular velocity of 100 rad/s, the total flux linking the coil is maximum when the plane of the coil is perpendicular to the magnetic flux. We can therefore write the expression of the magnetic flux passing

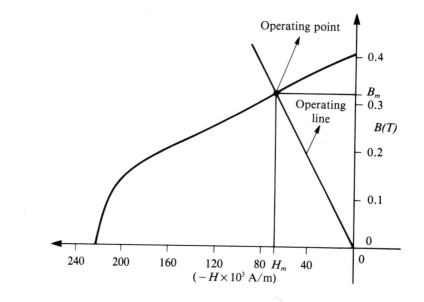

FIGURE 1.47
Demagnetization
Curve for
Example 1.18

through the area of the coil at any time t as

$$\phi_T(t) = 0.002488 \cos(100t)$$

From Faraday's law of induction, the induced voltage in the coil is

$$e(t) = -0.2488 \sin(100t) \text{ V}$$

1.9 PER UNIT SYSTEM

When an electric machine is analyzed using the exact magnitudes of its parameters, it is not immediately evident how its performance will compare with that of another similar type of machine designed for different voltage and power ratings. However, if the parameters and ratings of similar types of machines are given in terms of decimal fractions of prespecified base values of the respective quantities, the performance of one machine can be easily compared with that of another. These decimal fractions are known as the *per unit values*. For example, a statement that the primary winding resistance of a transformer is 10 Ω has no bearing on the primary winding resistances of other transformers. This value of resistance may be very high for one transformer and quite low for the other. However, if we say the per unit value of the primary winding resistance is 0.1, it means that the resistive drop with rated current will be 10% of the rated voltage. This statement has definite significance irrespective of the voltage and current ratings of the transformer.

To express the parameters of any machine in terms of its per unit values, the reference or base values must be selected first. It does not matter what values are used as the base values. However, the rated values of a machine are usually employed as the base values. For the fundamental base values, we can use the rated voltage and the rated current, the rated voltage and the rated power, or the rated current and the rated power. All other quantities can then be expressed in terms of the fundamental base quantities. In our discussion of the per unit system, we will use the rated voltage and the rated (apparent) power as the fundamental base values. If V_b and S_b are the fundamental voltage and power base values, the base values for the current and the impedance, respectively, are

$$I_b = \frac{S_b}{V_b} \tag{1.89}$$

and
$$Z_b = \frac{V_b}{I_b} \tag{1.90}$$

The parameters and the ratings of a machine can now be expressed as a decimal fraction of the base values. If Z is the impedance and V, I, and S are the voltage, current, and power ratings of a machine, we can represent them it terms of per unit quantities as follows:

$$Z_{pu} = \frac{Z}{Z_b} \qquad (1.91)$$

$$V_{pu} = \frac{V}{V_b} \qquad (1.92)$$

$$I_{pu} = \frac{I}{I_b} \qquad (1.93)$$

$$S_{pu} = \frac{S}{S_b} \qquad (1.94)$$

■ **EXAMPLE 1.19**

A capacitive load takes 2 A at a power factor of 0.9 when connected to a 600-V, 60-Hz source. Determine its per unit values when the base values for the voltage and power are 100 V and 1000 VA, respectively.

Solution

The fundamental base values are given as

$$V_b = 100 \text{ V}$$

and

$$S_b = 1000 \text{ VA}$$

The other base quantities are

$$I_b = \frac{1000}{100} = 10 \text{ A}$$

and

$$Z_b = \frac{100}{10} = 10 \text{ }\Omega$$

The exact quantities for the load, taking the applied voltage as the reference, are

$$\tilde{V} = 600\underline{/0°} \text{ V}$$

$$\theta = \cos^{-1}(0.9) = 25.84° \text{ (capacitive)}$$

$$\tilde{I} = 2\underline{/25.84°} \text{ A}$$

$$\hat{S} = \tilde{V}\tilde{I}^* = 1200\underline{/-25.84°} \text{ VA}$$

and
$$\hat{Z} = \frac{600\underline{/0°}}{2\underline{/25.84°}} = 300\underline{/-25.84°} \text{ } \Omega$$

In terms of the base quantities, the per unit values of the load are

$$\hat{S}_{pu} = \frac{1200\underline{/-25.84°}}{1000} = 1.2\underline{/-25.84°}$$

$$\tilde{V}_{pu} = \frac{600\underline{/0°}}{100} = 6\underline{/0°}$$

$$\tilde{I}_{pu} = \frac{2\underline{/25.84°}}{10} = 0.2\underline{/25.84°}$$

and
$$\hat{Z}_{pu} = \frac{300\underline{/-25.84°}}{10} = 30\underline{/-25.84°}$$

PROBLEMS

1.1. Determine the current I in the circuit shown in Fig. P1.1 using (a) mesh equations and (b) Thevenin's theorem.

1.2. Using Thevenin's theorem, determine the current as indicated in the circuit of Fig. P1.2.

1.3. Find the steady-state current $i(t)$ and the voltage $v(t)$ in the circuit given in Fig. P1.3 using (a) Thevenin's theorem and (b) mesh equations. Calculate the power supplied by the source and the power dissipated in the circuit.

1.4. A Y-connected load with an impedance of $12 - j15$ Ω per phase is connected to a balanced 230-V, 60-Hz, three-phase supply. Determine (a) the line and phase voltages, (b) the line and phase currents, and (c) the apparent, active and reactive powers in the load.

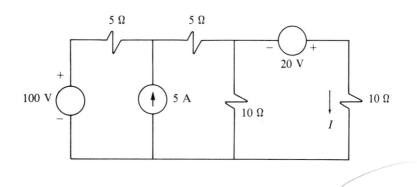

FIGURE P1.1
Electric Circuit for
Problem 1.1

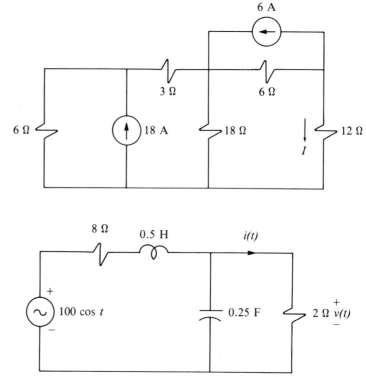

FIGURE P1.2
Electric Circuit for
Problem 1.2

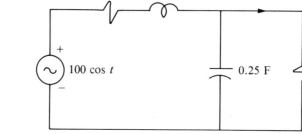

FIGURE P1.3
Electric Circuit for
Problem 1.3

1.5. A conductor 10 cm long is moving with a speed of 150 m/s in a magnetic-flux density of 0.5 T. Calculate the induced emf in the conductor if the magnetic field is being swept at an angle of 45°.

1.6. A conductor of length 50 cm carrying a current of 10 A is immersed in a magnetic-flux density of 5 kG at an angle of 30°. What is the force acting on the conductor? (*Note:* 10 kG = 1 Wb.)

1.7. A 10-cm square coil with 100 turns is held perpendicular to a uniform magnetic-flux density of 15 kG. The coil is then made to rotate. Calculate (a) the maximum flux passing through the coil, (b) the average voltage induced in the coil if it makes a quarter-turn in 1 s, and (c) the average induced emf between brushes if the coil rotates at 1200 rpm.

1.8. An armature conductor is rotating at 6000 rpm. The length of the conductor is 25 cm and the armature diameter is 65 cm. Determine (a) the voltage induced in the conductor while passing under a pole of uniform flux density of 1.8 T, and (b) the speed in rpm if the induced emf is 30 V.

1.9. A circular toroid is wound with 100 turns and has an air gap of 0.2 mm, and inner and outer diameters are 30 and 40 mm, respectively. The relative permeability of the magnetic material is 2000. Calculate the current in the

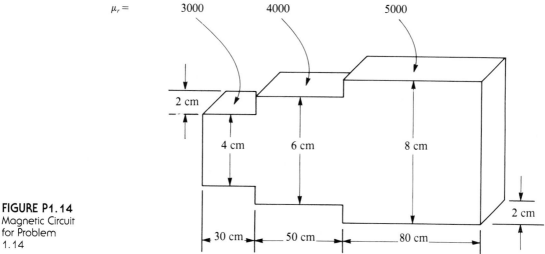

$\mu_r =$ 3000   4000   5000

2 cm

4 cm   6 cm   8 cm

2 cm

30 cm   50 cm   80 cm

FIGURE P1.14
Magnetic Circuit
for Problem
1.14

coil to establish a flux density in the air gap region of 1.2 T. What is the total flux in the toroid? Assume uniform flux distribution.

1.10. A coil sets up a flux of 0.5 Wb when wound over a magnetic material and a flux of 0.001 Wb when wound over a nonmagnetic material of the same dimensions. Under constant mmf conditions, what is the relative permeability of the magnetic material?

1.11. A 2000-turn coil is wound over a magnetic toroid having a diameter of 60 cm and mean length of 3 m. What must the current in the coil be to establish a flux of 0.5 Wb in the toroid? Assume $\mu_r = 500$.

1.12. A soft iron circular ring has a cross-sectional area of 1.45 cm², mean length of 542 mm, and an air gap of 4.75 mm. A coil with 300 turns is wrapped around the ring and carries a current of 2 A. Assume the relative permeability of iron to be 900. Determine the flux and the flux density in the ring. Neglect fringing.

1.13. A silicon steel ring has an air gap of 1.5 mm, cross-sectonal area of 1.5 cm², and mean magnetic length of 48 cm. It is wound with a coil of 250 turns and carrying a current of 4 A. Determine the flux and flux density in the ring.

1.14. A magnetic circuit is shown in Fig. P1.14. Compute the reluctance of each section. What is the total reluctance of the magnetic circuit? Assume $\mu_{r1} = 3000$, $\mu_{r2} = 4000$, and $\mu_{r3} = 5000$.

1.15. A magnetic circuit is made of silicon steel as shown in Fig. P1.15 with its pertinent dimensions in centimeters. The outer legs have 1000 turns each. It is required to produce a flux of 3.6 mWb in the air gap by applying equal current to both windings. What is the current in each winding?

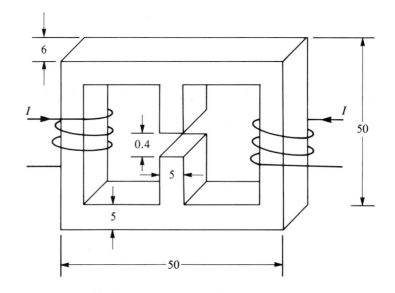

FIGURE P1.15
Magnetic Circuit
for Problem
1.15 (All
dimensions are
in centimeters)

1.16. The magnetic circuit of a 4-pole dc machine with active length of 56 mm is illustrated in Fig. P1.16. Determine the mmf per pole needed to establish a flux density of 1.0 T in the air gap under the pole. Assume that the effective arc of the air gap is the same as the width of the pole. All parts of the magnetic circuit are punched from silicon steel.

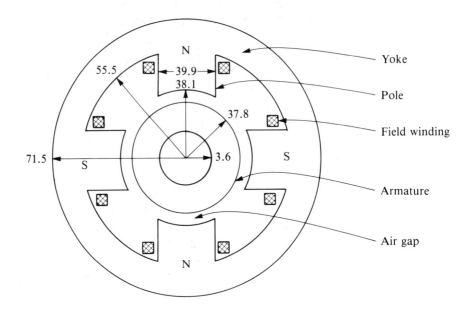

FIGURE P1.16
Magnetic Circuit
for Problem
1.16 (All
dimensions are
in millimeters)

1.17. An iron ring with a 50-cm mean diameter is made of square iron of 4-cm² cross section and is uniformly wound with two coils having 1000 and 400 turns. Calculate the self-inductance of each coil and the mutual inductance between them. Assume $\mu_r = 1500$.

1.18. Two coils having 100 and 500 turns are wound on a magnetic material with 40-cm² cross section and mean length of 100 cm. Only 60% of the flux produced by one coil links the other coil. If a current changes linearly from 0 to 100 A in 10 ms in one coil, determine the induced voltage in the other coil. Assume $\mu_r = 500$. What is the mutual inductance between them?

1.19. The area of a hysteresis loop for a magnetic material weighing 25 kg is 150 cm² when drawn to scales of 1 cm = 0.2 T and 1 cm = 200 A-t/m. Assuming the density of the material to be 7800 kg/m³, calculate the hysteresis loss at 60 Hz.

1.20. In a certain magnetic material, the hysteresis loss is 200 W when the frequency is 25 Hz and the maximum flux density is 1.2 T. What is the hysteresis loss when the maximum flux density is reduced to 0.8 T at 60 Hz? Assume the Steinmetz exponent is 1.6.

CHAPTER 2

Electromechanical Energy Conversion

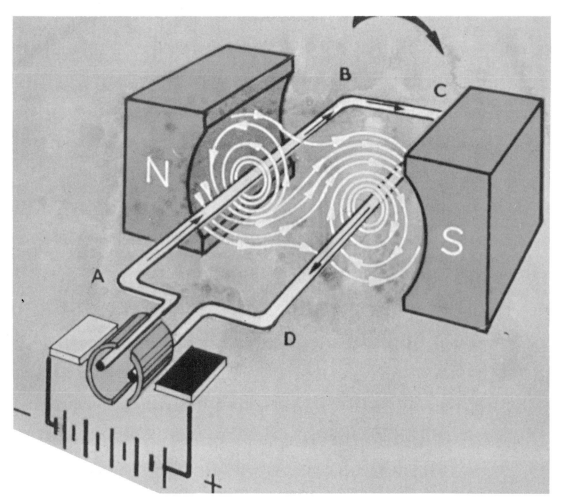

It is our good fortune that natural phenomena exists that provide several different means for the interchange of energy between a mechanical and an electrical system. Basically, the electromagnetic phenomenon is responsible for the electromechanical energy-conversion process. Electromechanical energy conversion involves either the conversion of electrical energy to mechanical energy or vice versa. Such an energy-conversion process is reversible except for the losses in the system. In other words, the energy can be transferred back and forth between the electrical and the mechanical systems. However, each time such a transfer is completed some of the energy is converted into heat and is lost from the systems forever. In this chapter, our aim is to explore the basic principles of electromechanical energy conversion and to develop simple models to determine either the force or the torque developed by various electromechanical devices.

Whenever a current-carrying conductor is immersed in a magnetic field, it experiences a force tending to move it in accordance with the Lorentz force equation. When the conductor moves in the direction of the electromagnetic force, the magnetic field aids in the conversion of electrical energy into mechanical energy. This is essentially the principle of operation of all motors. On the other hand, if the conductor is forced to move in the direction opposite to the electromagnetic force, mechanical energy is transferred into electrical energy through the associated magnetic field. Generator action is based on this principle. In these cases, the magnetic field acts as the medium for the energy transfer.

The energy-transfer process can also take place when the electric field acts as a medium. Consider the charged plates of a capacitor. A force of attraction tends to move the plates closer to each other in accordance with Coulomb's law. If one plate is allowed to move in the direction of the force, electrical energy will be converted into mechanical energy. However, any increase in the separation between the plates will cause the transfer of mechanical energy into electrical energy.

The conversion of energy from one form to another must, however, satisfy the principle of conservation of energy, because the mass of the materials used in the construction of electromechanical energy-conversion devices has to for practical purposes remain constant. Accordingly, the input energy must be equal to the summation of useful output energy, the energy dissipated as heat, and the change in the stored energy in the field. That is,

Input energy = output energy + energy dissipated as heat

$$+ \text{ change in the stored energy} \quad (2.1)$$

In this energy-balance equation, the output energy and the energy dissipated as heat are always positive. On the other hand, the change in the stored energy may be positive or negative depending on whether it is increasing or decreasing.

2.1   ELECTRIC-FIELD ENERGY CONVERSION

A lossless parallel-plate capacitor (Fig. 2.1) with surface area A and separation between the plates x is connected to a voltage source, resulting in a displacement current that can be expressed as

$$i = \frac{dQ}{dt} \qquad (2.2)$$

The amount of differential energy input to the capacitor in an infinitesimal time dt is

$$dW_i = vi\,dt$$

or

$$dW_i = v\,dQ \qquad (2.3)$$

If the capacitor is assumed to be lossless and the plates are held stationary, the input energy must increase the stored field energy of the capacitor. Thus, the amount of energy stored in the electric field in time dt is

$$dW_f = vdQ \qquad (2.4)$$

From Maxwell's equations as given in Chapter 1, it can be shown for a parallel-plate capacitor, neglecting the effect of fringing, that

$$v = Ex \qquad (2.5)$$

and

$$Q = DA \qquad (2.6)$$

where E and D are the electric-field intensity and electric-flux density within the region enclosed by the parallel plates.

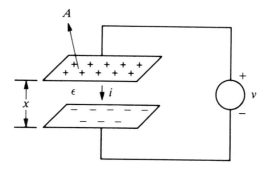

FIGURE 2.1
Parallel-Plate
Capacitor

Equation (2.4) can then be expressed as

$$dW_f = A \times E \, dD$$

or
$$\Delta W = \int_{W_{f1}}^{W_{f2}} dW_f = A \times \int_{D_1}^{D_2} E \, dD = V \int_{D_1}^{D_2} E \, dD \qquad (2.7)$$

where V is the volume bounded by the capacitor plates. For a homogeneous, isotropic, and linear medium, $D = \epsilon E$, where ϵ is the permittivity of the medium. For zero initial conditions, the change in field energy in time t can be expressed as

$$
\begin{aligned}
W_f &= \frac{1}{2} \epsilon E^2 V \\[1em]
&= \frac{1}{2} \epsilon A \frac{v^2}{x^2} \qquad (2.8) \\[1em]
&= \frac{1}{2} C v^2
\end{aligned}
$$

where
$$C = \frac{\epsilon A}{x} \qquad (2.9)$$

is the capacitance of the parallel-plate capacitor. Thus, in a lossless, fixed, parallel-plate capacitor, the electrical energy input translates into an increase in the stored energy in the capacitor in the form of electric field.

A force of attraction is exerted on the two plates of the parallel-plate capacitor by equal and opposite charge distribution on them. If one of the plates is free to move with respect to the other, the force of attraction will tend to impart motion. Therefore, some of the electrical energy will be converted into mechanical energy, which is also the developed energy. In this case, Eq. (2.1) reduces to

$$dW_i = dW_d + dW_f \qquad (2.10)$$

where dW_d is the developed energy. It is essentially the amount of work done by the system and can be expressed as

$$dW_d = F_d \, dx \qquad (2.11)$$

where F_d is the force developed and dx is the infinitesimal displacement in the direction of the force. With the help of Eqs. (2.3) and (2.10), the differential increase in the stored energy can be expressed as

$$dW_f = v \, dQ - F_d \, dx \qquad (2.12)$$

If we consider the applied voltage v and the displacement x as the independent variables, W_f, F_d, and Q must then be functions of v and x. Therefore,

$$dW_f(v, x) = \frac{\partial W_f(v, x)}{\partial v} dv + \frac{\partial W_f(v, x)}{\partial x} dx \tag{2.13}$$

and

$$dQ(v, x) = \frac{\partial Q(v, x)}{\partial v} dv + \frac{\partial Q(v, x)}{\partial x} dx \tag{2.14}$$

In carrying out the partial derivatives, say with respect to v, in these equations, it is implied that the other variable x must be considered as constant. Substituting these equations in Eq. (2.12), it can be shown that

$$F_d = v \frac{\partial Q}{\partial x} - \frac{\partial W_f}{\partial x} \tag{2.15}$$

Since $Q = Cv$ and W_f is known in terms of C and v from Eq. (2.8), Eq. (2.15) can be expressed in a simplified form as

$$F_d = \frac{1}{2} v^2 \frac{\partial C}{\partial x} \tag{2.16}$$

Substituting for C from Eq. (2.9) for a parallel-plate capacitor, we obtain

$$F_d = -\frac{1}{2} \epsilon A \frac{v^2}{x^2}$$

or

$$F_d = -\frac{1}{2} \epsilon A E^2 \tag{2.17}$$

Equation (2.17) represents the force acting on the plates of a parallel-plate capacitor charged to a potential difference of v volts or electric-field intensity of E V/m. The negative sign highlights the fact that it is a force of attraction.

■ EXAMPLE 2.1

Two parallel plates each having dimensions of 20 by 20 cm are held 2 mm apart in air. If the potential difference between the plates is 2 kilovolts (kV), determine (a) the energy stored by the capacitor, and (b) the force acting on the plates.

Solution

(a) The capacitance of the system is

$$C = \frac{\epsilon A}{x}$$

$$= \frac{8.85 \times 10^{-12} \times 20 \times 20 \times 10^{-4}}{2 \times 10^{-3}} = 177 \text{ pF}$$

The energy stored is

$$W_f = \frac{1}{2} CV^2 = 0.5 \times 177 \times 10^{-12} \times (2 \times 10^3)^2 = 354 \text{ microjoules } (\mu J)$$

(b) The electric field intensity is

$$E = \frac{V}{x}$$

$$= \frac{2000}{0.002} = 1.0 \text{ megavolts/m (MV/m)}$$

The force of attraction is

$$F_d = \frac{1}{2} \epsilon A E^2 = 0.5 \times 8.85 \times 10^{-12} \times 400 \times 10^{-4} \times 1.0 \times 10^6$$

$$= 177 \text{ nanonewtons (nN)}$$

■ EXAMPLE 2.2

The upper plate of a parallel-plate capacitor, as shown in Fig. 2.2, is held stationary while the lower plate is free to move. The surface area of each plate is 20 cm^2 and the separation is 5 mm. Determine the mass of an object suspended from the lower plate that will keep it stationary when the plates are subjected to a potential difference of 10 kV. What is the energy stored in the electric field?

Solution

To keep the lower plate stationary, the weight of the object must be equal to the magnitude of the force of attraction acting on the plates. That is,

$$mg = |F_d|$$

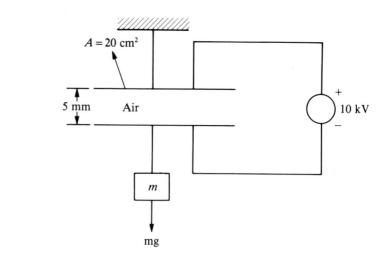

FIGURE 2.2
Air-Filled
Parallel-Plate
Capacitor for
Example 2.2

However,
$$|F_d| = \frac{1}{2} \epsilon A E^2$$

$$= 0.5 \times 8.85 \times 10^{-12} \times 20 \times 10^{-4}$$

$$\times \left(\frac{10 \times 10^{-3}}{5 \times 10^{-3}} \right)^2 = 35.4 \text{ mN}$$

Thus,
$$m = \frac{0.0354}{9.81} = 3.609 \text{ g}$$

The capacitance of the system is

$$C = \frac{\epsilon A}{x}$$

$$= \frac{8.85 \times 10^{-12} \times 20 \times 10^{-4}}{5 \times 10^{-3}} = 3.54 \text{ pF}$$

The field energy is

$$W_f = \frac{1}{2} C V^2$$

$$= 0.5 \times 3.54 \times 10^{-12} \times (10 \times 10^3)^2 = 177 \text{ } \mu\text{J}$$

The amount of force developed by the electrical systems in the preceding examples is very small even though the applied voltages are quite high. This fact clearly indicates that, to develop a substantial amount of force in such a system, either the applied voltage must be very high or the energy-conversion device must be quite large. However, in most applications neither of these conditions can be realized because of practical as well as economical reasons. This is why magnetic materials are widely used in designing such devices, as we will see in the subsequent section.

2.2 MAGNETIC-FIELD ENERGY CONVERSION

When a time-varying voltage is applied across the terminals of an N-turn coil of a magnetic circuit with mean length l and cross-sectional area A, as shown in Fig. 2.3, the current through the coil establishes a time-varying flux and thereby induces an opposing voltage in accordance with Faraday's law of induction. To sustain the current flow in the coil, the applied source must do work against the induced emf. Thus, the electrical input energy in time dt in a lossless, fixed magnetic circuit is

$$dW_i = vi\ dt = -ei\ dt \tag{2.18}$$

However, from Faraday's law of induction,

$$e = -\frac{d\lambda}{dt} = -N\frac{d\phi}{dt} \tag{2.19}$$

The energy input in time interval dt becomes

$$dW_i = i\ d\lambda = Ni\ d\phi \tag{2.20}$$

Substituting $Ni = Hl$ and $\phi = BA$, where B is the magnetic-flux density, in Eq. (2.20), we obtain

$$dW_i = lAH\ dB = VH\ dB \tag{2.21}$$

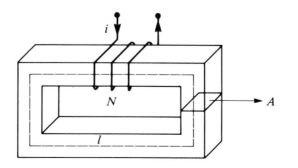

FIGURE 2.3
Magnetic Circuit with an *N*-Turn Coil

where V is the volume of the magnetic circuit. Note that $H\,dB$ represents the shaded area in Fig. 2.4a. The total energy input in one complete cycle is

$$W_i = V \oint H\,dB \tag{2.22}$$

where $\oint H\,dB$ stands for the integration over one complete cycle and represents the area enclosed by the loop in Fig. 2.4b. In time-invariant magnetic fields, Eq. (2.22) yields the energy dissipated as heat due to hysteresis.

The energy input in establishing the flux density initially from zero to B can be expressed as

$$W_i = V \int_0^B H\,dB \tag{2.23}$$

Since the magnetic system has been assumed to be fixed and lossless, the input electrical energy must represent the increase in magnetic field energy. Thus,

$$W_f = \int_0^\lambda i\,d\lambda$$

or
$$W_f = V \int_0^B H\,dB \tag{2.24}$$

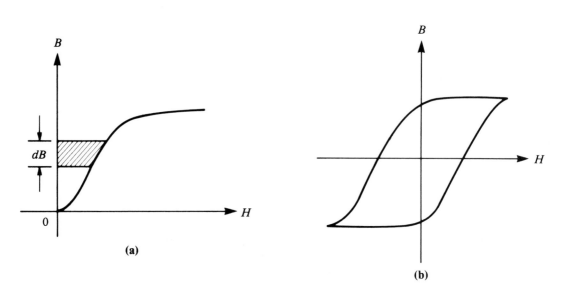

(a)

(b)

FIGURE 2.4
(a) Representation of Differential Flux Density on the B–H Curve; (b) Closed Loop Corresponding to the Integration of H over One Complete Cycle

For a homogeneous, isotropic, and linear medium, $B = \mu H$, where μ is the permeability of the magnetic material, the Eq. (2.24) reduces to

$$W_f = \frac{1}{2} \mu H^2 V$$

$$= \frac{1}{2} \frac{\mu A N^2}{l} i^2 \qquad (2.25)$$

$$= \frac{1}{2} L i^2$$

where
$$L = \frac{\mu A N^2}{l} = \frac{N^2}{\mathcal{R}} \qquad (2.26a)$$

is the inductance of the magnetic circuit and

$$\mathcal{R} = \frac{l}{\mu A} \qquad (2.26b)$$

is the corresponding reluctance.

■ EXAMPLE 2.3

The relation between the flux linkages and the current for the magnetic material in Fig. 2.3 is given by

$$\lambda = \frac{6i}{2i + 1} \text{ Weber-turns (Wb-t)}$$

Determine the energy stored in the magnetic field for $\lambda = 2$ Wb-t.

Solution

In terms of the flux linkages and the current, the energy stored in the magnetic field is given as

$$W_f = \int_0^\lambda i \, d\lambda$$

From the given relation between the flux linkages and the current,

$$i = \frac{\lambda}{6 - 2\lambda}$$

Thus,
$$W_f = \int_0^\lambda \frac{\lambda}{6 - 2\lambda}\, d\lambda$$

Setting $6 - 2\lambda = x$, this expression reduces to

$$W_f = \int_2^6 \left(\frac{1.5}{x} - 0.25\right) dx$$

$$= [1.5 \ln(x) - 0.25\, x]_2^6 = 0.648 \text{ J}$$

Let us now consider another magnetic circuit with one part free to move with respect to the other, as shown in Fig. 2.5. The energy conversion in a lossless magnetic system is governed by Eq. (2.1) and can be expressed as

$$dW_i = i\, d\lambda - F_d\, dx \tag{2.27}$$

If we choose λ and x as the independent variables in Eq. (2.27), dW_f, i, and F_d will then be multivariate functions of λ and x. Therefore,

$$dW_f(\lambda,\, x) = \frac{\partial W_f(\lambda,\, x)}{\partial \lambda}\, d\lambda + \frac{\partial W_f(\lambda,\, x)}{\partial x}\, dx \tag{2.28}$$

Comparing Eq. (2.27) with Eq. (2.28), it can be shown that

$$i = \frac{\partial W_f(\lambda,\, x)}{\partial \lambda} \tag{2.29}$$

and
$$F_d = -\frac{\partial W_f(\lambda,\, x)}{\partial x} \tag{2.30}$$

Equation (2.30) clearly indicates that the rate of decrease in the stored mag-

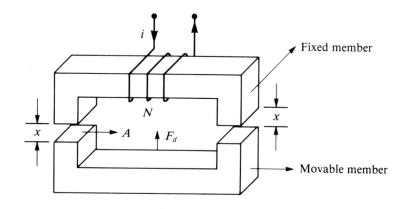

FIGURE 2.5
Magnetic System with Fixed and Movable Members

netic energy with respect to the displacement determines the force developed by the electromechanical device.

Since $Li = \lambda$, the stored magnetic energy can be expressed as

$$W_f = \frac{1}{2} Li^2 = \frac{1}{2} \frac{\lambda^2}{L} \qquad (2.31)$$

L is a function of λ and x and represents the effective inductance of the entire magnetic circuit. From Eq. (2.30), the developed force is

$$F_d = \frac{1}{2} \frac{\lambda^2}{L^2} \frac{\partial L}{\partial x} = \frac{1}{2} i^2 \frac{\partial L}{\partial x} \qquad (2.32)$$

Equation (2.32) is quite simple to use for linear magnetic circuits because the inductance can be determined using the reluctance concept introduced in Chapter 1. However, for nonlinear magnetic circuits we have to resort to the basic fields approach. For that purpose, consider only one of the air gaps shown in Fig. 2.5. The force acting in the other air-gap region can also be calculated in the same fashion. When one magnetic piece tends to move closer to the other, the depletion in field energy takes place in the air-gap region and is responsible for the development of the force.

The energy in the air-gap region, from Eq. (2.25), can be expressed as

$$W_f = \frac{1}{2} \mu_0 H_0^2 Ax = \frac{1}{2} A \frac{B_0^2}{\mu_0} x \qquad (2.33)$$

where B_0 is the flux density in the air gap.

The force developed per air gap, from Eq. (2.31), can now be expressed as

$$F_{dg} = -\frac{1}{2} A \frac{B_0^2}{\mu_0} \qquad (2.34)$$

or

$$F_{dg} = -\frac{1}{2} A \mu_0 H_0^2 \qquad (2.35)$$

This expression is very similar to the one obtained for electric-field energy conversion. Once again, the presence of the negative sign indicates that it is the force of attraction that tends to decrease x.

■ EXAMPLE 2.4

The magnetic circuit with the dimensions shown in Fig. 2.6 is excited by a 100-turn coil wound over the center leg. Determine the necessary excitation current

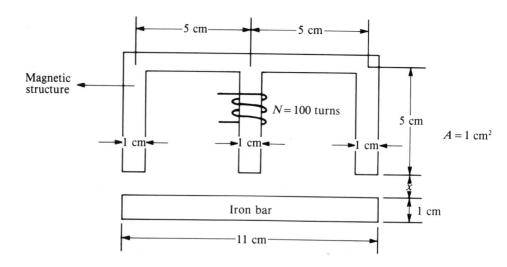

FIGURE 2.6
Magnetic
System for
Example 2.4

in the coil so that the iron bar remains suspended at a distance of 1 cm using (a) inductance and (b) the fields concept. Also calculate the energy stored in the magnetic field. The relative permeability and the specific gravity of the magnetic structure and the iron bar are 2000 and 7.85 g/cm³, respectively. Neglect leakage and fringing effects.

Solution

(a) Since the permeability of the magnetic circuit is given to be constant, the inductance concept for determining the force experienced by the iron bar can be employed. Let us assume that the distance between the iron bar and the remaining magnetic circuit is x meters. The mean length of either the left or the right leg of the magnetic circuit, l_1, including the part of the iron bar, can be computed as 0.16 m, and that of the center leg, l_c, is 0.055 m. The effective equivalent circuit in terms of the reluctances is shown in Fig. 2.7, where

$$\mathcal{R}_0 = \frac{0.16}{2000 \times 4\pi \times 10^{-7} \times 10^{-4}} = 636,620 \ \text{H}^{-1}$$

$$\mathcal{R}_g = \frac{x}{4\pi \times 10^{-7} \times 10^{-4}} = 7.958 \times 10^9 x \ \text{H}^{-1}$$

$$\mathcal{R}_c = \frac{0.055}{2000 \times 4\pi \times 10^{-7} \times 10^{-4}} = 218,838 \ \text{H}^{-1}$$

The effective reluctance of each of the outer legs is

$$\mathcal{R}_b = \mathcal{R}_0 + \mathcal{R}_g = 636,620 + 7.958 \times 10^9 x \ \text{H}^{-1}$$

The total reluctance is

$$\mathcal{R} = \mathcal{R}_c + \mathcal{R}_g + \frac{\mathcal{R}_b}{2} = 954,930 + 11.937 \times 10^9 x \ \text{H}^{-1}$$

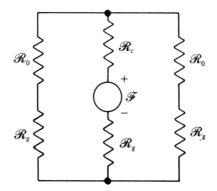

FIGURE 2.7
Electric Analog
Circuit for the
Magnetic Circuit
Given in Figure
2.6

The inductance is

$$L = \frac{N^2}{\mathcal{R}} = \frac{1}{1,193,700x + 95.49} \text{ H}^{-1}$$

From Eq. (2.32),

$$F_d = -\frac{1}{2} I^2 \frac{1,193,700}{(1,193,700x + 95.49)^2} \text{ N}$$

The negative sign only indicates that the force is acting in the upward direction. For $x = 1$ cm, the developed force becomes

$$F_d = -0.00412 I^2 \text{ N}$$

This force must be balanced by the force due to gravity acting in the downward direction on the iron bar. Thus,

$$F_d = mg = 7.85 \times 10^{-3} \times 11 \times 1 \times 9.81 = 0.847 \text{ N}$$

Therefore, $I = \sqrt{\dfrac{0.847}{0.00412}} = 14.34 \text{ A}$

The inductance of the circuit at $x = 1$ cm is $L = 83.11$ μH. The energy stored in the magnetic field is

$$W_f = \frac{1}{2} L I^2 = 0.5 \times 83.11 \times 10^{-6} \times 14.34 = 8.55 \text{ mJ}$$

(b) *Field concept:* The flux in the center leg must be twice as much as that in any one of the outer legs, as shown in Fig. 2.8. If B_0 is the flux density in the

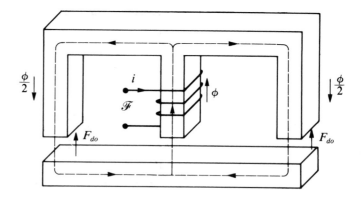

FIGURE 2.8
Magnetic-Flux
Densities and
Field-Intensities
for Example 2.4

outer leg, $2B_0$ will be the flux density in the center leg because of the same uniform area of the entire circuit. On the other hand, the force of attraction experienced by the central region of the iron bar will be four times as much as experienced by the outer regions. Thus, from Eq. (2.34),

$$F_d = \frac{1}{2} A \frac{B_0^2}{\mu_0} (1 + 1 + 4) = \frac{3AB_0^2}{\mu_0}$$

or

$$B_0 = \sqrt{\frac{0.847 \times 4\pi \times 10^{-7}}{3 \times 10^{-4}}} = 0.06 \text{ T}$$

Due to symmetry, the mmf requirements of each of the outer legs will be the same. In the outer air gap,

$$H_0 = \frac{0.06}{4\pi \times 10^{-7}} = 47746.48 \text{ A-t/m}$$

$$H_0 x = 47,746.48 \times 0.01 = 477.46 \text{ A-t}$$

In the outer magnetic region,

$$H_1 = \frac{0.06}{2000 \times 4\pi \times 10^{-7}} = 23.87 \text{ A-t/m}$$

and

$$H_1 l_1 = 23.87 \times 0.16 = 3.82 \text{ A-t}$$

Thus, the mmf needed for one of the outer legs is

$$\mathscr{F}_0 = 3.82 + 477.46 = 481.28 \text{ A-t}$$

For the center leg the air gap is

$$H_{0c} = \frac{2B_0}{\mu_0} = \frac{2 \times 0.06}{4\pi \times 10^{-7}} = 95{,}492.97 \text{ A-t/m}$$

$$H_{0c}x = 95{,}492.97 \times 0.01 = 954.93 \text{ A-t}$$

The magnetic part is

$$H_c = \frac{2 \times 0.06}{2000 \times 4\pi \times 10^{-7}} = 47.75 \text{ A-t/m}$$

$$H_c l_c = 47.75 \times 0.055 = 2.63 \text{ A-t}$$

Thus, the mmf needed for the center leg is

$$\mathscr{F}_c = 2.63 + 954.93 = 957.56 \text{ A-t}$$

and total mmf requirements are

$$\mathscr{F} = 957.56 + 481.28 = 1438.84 \text{ A-t}$$

Thus, the current in the coil is

$$I = \frac{1438.84}{100} = 14.39 \text{ A}$$

From the magnetic circuit analysis,

$$NI = \mathscr{R}\phi$$

Therefore,

$$\mathscr{R} = \frac{NI}{\phi} \quad \text{and} \quad L = \frac{N^2}{\mathscr{R}} = \frac{N\phi}{I}$$

From Eq. (2.25),

$$W_f = \frac{1}{2} LI^2 = \frac{1}{2} NI\phi = \frac{1}{2} NIBA$$

where ϕ is the flux linking the coil. Therefore, it is the flux in the center leg. Substituting the values of flux density and the cross-sectional area of the center leg in the preceding expression, we obtain

$$W_f = 0.5 \times 100 \times 14.39 \times (2 \times 0.06) \times 1 \times 10^{-4} = 8.6 \text{ mJ}$$

Thus far, we have developed the electromechanical energy-conversion model for a magnetic system considering λ and x as the independent variables. However, it is more suitable to choose i and x as the independent variables because it is the current that establishes the flux and thereby the total flux linkages. Consequently, the differential change in λ can be expressed in terms of partial derivatives as

$$d\lambda(i, x) = \frac{\partial \lambda(i, x)}{\partial i} di + \frac{\partial \lambda(i, x)}{\partial x} dx \qquad (2.36)$$

Replacing $d\lambda$ in Eq. (2.27) by its partial derivatives as given in Eq. (2.36), we obtain

$$dW_f(i, x) = i \frac{\partial \lambda(i, x)}{\partial i} di + \left(i \frac{\partial \lambda(i, x)}{\partial x} - F_d \right) dx \qquad (2.37)$$

The differential change in the stored magnetic field energy can be expressed as

$$dW_f(i, x) = \frac{\partial W_f(i, x)}{\partial i} di + \frac{\partial W_f(i, x)}{\partial x} dx \qquad (2.38)$$

Equating terms in Eqs. (2.37) and (2.38), we get

$$\frac{\partial W_f(i, x)}{\partial i} = i \frac{\partial \lambda(i, x)}{\partial i} \qquad (2.39)$$

and

$$\frac{\partial W_f(i, x)}{\partial x} = i \frac{\partial \lambda(i, x)}{\partial x} - F_d$$

$$F_d = \frac{\partial}{\partial x} [i\lambda(i, x) - W_f(i, x)] \qquad (2.40)$$

or

$$= \frac{\partial W_c(i, x)}{\partial x} \qquad (2.41)$$

where

$$W_c(i, x) = i\lambda(i, x) - W_f(i, x) \qquad (2.42)$$

is defined as the *co-energy* because it has the units of energy. It neither exists inside nor outside of the magnetic system but offers a convenient way to express the development of the force in a magnetic system. As is evident from Eq. (2.41), it is the rate of increase of the co-energy with respect to the displacement that results in the developed force. From the definition of co-energy as given by Eq. (2.42), it is obvious that co-energy is the area under the $\lambda - i$ characteristic as shown in Fig. 2.9. Thus, co-energy can also be expressed as

$$W_c(i, x) = \int_0^i \lambda \, di \tag{2.43}$$

Since $\lambda = N\phi$ and $i = \mathcal{F}/N$, the co-energy from Eq. (2.43) can be expressed as a function of $\mathcal{F}$ and x as

$$W_c(\mathcal{F}, x) = \int_0^{\mathcal{F}} \phi \, d\mathcal{F} \tag{2.44}$$

and
$$F_d = \frac{\partial W_c(\mathcal{F}, x)}{\partial x} \tag{2.45}$$

In a linear magnetic system where flux linkages are proportional to the current, the stored magnetic energy and the co-energy are equal, as illustrated in Fig. 2.10. On the other hand, for a saturated magnetic system, the stored magnetic energy is smaller than the co-energy.

All the preceding models are confined to the representation of a linearly operating electromechanical system. However, similar development can be carried out ab initio for rotating electromechanical devices, or the necessary relations can be deduced from the equations for the linear motion. The latter can be easily accomplished by expressing the developed energy in terms of the developed torque as

$$dW_d = T_d \, d\theta \tag{2.46}$$

where $d\theta$ is the differential change in angular displacement in the direction of the developed torque, T_d.

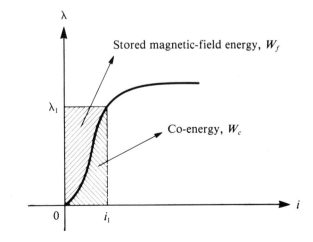

FIGURE 2.9
Graphical Interpretation of the Magnetic-Field Energy and Co-energy

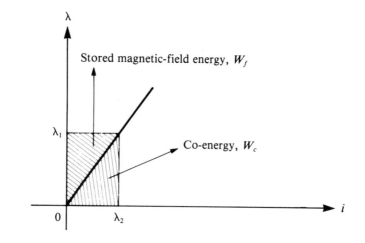

FIGURE 2.10
Magnetic-Field
Energy and Co-
energy in the
Case of a Linear
Magnetic
Medium

The developed torque can be expressed in terms of the decrease in stored magnetic field energy, from Eq. (2.30), as

$$T_d = -\frac{\partial W_f(\lambda, \theta)}{\partial \theta} \tag{2.47}$$

On the other hand, from Eq. (2.41), the developed torque can also be expressed in terms of the increase in co-energy as

$$T_d = \frac{\partial W_c(i, \theta)}{\partial \theta} \tag{2.48}$$

■ **EXAMPLE 2.5**

A magnetic circuit of a plunger has a cross-sectional area of 5 cm² and the dimensions shown in Fig. 2.11a. Calculate the force exerted on the plunger and the co-energy when the distance x is 2 cm and the current in the 100-turn coil is 5 A. The relative permeabilities of magnetic material and nonmagnetic bushing are 2000 and 1, respectively. Assume the fringing and leakage effects are negligible.

Solution

The electrical analog of the magnetic circuit given in Fig. 2.11b allows the calculation of total inductance of the system in terms of the individual reluctances.

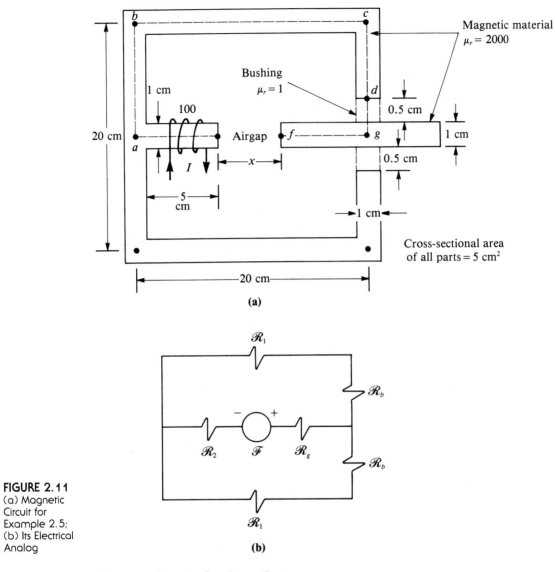

FIGURE 2.11
(a) Magnetic
Circuit for
Example 2.5;
(b) Its Electrical
Analog

The mean lengths for the paths are:

$abcd$: $\quad l_1 = 10 + 20 + 9 = 39 \text{ cm} = 0.39 \text{ m}$

$ae + fg$: $\quad l_2 = 20.5 \text{ cm} - x = 0.205 - x \text{ m}$

dg $\quad l_b = 0.5 \text{ cm} \quad = 0.005 \text{ m}$

ef: $\quad l_g = x \text{ m}$

The cross-sectional area of each path is $A = 5$ cm$^2 = 0.0005$ m^2. Thus, the reluctances are

$$\mathcal{R}_1 = \frac{0.39}{2000 \times 4\pi \times 10^{-7} \times 0.0005} = 310352 \text{ H}^{-1}$$

$$\mathcal{R}_2 = \frac{0.205 - x}{2000 \times 4\pi \times 10^{-7} \times 0.0005} = 163,134 - 795,775x \text{ H}^{-1}$$

$$\mathcal{R}_b = \frac{0.005}{4\pi \times 10^{-7} \times 0.0005} = 7,957,747 \text{ H}^{-1}$$

$$\mathcal{R}_g = \frac{x}{4\pi \times 10^{-7} \times 0.0005} = 1.59155 \times 10^9 x \text{ H}^{-1}$$

The total reluctance is

$$\mathcal{R} = \mathcal{R}_2 + \mathcal{R}_g + 0.5(\mathcal{R}_1 + \mathcal{R}_b)$$

$$= 4,297,184 + 1.590754 \times 10^9 x \text{ H}^{-1}$$

The total inductance is

$$L = \frac{N^2}{\mathcal{R}} = \frac{1}{159,075x + 430} \text{ H}$$

The force developed is

$$F_d = \frac{1}{2} I^2 \frac{\partial L}{\partial x} = -\frac{1}{2} I^2 \frac{159,075}{(159,075x + 430)^2} \text{ N}$$

For $x = 2$ cm $= 0.02$ m and $I = 5$ A,

$$L = 276.89 \text{ } \mu\text{H}$$

and

$$F_d = -0.1525 \text{ N (force of attraction)}$$

The energy stored is

$$W = \frac{1}{2} LI^2 = 0.5 \times 2.76.89 \times 10^{-6} \times 5^2 = 3.46 \text{ mJ}$$

Since the permeability is assumed to be constant, the energy stored in the magnetic field must be equal to the co-energy. The co-energy from Eq. (2.42) with $\lambda = LI$

for a linear magnetic system is

$$W_c = LI^2 - \frac{1}{2} LI^2 = \frac{1}{2} LI^2 = 3.46 \text{ mJ}$$

■ EXAMPLE 2.6

Determine the minimum amount of current required to keep the ferromagnetic plate at a distance of 1 mm from the pole faces of an electromagnet having 1000 turns when the torque exerted by the spring at an effective radius of 20 cm is 20 Newton-meters (N-m) as shown in Fig. 2.12. Assume that each pole face is 3 cm square and the ferromagnetic material is infinitely permeable. Ignore leakage and fringing effects.

Solution

The force exerted on the ferromagnetic plate by a torque of 20 N-m at a radius of 20 cm is $F = 20/0.2 = 100$ N. To balance the exerted force, the magnitude of the force developed by the electromagnet must be 100 N. The total reluctance of the magnetic system is essentially due to the two air gaps, each x meters long and 9 cm^2 in cross section. Thus,

$$\mathcal{R} = \frac{2x}{4\pi \times 10^{-7} \times 9 \times 10^{-4}} = 1.768 \times 10x \text{ H}^{-1}$$

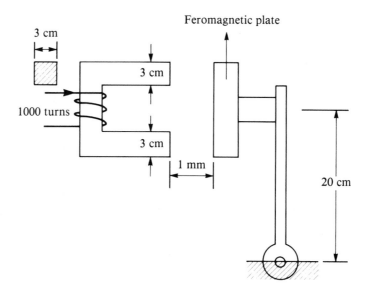

FIGURE 2.12
Electromagnetic
Device for
Example 2.6

The inductance is

$$L = \frac{N^2}{\mathcal{R}} = \frac{1000^2}{1.768 \times 10^9 x} = \frac{5.655 \times 10^{-4}}{x} \text{ H}$$

$$F_d = \frac{1}{2} I^2 \frac{\partial L}{\partial x} = -\frac{1}{2} I^2 \frac{5.655 \times 10^{-4}}{x^2} \text{ N}$$

At $x = 1$ mm $= 0.001$ m, $F_d = -282.74I^2$. Therefore, $282.74I^2 = 100$ or $I = 0.595$ A.

■ EXAMPLE 2.7

An electromechanical system is shown in Fig. 2.13. Develop a set of differential equations governing its behavior after the switch is closed.

Solution

After the switch is closed, the equation of motion can be written as

$$m \frac{du}{dt} = F_d - vu - k(l - x)$$

where m, u, v, and x are the mass, velocity, coefficient of friction, and distance of the movable part from the electromagnet, respectively. The spring constant is denoted by k. However,

$$u = -\frac{dx}{dt}$$

where the negative sign accounts for the decrease in x.

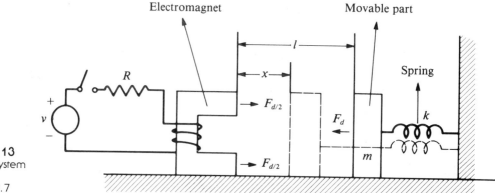

FIGURE 2.13
Dynamic System
Given in
Example 2.7

The equation satisfying the electric circuit is

$$v = \frac{d\lambda}{dt} + Ri$$

where

$$\lambda = Li.$$

Neglecting fringing and leakage, the inductance of the magnetic circuit can be expressed as

$$L = \frac{\mu_0 N^2 A}{\mu_0 A \mathscr{R}_m + 2x}$$

where $\mathscr{R}_m$ is the reluctance of the magnetic material. The force developed is

$$F = -\frac{\partial W_f(\lambda, x)}{\partial x} = -\frac{1}{2}\frac{\partial}{\partial x}\left(\frac{\lambda^2}{L}\right) = -\frac{\lambda^2}{\mu_0 A N^2}$$

Finally, the set of differential equations that govern the behavior of the given system can be summarized as follows:

$$\frac{du}{dt} = -\frac{1}{m\mu_0 A N^2}\lambda^2 - \frac{v}{m}u + \frac{k}{m}x - \frac{k}{m}l$$

$$\frac{dx}{dt} = -u$$

$$\frac{d\lambda}{dt} = -\frac{R\mathscr{R}_m}{N^2}\lambda - \frac{2R}{N^2\mu_0 A}x\lambda + v$$

These differential equations are nonlinear in terms of the independent variables x and λ and can be solved numerically.

2.3   RELUCTANCE MOTOR

To illustrate the application of electromechanical energy-conversion principles to rotating machines, consider an elementary, singly excited, 2-pole reluctance motor as shown in Fig. 2.14. It consists of two parts, a freely rotating member (rotor) revolving under the poles of a stationary magnetic circuit (stator), which carries the excitation winding as depicted.

If the angular displacement θ between the rotor and the stator magnetic axes is zero (which is referred to as the direct or d-axis position), the effective air gap and thereby the reluctance of the magnetic system will be minimum, as is evident from Fig. 2.14. To simplify the theoretical development, assume that the stator

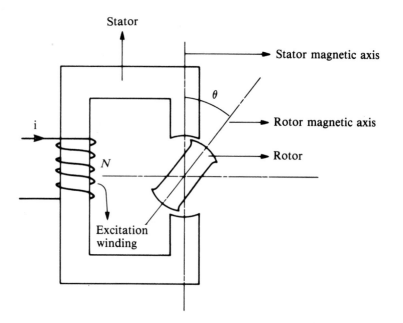

FIGURE 2.14
Reluctance
Motor

and rotor offer negligible reluctances in comparison with the reluctance of the air gap. Thus, at $\theta = 0$, the effective reluctance of the motor is

$$\mathcal{R}(0) = \frac{2g}{\mu_0 A} \tag{2.49}$$

where g and A are the effective air-gap length and area per pole, respectively.

Since the inductance is inversely proportional to the reluctance, the inductance corresponding the previous position is maximum. However, when the stator and the rotor magnetic axes are at right angles to each other (which is the quadrature or q-axis position), the reluctance is maximum, leading to a minimum inductance. As the rotor revolves with a uniform speed, ω_m, the inductance will go through maxima and minima as shown in Fig. 2.15. The variation in inductance

FIGURE 2.15
Variation of the
Inductance as
the Rotor
Rotates in a
Reluctance
Motor

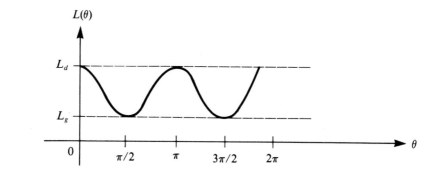

as a function of angular displacement can be obtained by Fourier series analysis as

$$L(\theta) = 0.5(L_d + L_q) + 0.5(L_d - L_q) \cos 2\theta \qquad (2.50)$$

The torque developed by the reluctance motor can be obtained from the co-energy expression given in Eq. (2.48) as

$$T_d = \frac{\partial W_c(i, \theta)}{\partial \theta} = \frac{1}{2} i^2 \frac{\partial L}{\partial \theta}$$

$$= -\frac{1}{2} i^2 (L_d - L_q) \sin 2\theta \qquad (2.51)$$

However, θ can be expressed as

$$\theta = \omega_m t + \delta \qquad (2.52)$$

where δ is the initial position of the rotor magnetic axis with respect to the stator magnetic axis. Thus, the torque developed can be rewritten as

$$T = -\frac{1}{2} i^2 (L_d - L_q) \sin[2(\omega_m t + \delta)] \qquad (2.53)$$

It is evident that the initial torque developed by the motor will be zero if $\delta = 0°$ and will be maximum when $\delta = 45°$. The presence of the negative sign in the torque expression highlights the fact that the torque tends to align the rotor under the nearest pole of the stator and thereby defines the direction of rotation. As is evident from Eq. (2.53), the torque developed by the motor will vanish for a round rotor because L_d will be equal to L_q.

■ EXAMPLE 2.8

The current intake of a 2-pole reluctance motor at 60 Hz is 10 A (rms), and its inductances are $L_d = 2$ H and $L_q = 1$ H. Develop a criterion for the motor speed such that the average torque developed is maximum.

Solution

Substituting the given values in Eq. (2.53), the torque developed by the motor is

$$T_d = -100 \cos^2 377t \sin[2(\omega_m t + \delta)]$$

Using the following trigonometric identities,

$$2 \cos^2 x = 1 + \cos 2x$$

and
$$2 \sin x \cos y = \sin(x + y) + \sin(x - y)$$

the torque expression can be rewritten as

$$T_d = 50\{\sin[2(\omega_m t + \delta)] + 0.5 \sin[2(377 + \omega_m)t + 2\delta]$$
$$+ 0.5 \sin[2(\omega_m - 377)t + 2\delta]\}$$

It is apparent from this expression that the average torque developed by the motor will be zero unless $\omega_m = 377$ rad/s and $\delta \neq n\pi$, where $n = 0, 1, 2, \ldots$. Therefore, for the development of an average torque the rotor must rotate at an angular velocity equal to the angular frequency of the supply. This, in fact, is known as the synchronous speed of the motor. Thus, the torque developed at the synchronous speed is

$$T_d = -25 \sin 2\delta$$

which is maximum when $\delta = 45°$.

PROBLEMS

2.1. The energy stored in a fixed, air-filled, parallel-plate capacitor is 3 J. The potential difference between the plates is 100 kV and the separation is 2 cm. Calculate the surface area of each plate and the force of attraction between them.

2.2. The upper plate of a parallel-plate capacitor is stationary, while the lower plate with a mass of 10 g is free to move and is held at a potential of 50 kV with respect to ground. The surface area of each plate is 50 cm². For an air-gap spacing of 2 cm, determine the necessary voltage that must be applied to the upper plate with respect to ground to keep the plates stationary.

2.3. The energy stored in a magnetic circuit is expressed as $W_f = -5 \ln \lambda - \lambda$. What is the relationship between the applied current and the flux linkages? Calculate the inductance of the magnetic circuit for a current of 2 A.

2.4. Calculate the force of attraction acting on the movable part shown in Fig. P2.4. The current applied to the 100-turn coil is 20 A and the relative permeability of the magnetic material is 1000.

2.5. An electromagnet with a relative permeability of 1000 and a uniform cross-sectional area of 150 cm² has the dimensions given in Fig. P2.5. If the current applied to the series-connected coils is 100 A, determine whether the elec-

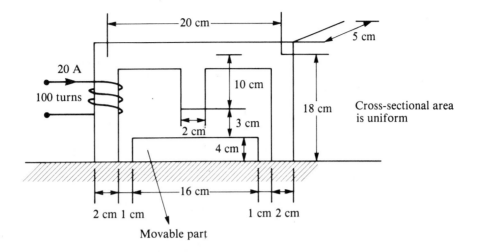

FIGURE P2.4
Magnetic Circuit
for Problem 2.4

Movable part

tromagnet is capable of lifting a ferromagnetic bar having a relative permeability of 300 and a mass of 100 kg.

2.6. The magnetic circuit of a plunger has the dimensions shown in Fig. P2.6. Calculate (a) the total flux linkages, (b) the stored energy in the magnetic circuit, and (c) the force of attraction acting on the plunger. The relative permeability of the magnetic material is 1500.

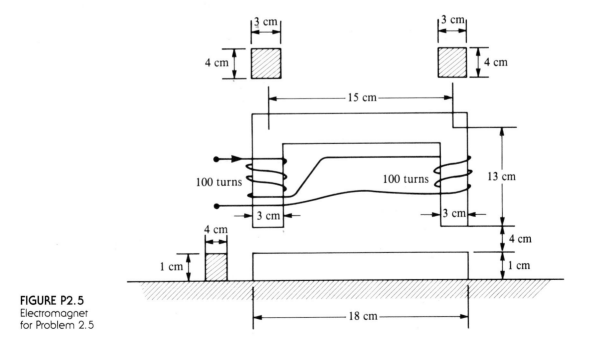

FIGURE P2.5
Electromagnet
for Problem 2.5

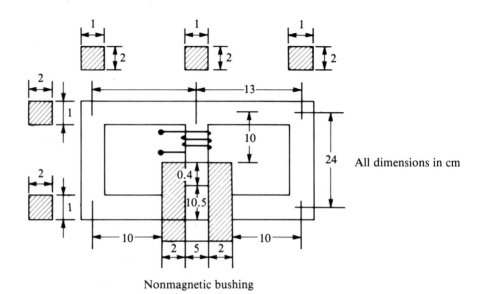

FIGURE P2.6
Magnetic Circuit
of the Plunger
for Problem 2.6

All dimensions in cm

Nonmagnetic bushing

2.7. The 100-turn coil of the electromagnetic device shown in Fig. P2.7 carries a current of $i(t) = 15 \cos 377t$ A. Calculate the average value of the force of attraction exerted on a fixed magnetic bar. The relative permeability of the magnetic material is 800.

2.8. A U-shaped electromagnet is required to lift an iron bar that is at a distance of 1 mm. The cross-sectional area of each pole is 12 cm^2 and the flux density in the air gap is 0.8 T. Determine the pulling force developed by the magnet. Assume that the electromagnet and iron have no reluctances.

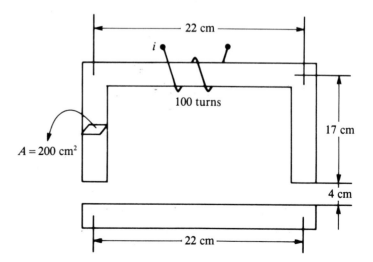

FIGURE P2.7
Electromagnet
for Problem 2.7

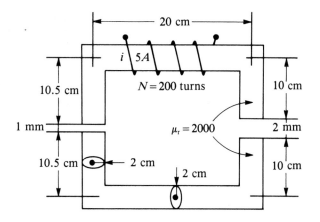

FIGURE P2.9
Magnet Structure
for Problem 2.9

2.9. The dimensions of a magnetic circuit with a uniform circular cross section having a radius of 2 cm are shown in Fig. P2.9. What is the force acting on each pole? Determine the stored energy in each nonmagnetic region.

2.10. The necessary dimensions of a cylindrical electromagnet in close contact with a 2-cm-thick iron disc are shown in Fig. P2.10. The specific gravity of iron is 7.85 g/cm³ and its relative permeability is 500. Determine the current in the 500-turn coil. The relative permeability of the electromagnet is 2100.

2.11. The effective reluctance of a 2-pole reluctance motor is given by

$$\mathcal{R}(\theta) = 1500 - 850 \cos 2\theta$$

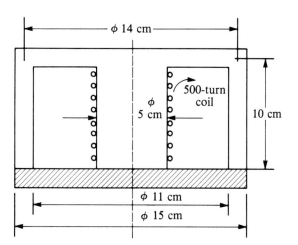

FIGURE P2.10
Cylindrical
Electromagnet
for Problem
2.10

Determine the developed torque by the motor when the current in the 150-turn coil is 5 A and the rotor magnetic axis is at 25° with respect to the stator magnetic axis. What must the initial rotor position be for the developed torque to be maximum?

2.12. If the reluctance motor given in Problem 2.11 is operated from a current source of 5 sin 314t A, determine (a) the speed, and (b) the average torque developed by the motor.

CHAPTER 3

Transformers

In its simplest form, a transformer consists of two insulated windings usually wound over a magnetic material with relatively high permeability. Highly permeable magnetic materials are used to ensure the most effective flux linkages between the windings on one hand and a low reluctance path on the other. To keep the core loss to minimum levels, thin laminations are used for the magnetic core of the transformer. For special applications such as radio and communication equipment, however, the core of the transformer may be made of nonmagnetic materials. In such cases, the transformer is referred to as an *air-core* transformer.

When a current flows through one of the windings, it produces a magnetic flux in the magnetic material, which links more or less completely with the turns of the other winding, thereby inducing a voltage in it. The frequency of the induced voltage in the second winding is the same as that of the current in the first winding. The winding that receives energy from the power source is defined as the *primary*, while the winding that delivers energy to the load is known as the *secondary*. Either winding may be referred to as the primary, while the other acts as the secondary. Since the magnitude of the induced voltage in the secondary depends on the number of turns in that winding, the induced voltage may be less than, equal to, or greater than the voltage applied to the primary winding.

When the applied voltage on the primary is equal to the induced voltage in the secondary, the transformer is said to have a *one-to-one ratio* and is employed basically for the purpose of electrically isolating the secondary side of the circuit from its primary side. Such a transformer is usually called an isolation transformer.

Since the voltage induced in the secondary winding is due to the changes in the magnetic flux, a transformer can be utilized for dc isolation. That is, if the input voltage or current in the primary side consists of both dc and ac components, the voltage and current on the secondary side will be purely alternating in nature.

When the voltage on the secondary side of the transformer is greater than that on the primary side, the transformer is referred to as a *step-up* transformer. On the other hand, a *step-down* transformer has low voltage on the secondary side. To minimize the copper losses of a transmission line, electric energy is transmitted at high voltage levels over long distances until it reaches the consumer. Then, at the premises of the consumer, the voltage is reduced to a level that is the standard for the community. Thus, a step-up transformer is used at the generation site and a step-down transformer is installed at the consumer's premises.

Transformers are also used for impedance matching. A known impedance can be raised or lowered to match the rest of the circuit for maximum power transfer.

3.1 AN IDEAL TRANSFORMER

For the sake of simplicity, let us consider a two-winding transformer as shown in Fig. 3.1, where the primary winding with N_1 turns is connected to a time-varying voltage source v_1, and the secondary winding with N_2 turns is left open.

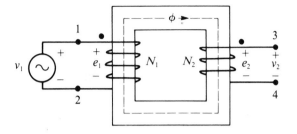

FIGURE 3.1
Elementary
Two-Winding
Transformer

In the case of an ideal transformer, let us postulate that:

1. The magnetic material is infinitely permeable and does not saturate.
2. The magnetic core does not exhibit any core loss.
3. All the flux is confined within the magnetic core.
4. Each winding has no resistance.

According to Faraday's law of induction, the magnetic flux ϕ in the core induces a voltage, e_1, in the primary winding, which opposes the applied voltage. For the polarities of the applied and induced voltages as shown in the figure, we can write

$$v_1 = e_1 = N_1 \frac{d\phi}{dt} \tag{3.1}$$

Similarly, the induced voltage in the secondary winding is

$$e_2 = v_2 = N_2 \frac{d\phi}{dt} \tag{3.2}$$

It is obvious from Fig. 3.1 that terminal 1 of the primary winding is positive with respect to its terminal 2, while terminal 3 of the secondary is positive with respect to its terminal 4 at any given instant. Since terminal 3 is of the same polarity as terminal 1, they are said to follow each other. Similarly, we can say that terminal 4 of the secondary winding follows terminal 2 of the primary winding and, therefore, they are like polarity terminals. To indicate the like polarity relationship, either terminals 1 and 3 or terminals 2 and 4 may be indicated by dots. In Fig. 3.1, terminals 1 and 3 are dotted.

Dividing Eq. (3.1) by Eq. (3.2), we obtain

$$\frac{v_1}{v_2} = \frac{e_1}{e_2} = \frac{N_1}{N_2} \tag{3.3}$$

The ratio of the number of turns in the primary to that of the secondary is defined

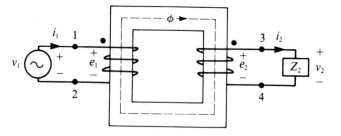

FIGURE 3.2
Ideal
transformer
under Load
Conditions

as the ratio of transformation or simply the a ratio. That is,

$$\frac{N_1}{N_2} = a \tag{3.4}$$

Let i_1 and i_2 be the currents through the primary and secondary windings, respectively, when the latter is connected to a load as shown in Fig. 3.2. The mmfs of the primary and the secondary windings are

$$\mathscr{F}_1 = N_1 i_1 \tag{3.5}$$

and

$$\mathscr{F}_2 = N_2 i_2 \tag{3.6}$$

respectively. In an ideal transformer, the net mmf must be zero due to its infinite permeability. Therefore,

$$N_1 i_1 - N_2 i_2 = 0$$

or

$$\frac{i_2}{i_1} = \frac{N_1}{N_2} = a \tag{3.7}$$

From Eqs. (3.3) and (3.7),

$$\frac{v_1}{v_2} = \frac{i_2}{i_1}$$

or

$$v_1 i_1 = v_2 i_2 \tag{3.8}$$

Equation (3.8) highlights the fact that at any instant t the output power is the same as the input power. This result is a natural consequence of the assumed conditions that there are no ohmic or core losses in the ideal transformer.

For sinusoidal variations in the impressed voltage, the magnetic flux in the core will also vary sinuusoidally under ideal conditions. Therefore, the magnetic flux at any instant t may be represented by the equation

$$\phi = \phi_m \sin \omega t \tag{3.9}$$

where ϕ_m is the maximum value of the magnetic flux and ω is the angular frequency.

Equation (3.1) can now be written as

$$e_1 = N_1 \omega \phi_m \cos \omega t \tag{3.10}$$

Equation (3.10) can be expressed in phasor form in terms of its rms or effective value, as

$$\tilde{E}_1 = \frac{1}{\sqrt{2}} N_1 \omega \phi_m \underline{/0^\circ} = 4.44 f N_1 \phi_m \underline{/0^\circ} \tag{3.11}$$

In the same way, we can derive the following equation for the secondary side:

$$\tilde{E}_2 = 4.44 f N_2 \phi_m \underline{/0^\circ} \tag{3.12}$$

From Eqs. (3.11) and (3.12), we get

$$\frac{\tilde{V}_1}{\tilde{V}_2} = \frac{\tilde{E}_1}{\tilde{E}_2} = \frac{N_1}{N_2} = a \tag{3.13}$$

where $\tilde{V}_1 = \tilde{E}_1$ and $\tilde{V}_2 = \tilde{E}_2$ under ideal conditions. Since $\tilde{E}_1$ and $\tilde{E}_2$ are in phase, this relation can also be written in terms of their magnitudes.

It can also be shown that

$$\frac{\tilde{I}_2}{\tilde{I}_1} = \frac{N_1}{N_2} = a \tag{3.14}$$

where $\tilde{I}_1$ and $\tilde{I}_2$ are the currents in the primary and secondary windings, respectively. Once again, $\tilde{I}_1$ and $\tilde{I}_2$ are in phase for an ideal transformer.

Equation (3.8) can be expressed in terms of phasor quantities as

$$\tilde{V}_1 \tilde{I}_1^* = \tilde{V}_2 \tilde{I}_2^* \tag{3.15}$$

That is, the complex power supplied to the primary is equal to the complex power delivered by the secondary to the load. In terms of the apparent powers, Eq. (3.15) can be written as

$$V_1 I_1 = V_2 I_2 \tag{3.16}$$

Let $\tilde{Z}_2$ be the load impedance on the secondary side such that

$$\hat{Z}_2 = \frac{\tilde{V}_2}{\tilde{I}_2} \tag{3.17}$$

$$= \frac{1}{a^2} \frac{\tilde{V}_1}{\tilde{I}_1}$$

$$= \frac{1}{a^2} \hat{Z}_1$$

or
$$\hat{Z}_1 = a^2 \hat{Z}_2 \tag{3.18}$$

where $\hat{Z}_1 = \tilde{V}_1/\tilde{I}_1$ is the load impedance as referred to the primary side. Thus, any impedance on the secondary side can be referred to the primary side by multiplying it by the square of the ratio of transformation. This forms the basis to utilize transformers for impedance matching.

■ EXAMPLE 3.1

The maximum flux density in the core of a 440/4000-V, 60-Hz, single-phase transformer is 0.8 T. If the induced voltage per turn is 10 V, determine (a) the primary and the secondary turns, and (b) the cross-sectional area of the core.

Solution

The induced voltage in the primary is

$$E_1 = N_1 \times \text{(emf induced per turn)}$$

$$440 = N_1 \times 10$$

Thus,
$$N_1 = 44 \text{ turns}$$

Similarly,
$$N_2 = \frac{4400}{10} = 440 \text{ turns}$$

Since
$$E_1 = 4.44 f N_1 \phi_m = 4.44 f N_1 B_m A,$$

$$440 = 4.44 \times 60 \times 44 \times 0.8 \times A$$

Thus, the cross-sectional area is: $A = 0.047 \text{ m}^2$

■ EXAMPLE 3.2

A coil having 200 turns is immersed in a 60-Hz magnetic flux with a peak value of 4 mWb. Obtain the expression for the instantaneous value of the induced voltage.

Solution

The angular frequency is

$$\omega = 2\pi f = 2 \times \pi \times 60 = 377 \text{ rad/s}$$

The rms value of the induced voltage is

$$E = 4.44 f N \phi_m = 4.44 \times 60 \times 200 \times 0.004 = 213 \text{ V}$$

The peak value is

$$E_m = \sqrt{2}\, E = 1.414 \times 213 = 301 \text{ V}$$

Thus, the expression for the instantaneous value of the induced voltage is

$$e = 301 \cos(377t) \text{ V}$$

■ EXAMPLE 3.3

The core of a two-winding transformer shown in Fig. 3.3a is subjected to the magnetic-flux variations as shown in Fig. 3.3b. Calculate the induced voltages in the winding.

Solution

Since the polarities are already marked on the windings, when terminal b becomes positive with respect to terminal a, terminal c is also positive with respect to terminal d.

For the time interval from 0 to 0.06 s, the magnetic flux increases linearly and is given by the following equation:

$$\phi = \frac{0.009}{0.06} t = 0.15t$$

Thus the voltage between terminals a, b is

$$e_{ab} = -e_{ba} = -N_{ab} \frac{d\phi}{dt} = -200 \times 0.15 = -30 \text{ V}$$

and the voltage between terminals c, d is

$$e_{cd} = N_{cd} \frac{d\phi}{dt} = 500 \times 0.15 = 75 \text{ V}$$

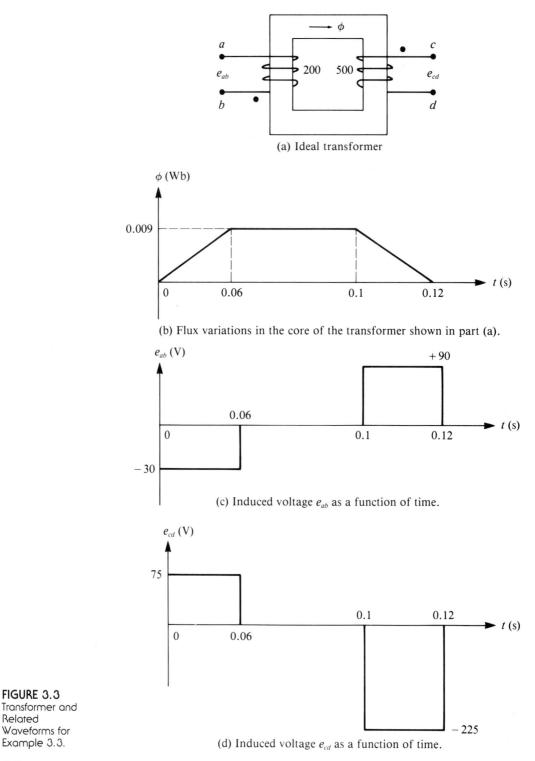

(a) Ideal transformer

(b) Flux variations in the core of the transformer shown in part (a).

(c) Induced voltage e_{ab} as a function of time.

(d) Induced voltage e_{cd} as a function of time.

FIGURE 3.3
Transformer and
Related
Waveforms for
Example 3.3.

From 0.06 to 0.1 s, the voltage induced is zero as there is no variation in the flux. In the time interval from 0.1 to 0.12 s, it can be shown that the rate of change of magnetic flux in the core is -0.45 Wb/s. The induced voltages are

$$e_{ab} = -200 \times (-0.45) = 90 \text{ V}$$

and

$$e_{cd} = 500 \times (-0.45) = -225 \text{ V}$$

The voltage waveforms are shown in Figs. 3.3c and 3.3d for the ab and cd windings, respectively.

■ EXAMPLE 3.4

An ideal transformer has 30 turns on the primary and 750 turns on the secondary. The primary winding is connected to a 240-V, 50-Hz source. The secondary winding supplies a load of 4 A at a lagging power factor of 80%. Determine (a) the ratio of transformation, (b) the current in the primary winding at rated load, (c) the magnetic flux in the core. (d) Draw the phasor diagram.

Solution

(a) The ratio of transformation is $a = 30/750 = 0.04$.

(b) Since $I_2 = 4$ A, $I_1 = I_2/a = 4/0.04 = 100$ A.

(c) For an ideal transformer, the maximum value of the magnetic flux in the core is given by

$$\phi_m = \frac{E_1}{4.44 f N_1} = \frac{240}{4.44 \times 50 \times 30} = 0.036 \text{ Wb}$$

(d) Since the power factor is 80% lagging, the power factor angle is

$$\theta = \cos^{-1}(0.8) = 36.9° \text{ (lag)}$$

Lagging angles are usually written with a negative sign. Thus, $\theta = -36.9°$. The phasor diagram is shown in Fig. 3.4, where $V_2 = V_1/a = 240/0.04 = 6000$ V.

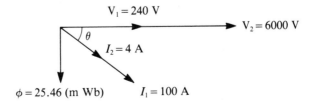

FIGURE 3.4
Phasor Diagram
for Example 3.4

3.2 IRON-CORE TRANSFORMER

In the discussion of an ideal transformer, it was assumed that all the flux produced by the primary winding was linking the secondary winding. In practice, such a condition is very difficult to realize. To maximize the flux linkages between the primary and the secondary, the two windings are usually wound over the same leg of the transformer as in the construction of a shell-type transformer, shown in Fig. 3.5. In the core-type transformer, shown in Fig. 3.6, each winding is evenly split and wound one upon the other on both legs of the rectangular core. The nomenclature is derived from the fact that in a shell-type transformer the core encircles the windings, whereas the windings encircle the core in a core-type transformer. Placing one winding over the other is an attempt to ensure that almost all the flux links both windings. However, a part of the magnetic flux that links only one of the windings and completes its path largely in air is known as the *leakage flux*. Therefore, when both windings are carrying currents, each produces its own leakage flux, as illustrated in Fig. 3.7. The flux that remains in the magnetic core and links both the windings is termed the *mutual flux*.

Although it is a small fraction of the total flux produced by a winding, the leakage flux does affect the behavior of the transformer. The total flux produced by a winding is the sum of the mutual flux and the leakage flux for that winding. Thus, we can model a winding by two windings in such a way that one part is responsible for the production of mutual flux, while the other accounts for the leakage flux. If we redefine the leakage flux as that value of the flux that links all the turns of the winding producing it, the two windings will have the same number of turns as the original winding. Such hypothetical winding arrangements are shown in Fig. 3.8 for a two-winding transformer. The winding surrounding the core sets up the mutual flux, while the other accounts for the leakage flux. The two windings encircling the core now satisfy the conditions for an ideal transformer.

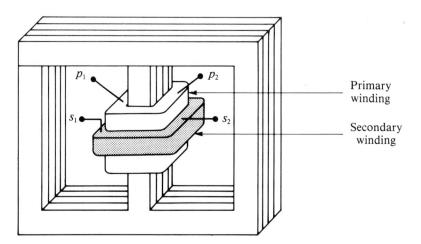

FIGURE 3.5
Shell-Type
Transformer

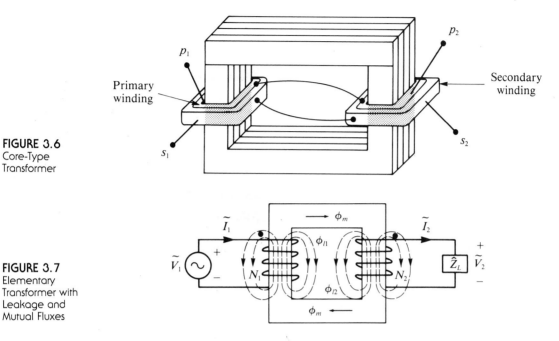

FIGURE 3.6
Core-Type
Transformer

FIGURE 3.7
Elementary
Transformer with
Leakage and
Mutual Fluxes

We can now determine the induced voltage by the leakage flux in the primary winding as

$$e_{l1} = N_1 \frac{d\phi_{l1}}{dt}$$

$$= N_1 \frac{d\phi_{l1}}{di_1} \frac{di_1}{dt}$$

$$= L_{l1} \frac{di_1}{dt} \tag{3.19}$$

where L_{l1} is the leakage inductance as defined in Chapter 1, and i_1 is the current in the primary winding.

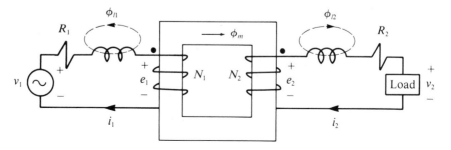

FIGURE 3.8
Hypothetical
Windings
Showing Mutual
and Leakage-
Flux Linkages
Separately

Similarly, the voltage induced in the secondary winding by its leakage flux will be

$$e_{l2} = L_{l2} \frac{di_2}{dt} \tag{3.20}$$

where L_{l2} is the leakage inductance of the secondary winding and i_2 is the corresponding current.

For the sinusoidal variations in the currents, Eqs. (3.19) and (3.20) can be expressed in phasor form as

$$\tilde{E}_{l1} = j\tilde{I}_1 X_1 \tag{3.21}$$

and

$$\tilde{E}_{l2} = j\tilde{I}_2 X_2 \tag{3.22}$$

where X_1 and X_2 are the leakage reactances and $\tilde{I}_1$ and $\tilde{I}_2$ are the currents in the primary and secondary windings, respectively.

The equivalent circuit of the transformer, including the winding resistance and leakage reactance, is given in Fig. 3.9.

■ EXAMPLE 3.5

The secondary of a transformer has 180 turns and carries a current of 18 A, 60 Hz at full load. The peak value of the mutual flux is 20 mWb, and the corresponding leakage flux associated with the secondary is 3 mWb. Calculate: (a) the voltage induced in the secondary by the leakage flux, (b) the voltage induced in the secondary by the mutual flux, (c) the leakage reactance of the secondary winding.

Solution

(a) The voltage induced by the secondary leakage flux is

$$E_{l2} = 4.44 f N_2 \phi_{l2}$$

$$= 4.44 \times 60 \times 180 \times 0.003 = 143.9 \text{ V}$$

(b) The voltage induced in the secondary winding by the mutual flux is

$$E_2 = 4.44 \times 60 \times 180 \times 0.02 = 959 \text{ V}$$

(c) The leakage reactance of the secondary winding is

$$X_2 = \frac{E_{l2}}{I_2} = 143.9/18 = 8 \text{ } \Omega$$

Since the core of a practical transformer is not ideal, it will have finite

FIGURE 3.9
Equivalent
Circuit for a
Transformer
Including
Winding
Resistance and
Leakage
Reactance

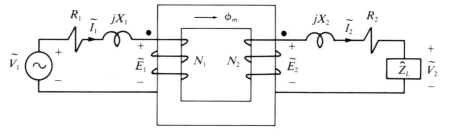

permeability and exhibit core loss. Therefore, even when the secondary is left open (no-load condition) and voltage is impresed on the primary winding, there will be some current, known as the excitation current $\tilde{I}_\phi$, in the primary winding. Under no-load condition, the induced emf $\tilde{E}_1$ by the mutual flux ϕ_M differs from the applied voltage $\tilde{V}_1$ by the impedance drop $\tilde{I}_1 \hat{Z}_1$, where $\hat{Z}_1 = R_1 + jX_1$ as is evident from Fig. 3.9. If the secondary winding supplies the current $\tilde{I}_2$ to the load, it will cause the current in the primary winding to increase from its no-load value. Such an increase in the primary winding current results in larger voltage drop across the leakage impedance. If the impressed voltage is held constant, the induced emf $\tilde{E}_1$ will drop. However, in the normal operating range of most transformers the leakage impedance drop is within 3% to 5% of $\tilde{V}_1$. It can therefore be said that $\tilde{E}_1$ remains substantially constant. In other words, the mutual flux, ϕ_M, is essentially the same under normal loading conditions. Since most transformers are designed to operate near the knee of the magnetization curve, the magnetization current $\tilde{I}_m$, that is, the current required to establish ϕ_M, is also constant. Since core loss depends on ϕ_M, it can also be considered as constant. Therefore, the core-loss component of the excitation current, $\tilde{I}_c$, remains nearly constant. For all practical purposes, we can say that the excitation current is proportional to $\tilde{E}_1$ and is given by

$$\tilde{I}_\phi = \frac{\tilde{E}_1}{\hat{Z}_\phi} \tag{3.23}$$

where $\hat{Z}_\phi$ is the equivalent exciting impedance.

It is a common practice to represent $\hat{Z}_\phi$ by a parallel combination of an equivalent resistance R_c, which accounts for the core loss and a magnetizing reactance X_m to account for the mmf drop in the core of the transformer. The equivalent circuit of the transformer, including R_c and X_m, is given in Fig. 3.10. It is obvious from the figure that

$$\tilde{I}_\phi = \tilde{I}_c + \tilde{I}_m$$

$$= \frac{\tilde{E}_1}{R_c} + \frac{\tilde{E}_1}{jX_m} \tag{3.24}$$

$\tilde{I}_c$ is in phase with $\tilde{E}_1$, while $\tilde{I}_m$ lags $\tilde{E}_1$ by 90°.

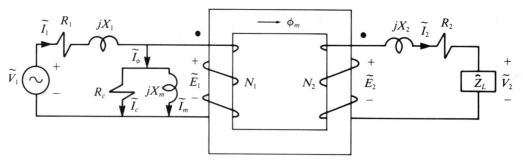

FIGURE 3.10
Equivalent Circuit for a Transformer, Including Winding and Exciting Impedances

In the equivalent-circuit representation of a transformer, the magnetic core is rarely shown. Sometimes, parallel lines are drawn between the two windings to highlight the presence of a magnetic core. We will use such an equivalent-circuit representation. If the parallel lines between the two windings are missing, our interpretation would be that a nonmagnetic core is being used for the transformer. With that understanding, the exact equivalent circuit of our iron-core transformer is given in Fig. 3.11. In this figure, a dashed box is also drawn to show that the circuit enclosed by it is none other than our so-called ideal transformer. All the ideal-transformer relationships can be applied to this circuit. The load current $\tilde{I}_2$ on the secondary side is represented on the primary side as $\tilde{I}_2'$.

Since the excitation current is supplied by the primary circuit, the difference between the mmfs of the primary and secondary windings must be equal to the mmf required for exciting the transformer. That is,

$$\tilde{I}_\phi N_1 = \tilde{I}_1 N_1 - \tilde{I}_2 N_2$$

$$\tilde{I}_\phi = \tilde{I}_1 - \frac{\tilde{I}_2}{a}$$

or
$$\tilde{I}_\phi = \tilde{I}_1 - \tilde{I}_2' \qquad\qquad (3.25)$$

FIGURE 3.11
Exact Equivalent Circuit for a Transformer with Parallel Lines for the Magnetic Core

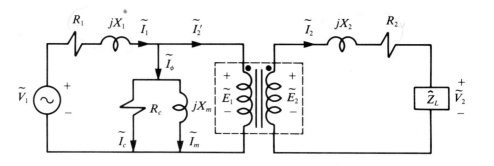

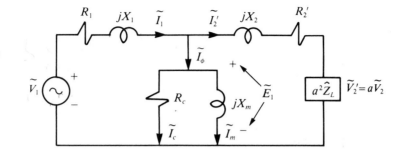

FIGURE 3.12
Equivalent
Circuit for a
Transformer as
Referred to the
Primary Side

It is possible to represent a transformer with an equivalent circuit that does not require an ideal transformer. Such equivalent circuits are drawn with reference to a given winding. Figure 3.12 shows such an equivalent circuit referred to the primary winding of the transformer. Note that the circuit elements that were on the primary side remain unchanged. Only the circuit elements on the secondary side have been transformed to the primary side. Figure 3.13 shows another equivalent-circuit representation, but referred to the secondary side.

■ EXAMPLE 3.6

The primary of a 2400/120-V, 60-Hz transformer has a resistance of 0.5 Ω and a leakage reactance of 1.2 Ω. What is the induced emf in the primary when the primary current is 80 A at a lagging power factor of 0.707?

Solution

The impedance of the primary winding is

$$\hat{Z}_1 = 0.5 + j1.2 = 1.3\underline{/67.38°}\ \Omega$$

The primary winding current is

$$\tilde{I}_1 = 80\ \underline{/-45°}\ \text{A}$$

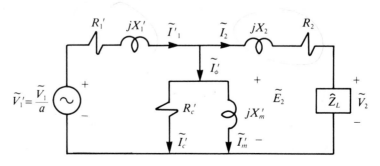

FIGURE 3.13
Equivalent
Circuit for a
Transformer as
Referred to the
Secondary Side

Since $\tilde{V}_1 = \tilde{E}_1 + \tilde{I}_1 \hat{Z}_1$,

$$\tilde{E}_1 = \tilde{V}_1 - \tilde{I}_1 \hat{Z}_1 = 2400 - (80\underline{/-45°}) \times (1.3\underline{/67.38°})$$

$$= 2400 - 104\underline{/22.38°}$$

$$= 2304.2\underline{/-0.99°} \text{ V}$$

■ EXAMPLE 3.7

A 2-kVA, 500/100-V, 60-Hz single-phase transformer draws a no-load current of 1.0 A at a lagging power factor of 0.16. If the power factor (pf) of the load is 0.8 lagging, determine the secondary and primary winding currents at full load.

$P = VI$

$I = P|V$

Solution

At no-load, $\tilde{I}_\phi = 1.0\underline{/-80.8°}$ A. At full load the current in the secondary winding is $I_2 = 2000/100 = 20$ A at 0.8 pf (lag).

Thus, $\tilde{I}_2 = 20\underline{/-36.87°}$ A

The nominal turns ratio is $a = 500/100 = 5$.

$$\tilde{I}_2' = \frac{\tilde{I}_2}{a} = 4\underline{/-36.87°} \text{ A}$$

The current in the primary winding is

$$\tilde{I}_1 = \tilde{I}_2' + \tilde{I}_\phi$$

$$= 4\underline{/-36.87°} + 1\underline{/-80.8°}$$

$$= 4.77\underline{/-45.23°} \text{ A}$$

Thus, at full load the current in the primary winding is 4.77 A at a lagging power factor of 0.704.

■ EXAMPLE 3.8

A 2500/500-V transformer has $R_1 = 2.5 \ \Omega$, $X_1 = 7.5 \ \Omega$, $R_2 = 0.05 \ \Omega$, $X_2 = 0.25 \ \Omega$, and no-load current of 2 A at a lagging power factor of 0.16. The secondary supplies a rated load of 50 A at a 0.8 lagging power factor. What is the applied voltage?

Solution

The transformation ratio is $a = 2500/500 = 5$.

$$\tilde{I}_2 = 50\underline{/-36.87°} \text{ A}$$

$$\hat{Z}_2 = 0.05 + j0.25 = 0.255\underline{/78.69°} \ \Omega$$

$$\tilde{E}_2 = \tilde{V}_2 + \tilde{I}_2\hat{Z}_2$$

$$= 500 + (50\underline{/-36.87°}) \times (0.255\underline{/78.69°})$$

$$= 509.57\underline{/0.96°} \text{ V}$$

$$\tilde{E}_1 = a\tilde{E}_2 = 2547.85\underline{/0.96°} \text{ V}$$

$$\tilde{I}'_2 = \frac{50}{5}\underline{/-36.87°} = 10\underline{/-36.87°} \text{ A}$$

$$\tilde{I}_1 = \tilde{I}'_2 + \tilde{I}_\phi = 10\underline{/-36.87°} + 2\underline{/-80.8°} = 11.52\underline{/-43.77°} \text{ A}$$

$$\hat{Z}_1 = 2.5 + j7.5 = 7.906\underline{/71.57°} \ \Omega$$

$$\tilde{V}_1 = \tilde{E}_1 + \tilde{I}_1\hat{Z}_1$$

$$= 2547.85\underline{/0.96°} + (11.52\underline{/-43.77°}) \times (7.906\underline{/71.57°})$$

$$= 2629.4\underline{/1.86°} \text{ V}$$

When a transformer, as shown in Fig. 3.11, operates under steady-state conditions, an insight into its currents, voltages, and phase relationships can be easily obtained with the help of its phasor diagram. Even though a phasor diagram can be developed by using any phasor quantity as a reference, use of the load voltage as a reference in analyzing transformers is highly recommended.

Let $\tilde{V}_2$ be the voltage across the load impedance $\hat{Z}_L$ and $\tilde{I}_2$ be the load current. Depending on $\hat{Z}_L$, the load current may be in phase, leading, or lagging the load voltage $\tilde{V}_2$. In this particular case, let us assume that $\tilde{I}_2$ lags $\tilde{V}_2$ by an angle of θ_2. Let us draw a horizontal line to represent the magnitude of the phasor $\tilde{V}_2$, being a reference phasor, as shown in Fig. 3.14. The current phasor, $\tilde{I}_2$, is now drawn lagging $\tilde{V}_2$ by an angle of θ_2. From the equivalent circuit,

$$\tilde{E}_2 = \tilde{V}_2 + \tilde{I}_2R_2 + j\tilde{I}_2X_2 \tag{3.26}$$

Let us now proceed to construct the phasor $\tilde{E}_2$, where $\tilde{E}_2$ is the induced voltage in the secondary winding as given by Eq. (3.26). The voltage drop across the secondary winding resistance R_2 is in phase with $\tilde{I}_2$. Since it must be added to $\tilde{V}_2$, the phasor $\tilde{I}_2R_2$ can then be drawn starting at the tip of $\tilde{V}_2$ and parallel to $\tilde{I}_2$. The tip of $\tilde{I}_2R_2$ now represents the phasor sum of $\tilde{V}_2$ and $\tilde{I}_2R_2$. The voltage

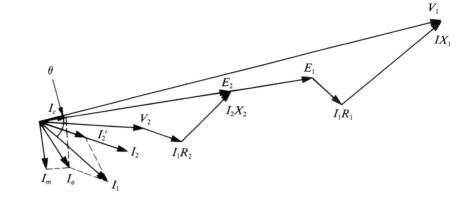

FIGURE 3.14
Phasor Diagram for the Equivalent Circuit of the Transformer Given in Figure 3.11

drop across the secondary leakage reactance can now be added at the tip of phasor $\tilde{I}_2 R_2$ by drawing it perpendicular to the current $\tilde{I}_2$. A line drawn from the origin to the tip of $\tilde{I}_2 X_2$ then represents the phasor $\tilde{E}_2$. This completes the drawing of phasors for the secondary winding.

The induced voltage in the primary winding is equal to the induced voltage on the secondary side times the transformation ratio (i.e., $\tilde{E}_1 = a\tilde{E}_2$). For our purpose, let us assume that the ratio of transformation is greater than unity. In that case, $\tilde{E}_1$ is greater than $\tilde{E}_2$ and can be represented by extending the phasor $\tilde{E}_2$ as shown.

The current $\tilde{I}_c$ that supplies the core losses is in phase with $\tilde{E}_1$, while the magnetization current $\tilde{I}_m$ lags $\tilde{E}_2$ by 90°. These current phasors can then be drawn from the origin. The phasor sum of these currents yields the excitation current phasor, $\tilde{I}_\phi$.

Since the secondary current as referred to the primary is $\tilde{I}_2' = \tilde{I}_2/a$, the sum of $\tilde{I}_2'$ and $\tilde{I}_\phi$ yields the current $\tilde{I}_1$ that flows through the power source. The voltage drops across the primary impedance, $\hat{Z}_1 = R_1 + jX_1$, can now be added to the phasor diagram to obtain voltage phasor $\tilde{V}_1$, as shown in Fig. 3.14. The phasor diagram is now complete. The angle between the phasors $\tilde{V}_1$ and $\tilde{I}_1$ is the power factor angle, θ_1, for the input power.

We can also draw the phasor diagram for the exact equivalent circuits as viewed from the primary and secondary sides. The exact equivalent circuit involves a great deal of calculations, which are not only laborious and time consuming but also prone to an error. In cases where the accuracy of the calculations is not as important, the exact equivalent circuit can be replaced with an approximate equivalent circuit. One such circuit as viewed from the primary side is shown in Fig. 3.15. Note that the parallel branch, which consists of the core-loss resistance R_c and magnetizing reactance X_m, has been moved and connected directly across the applied voltage. By doing so, an error has been introduced in calculating the voltage drop across the primary winding impedance $\hat{Z}_1$ because the current now flowing through it does not include the excitation current. However, under

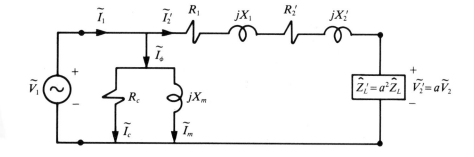

FIGURE 3.15
Approximate
Equivalent
Circuit of the
Transformer as
Referred to the
Primary

full-load conditions, the excitation current is a very small fraction of the primary current and, therefore, the amount of error introduced is not significant.

The currents in R_c and X_m can now be determined as

$$\tilde{I}_c = \frac{\tilde{V}_1}{R_c} \tag{3.27}$$

and

$$\tilde{I}_m = \frac{\tilde{V}_1}{jX_m} \tag{3.28}$$

The resistances and the leakage reactances of the two windings can be combined to form an effective impedance, as shown in Fig. 3.16, where

$$\hat{Z}_{e1} = R_{e1} + jX_{e1} \tag{3.29}$$

$$R_{e1} = R_1 + R_2a^2 \tag{3.30}$$

and

$$X_{e1} = X_1 + X_2a^2 \tag{3.31}$$

The phasor diagram for the approximate equivalent circuit of the transformer as referred to the primary side is given in Fig. 3.17. Similarly, we can draw the

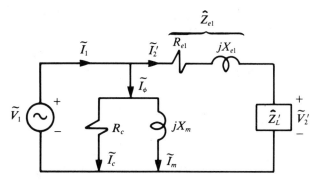

FIGURE 3.16
Simplified
Version of
Figure 3.15

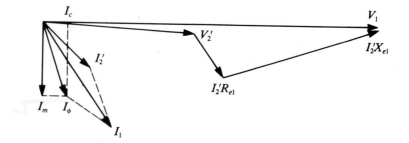

FIGURE 3.17
Phasor Diagram
for the Circuit
Shown in Figure
3.16

approximate equivalent circuit (Fig. 3.18) and the corresponding phasor diagram (Fig. 3.19) for the transformer as referred to the secondary side.

■ **EXAMPLE 3.9**

A 10-kVA, 2400/120-V, 50-Hz transformer has a high-voltage winding resistance of 0.1 Ω and a leakage reactance of 0.22 Ω. The low-voltage winding resistance is 0.035 Ω and leakage reactance is 0.012 Ω. Find the equivalent winding resistance, reactance, and impedance referred to the high-voltage side by using the approximate equivalent circuit.

Solution

The transformation ratio is $a = 2400/120 = 20$. In this transformer, the high-voltage side is, in fact, the primary side. Hence the values as referred to the high-voltage (primary) side are

$$R_{e1} = R_1 + a^2 R_2 = 0.1 + 20^2 \times 0.035 = 14.1 \ \Omega$$

and $\qquad X_{e1} = X_1 + a^2 X_2 = 0.22 + 20^2 \times 0.012 = 5.02 \ \Omega$

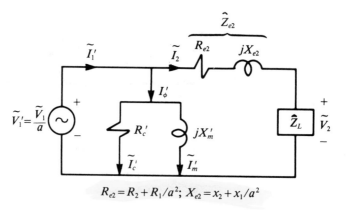

FIGURE 3.18
Approximate
Equivalent
Circuit of the
Transformer as
Referred to the
Secondary Side

$$R_{e2} = R_2 + R_1/a^2; \ X_{e2} = x_2 + x_1/a^2$$

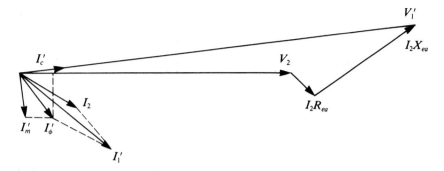

FIGURE 3.19
Phasor Diagram
for the Circuit
Shown in Figure
3.18

The impedance as referred to the primary side is

$$\hat{Z}_{e1} = R_{e1} + jX_{e1} = 14.1 + j5.02 = 15\underline{/19.6°} \; \Omega$$

■ EXAMPLE 3.10

A 10-kVA, 3300/600-V, 60-Hz, single-phase step-down transformer has 550 turns
on the primary and 100 turns on the secondary. The mean length of the magnetic
path in iron core is 2 m. Under no-load conditions with a rated voltage applied
to the primary winding, the maximum flux density in the core is 0.8 T, and the
corresponding magnetic-field intensity is 500 A-t/m. Determine the no-load current
and the power factor. Assume the core loss is 1.5 W/kg and the specific gravity
of iron is 7.8 g/cm³.

Solution

The mmf of the core is $\mathcal{F} = 500 \times 2 = 1000$ A-t. The maximum value of the
magnetizing current is

$$I_m \mid_{\max} = \frac{1000}{550} = 1.818 \text{ A}$$

The rms value of the magnetizing current is $I_m = 1.818/\sqrt{2} = 1.286$ A. The length
of the core is given. We must calculate its area in order to obtain its volume.
Since $E_1 = 4.44fN_1B_mA,$

$$A = \frac{3300}{4.44 \times 60 \times 550 \times 0.8} = 280 \text{ cm}^2$$

The volume of the core is

$$\text{Volume} = 280 \times 200 = 56,000 \text{ cm}^3$$

The mass of the iron core is

$$7.8 \times 56,000 = 436,800 \text{ g} = 436.8 \text{ kg}$$

Thus, the core loss

$$P_c = 1.5 \times 436.8 = 655.2 \text{ W}$$

The core-loss current is

$$I_c = 655.2/3300 = 0.199 \text{ A}$$

The excitation current is

$$I_\phi = \sqrt{I_c^2 + I_m^2} = \sqrt{0.199^2 + 1.286^2} = 1.301 \text{ A}$$

Since $I_c = I_\phi \cos \theta$, the power factor is

$$\cos \theta = \frac{0.199}{1.301} = 0.153$$

or

$$\theta = 81.2° \text{ (lag)}$$

3.3 VOLTAGE REGULATION

When the impressed voltage on the primary of the transformer is held constant, the secondary terminal voltage decreases with the load because of the primary and secondary winding impedances, the equivalent core-loss resistance, and the magnetizing reactance, as is evident from the equivalent circuit of the transformer given in Fig. 3.13. A quantity of particular interest is the net change in the magnitude of the secondary terminal voltage when the load is gradually decreased from full load to no load, with the primary winding voltage and temperature of the transformer maintained constant. When the change is expressed as a percentage of full-load (rated) voltage, it is usually referred to as the *voltage regulation* (VR) of the transformer.

$$\text{VR\%} = 100 \times \frac{E_2 - V_2}{V_2} \tag{3.32}$$

where V_2 is the full-load voltage and E_2 is the no-load voltage. Voltage regulation can also be expressed as

$$\text{VR\%} = 100 \times \frac{aE_2 - aV_2}{aV_2} \tag{3.33}$$

Since $E_1 = aE_2$, the voltage regulation becomes

$$\text{VR\%} = 100 \times \frac{E_1 - aV_2}{aV_2} \qquad (3.34)$$

For approximate calculations, V_1 can be substituted for E_1, as is evident from the approximate equivalent circuit of the transformer shown in Fig. 3.16. Thus,

$$\text{VR\%} = 100 \times \frac{V_1 - aV_2}{aV_2} \qquad (3.35)$$

However, if we divide the numerator and the denominator of Eq. (3.35) by the turns ratio, we get

$$\text{VR\%} = 100 \times \frac{(V_1/a) - V_2}{V_2} \qquad (3.36)$$

This becomes the definition of voltage regulation based on the approximate equivalent circuit of the transformer, as shown in Fig. 3.18.

■ **EXAMPLE 3.11**

A 1.1-kVA, 440/110-V, 60-Hz transformer has the following parameters as referred to the primary side: $R_{e1} = 1.5\ \Omega$, $X_{e1} = 2.5\ \Omega$, $R_c = 3000\ \Omega$, and $X_m = 2500\ \Omega$. The transformer is delivering full load at rated voltage when the power factor is 0.707 (lagging). Determine the primary voltage and the voltage regulation of the transformer.

Solution

The transformation ratio is $a = 440/110 = 4$. Referring to the approximate equivalent circuit of the transformer given in Fig. 3.16, the secondary voltage as referred to the primary is

$$V_2' = aV_2 = 4 \times 110 = 440\ \text{V}$$

The rated current on the secondary side is

$$I_2 = \frac{1.1 \times 1000}{110} = 10\ \text{A}$$

Since I_2 lags by $\theta_2 = \cos^{-1}(0.707) = 45°$, $\tilde{I}_2 = 10\underline{/-45°}$ A. Thus,

$$\tilde{I}_2' = \frac{\tilde{I}_2}{a} = \frac{10\underline{/-45°}}{4} = 2.5\underline{/-45°} \text{ A}$$

The voltage that must be applied to the primary side is

$$\tilde{V}_1 = 440 + (2.5\underline{/-45°} \times (1.5 + j2.5)$$

$$= 440 + 7.288\underline{/14.04°}$$

$$= 447.073\underline{/0.23°}$$

The percent of voltage regulation is

$$\text{VR\%} = 100 \times \frac{447.073 - 440}{440} = 1.61\%$$

3.4   TRANSFORMER EFFICIENCY

Similar to other machines in which the energy-conversion process is involved, the efficiency of a transformer is also defined as the ratio of the output power to the input power. In an ideal transformer, the output power was equal to the input power, as there were assumed to be no losses in the energy-transfer process. However, each winding of a transformer has some finite resistance and therefore has copper losses. In addition, there are core losses in the transformer. All these power losses contribute to the heating of the transformer. Furthermore, the presence of these losses indicates that the power input has to be greater than the power output by the amount of total losses in the transformer.

If the current $\tilde{I}_1$ and the applied voltage $\tilde{V}_1$ have a power factor angle of θ_1, the power input will be

$$P_{\text{in}} = V_1 I_1 \cos \theta_1 \qquad (3.37)$$

Similarly, if $\tilde{I}_2$ and $\tilde{V}_2$ are the load current and voltage, respectively, and θ_2 is the power factor angle between them, the power output of the transformer will be

$$P_o = V_2 I_2 \cos \theta_2 \qquad (3.38)$$

The efficiency of the transformer is

$$\eta = \frac{P_o}{P_{\text{in}}} = \frac{V_2 I_2 \cos \theta_2}{V_1 I_1 \cos \theta_1} \qquad (3.39)$$

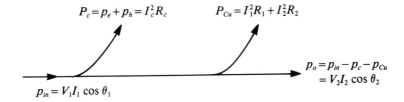

FIGURE 3.20
Power-Flow
Diagram for the
Transformer

However, the efficiency can be expressed in terms of losses as

$$\eta = \frac{P_{in} - P_L}{P_{in}} = \frac{P_o}{P_o + P_L} \tag{3.40}$$

where

$$P_L = P_{Cu} + P_c \tag{3.41}$$

and

$$P_{Cu} = I_1^2 R_1 + I_2^2 R_2 \tag{3.42}$$

and

$$P_c = I_c^2 R_c \tag{3.43}$$

In calculating the efficiency at different load conditions, the core loss is usually assumed to be constant. The power-flow relation between the input power, copper, and core losses and output power is known as the power-flow diagram and is illustrated in Fig. 3.20.

■ EXAMPLE 3.12

A 30-kVA, 1000/100-V, 60-Hz transformer has the following parameters as referred to the high-voltage side: $R_1 = 0.1 \; \Omega$, $X_1 = 0.3 \; \Omega$, $R_2' = 0.08 \; \Omega$, $X_2' = 0.27 \; \Omega$, $R_c = 900 \; \Omega$, and $X_m = 500 \; \Omega$. Determine the efficiency of the transformer when the load takes the rated current at a lagging power factor of 0.7.

Solution

The transformation ratio is $a = 1000/100 = 10$. Referring to the exact equivalent circuit of the transformer as given in Fig. 3.12 and considering the secondary rated voltage as a reference, we have

$$\tilde{V}_2' = a\tilde{V}_2 = 10 \times 100\underline{/0°} = 1000\underline{/0°} \text{ V}$$

From the lagging power factor of 0.7, $\theta = 45.57°$. The rated current is $I_2' = 30,000/1000 = 30$ A. Thus,

$$\tilde{I}_2' = 30\underline{/-45.57°}$$

$$\tilde{E}_1 = \tilde{V}_2' + \tilde{I}_2'(R_2' + jX_2')$$

$$= 1000 + (30\underline{/-45.57°})(0.08 + j0.27)$$

$$= 1007.47\underline{/0.22°} \text{ V}$$

$$\tilde{I}_c = \frac{1007.47\underline{/0.22°}}{900} = 1.12\underline{/0.22°} \text{ A}$$

$$\tilde{I}_m = \frac{1007.47\underline{/0.22}}{500\underline{/90°}} = 2.02\underline{/-89.78°} \text{ A}$$

$$\tilde{I}_\phi = \tilde{I}_c + \tilde{I}_m$$

$$= 1.12\underline{/0.22°} + 2.02\underline{/-89.78°} = 1.128 - j2.016 \text{ A}$$

$$\tilde{I}_1 = \tilde{I}_2' + \tilde{I}_\phi$$

$$= 30\underline{/-45.57°} + 1.128 - j2.016$$

$$= 32.235\underline{/-46.65°} \text{ A}$$

$$\tilde{V} = \tilde{E}_1 + \tilde{I}_1(R_1 + jX_1)$$

$$= 1007.47\underline{/0.22°} + (32.235\underline{/-46.65°})(0.1 + j0.3)$$

$$= 1016.74\underline{/0.46°} \text{ V}$$

The power output is

$$P_o = R_e[\tilde{V}_2'\tilde{I}_2'^*]$$

$$= R_e[(1000\underline{/0°})(30\underline{/45.57°})]$$

$$= 21,000 \text{ W}$$

The power input is

$$P_{in} = R_e[\tilde{V}_1\tilde{I}_1^*]$$

$$= R_e[(1016.74\underline{/0.46°})(32.235\underline{/46.65°})]$$

$$= 22,306.17 \text{ W}$$

The efficiency is

$$\eta = \frac{21,000}{22,306.17} = 0.9414 \quad \text{or} \quad 94.14\%$$

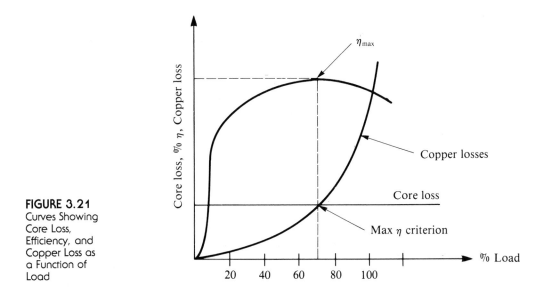

FIGURE 3.21
Curves Showing
Core Loss,
Efficiency, and
Copper Loss as
a Function of
Load

3.4.1 Maximum Efficiency Criterion

The efficiency of a transformer can be improved by improving the power factor of the load. For any load the transformer will operate at its peak efficiency when the power factor of the load is unity. On the other hand, there exists a definite load current for which the efficiency of the transformer is maximum.

Let us consider the equivalent circuit of the transformer as referred to the primary side (Fig. 3.16). The power input is $V_1 I_1 \cos \theta_1$. Let us assume that the core loss, P_c, for the transformer is constant at any load, as shown in Fig. 3.21. The copper losses are $I_1^2 R_{e1}$. The efficiency of the transformer from Eq. (3.40) is

$$\eta = \frac{V_1 I_2' \cos \theta_1 - P_c - I_2'^2 R_{e1}}{V_1 I_2' \cos \theta_1}$$

$$= 1 - \frac{P_c}{V_1 I_2' \cos \theta_1} - \frac{I_2' R_{e1}}{V_1 \cos \theta_1}$$

Differentiating both sides with respect to I_2', we get

$$\frac{d\eta}{dI_2'} = \frac{P_c}{V_1 I_2'^2 \cos \theta_1} - \frac{R_{e1}}{V_1 \cos \theta_1}$$

For efficiency to be maximum, $d\eta/dI_2' = 0$. Hence, the preceding equation yields

$$\frac{P_c}{V_1 I_2'^2 \cos \theta_1} = \frac{R_{e1}}{V_1 \cos \theta_1}$$

or

$$I_2'^2 R_{e1} = P_c \tag{3.44}$$

In accordance with Eq. (3.44), the intersection of the core-loss and copper-loss curves yields the load current for the maximum efficiency of the transformer. That is,

$$I_2' = \sqrt{\frac{P_c}{R_{e1}}} \tag{3.45}$$

or

$$I_2 = a I_2' = \sqrt{\frac{P_c}{R_{e2}}} \tag{3.46}$$

It is possible to design a transformer so that it operates at its maximum efficiency at full load by properly selecting its parameters. If I_{fl} is the full-load current, the full-load copper loss is $I_{fl}^2 R_{e1}$. Let I_η be the current at maximum efficiency of the transformer; then

$$I_\eta^2 R_{e1} = P_c$$

or

$$I_\eta = \sqrt{\frac{P_c}{R_{e1}}}$$

$$= I_{fl} \sqrt{\frac{P_c}{I_{fl}^2 R_{e1}}} \tag{3.47}$$

$$= I_{fl} \sqrt{\frac{\text{core loss}}{\text{full-load copper loss}}}$$

Likewise, the rating of the transformer at maximum efficiency would be

$$\text{kVA at max. efficiency} = \text{full-load kVA} \sqrt{\frac{\text{core loss}}{\text{full-load copper loss}}} \tag{3.48}$$

■ EXAMPLE 3.13

A 150-kVA transformer has a core loss of 16 kW and a copper loss of 25 kW. Calculate the kVA loading at which the efficiency is maximum. At this loading, what would be its efficiency at (1) unity power factor, and (2) 0.8 power factor lagging?

Solution

The load kVA for maximum efficiency is $150\sqrt{16/25} = 120$ kVA. Total loss at unity power factor is $16 \times 2 = 32$ kW. The efficiency is

$$\eta_{max} = \frac{120}{120 + 32} = 0.789 \quad \text{or} \quad 78.9\%$$

At 0.8 pf lagging, the total loss is the same. Thus, the efficiency is

$$\eta = \frac{120 \times 0.8}{120 \times 0.8 + 32} = \frac{96}{128} = 0.75 \quad \text{or} \quad 75\%$$

■ EXAMPLE 3.14

The maximum efficiency of a 10-kVA transformer occurs when it operates at 80% of its rated load with unity power factor. What is the rated efficiency of the transformer at a lagging power factor of 0.85? Assume the no-load loss is 400 W and the load voltage is the same at all load levels.

Solution

At maximum efficiency:

Power output: $P_o = 10,000 \times 0.8 = 8000$ W

Core-loss (given): $P_c = 400$ W

Copper loss = core loss: $P_{Cu} = 400$ W

Power input: $P_{in} = 8000 + 400 + 400 = 8800$ W

Maximum efficiency: $\eta_{max} = 8000/8800 = 0.909$ or 90.9%

At full-load with 0.85 pf lagging:

Power output: $P_o = 10,000 \times 0.85 = 8500$ W

Core loss: $P_c = 400$ W (assumed constant)

From the approximate equivalent circuit of the transformer (Fig. 3.16), the copper loss is $I_2'^2 R_{e1}$. Let V_2' be the load voltage. The load current at maximum efficiency is

$$I_{2\,max}' = \frac{8000}{V_2'}$$

and at full-load

$$I_2' = \frac{10,000}{V_2'}$$

Thus, $$I_2' = I_{2\,\text{max}}' \frac{10,000}{8000} = 1.25 I_{2\,\text{max}}'$$

The copper loss at full-load is

$$I_2'^2 R_{e1} = 1.25^2 I_{2\text{max}}'^2 R_{e1} = 1.5625 \times 400 = 625 \text{ W}$$

The power input is

$$P_{\text{in}} = 8500 + 400 + 625 = 9525 \text{ W}$$

The full-load efficiency is

$$\eta = \frac{8500}{9525} = 0.8399 \quad \text{or} \quad 83.99\%$$

3.5 TRANSFORMER ANALYSIS ON A PER UNIT BASIS

As explained in Chapter 1, the two different transformer designs can be compared on a per unit (pu) basis. Irrespective of the voltage and power ratings of the transformers, the per unit quantities must be similarly related to obtain similar performance. To employ the per unit system for transformers, the rated voltage and apparent power will be taken as the fundamental base quantities, as illustrated by the following examples.

■ EXAMPLE 3.15

Determine the per unit parameters, the voltage regulation, and the efficiency of the transformer discussed in Example 3.11.

Solution

The base values of the fundamental quantities for the transformer and its approximate equivalent circuit, as given in Fig. 3.16, are

$$S_b = 1100 \text{ VA}$$

and $$V_{2b}' = 440 \text{ V}$$

The other base quantities are

$$I'_{2b} = 1100/440 = 2.5 \text{ A}$$

and
$$Z_b = 440/2.5 = 176 \ \Omega$$

In terms of the base quantities, the equivalent-circuit parameters of the transformer are

$$R_{e1 \, pu} = \frac{1.5}{176} = 0.0085$$

$$X_{e1 \, pu} = \frac{2.5}{176} = 0.0142$$

$$R_{c \, pu} = \frac{3000}{176} = 17.0455$$

$$X_{m \, pu} = \frac{2500}{176} = 14.2045$$

Since the rated voltage and the current are the base values, the per unit load voltage and currents are

$$\hat{V}'_{2 \, pu} = 1\underline{/0^\circ}$$

and
$$\hat{I}'_{2 \, pu} = 1\underline{/-45^\circ}$$

where -45° corresponds to a lagging power factor of 0.707. The equivalent circuit of the transformer as referred to the primary side on a per unit basis is shown in Fig. 3.22.

The per unit applied voltage is

$$\hat{V}_{1 \, pu} = 1\underline{/0^\circ} + (1\underline{/-45^\circ})(0.0085 + j0.0142)$$

$$= 1.0161 + j0.004 = 1.0161\underline{/0.23^\circ}$$

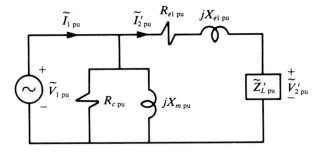

The primary voltage is

$$\tilde{V}_1 = V'_{2b}\tilde{V}_{1\,pu} = 447.08\underline{/0.23°}$$

The voltage regulation is

$$VR = 1.0161 - 1 = 0.0161$$

or

$$VR\% = 1.61\%$$

$$\tilde{I}_{c\,pu} = \frac{1.0161 + j0.004}{17.0455} = 0.0596 + j0.0002$$

$$\tilde{I}_{m\,pu} = \frac{1.0161 + j0.004}{j14.2045} = 0.0003 - j0.0715$$

$$\tilde{I}_{1\,pu} = \tilde{I}'_{2\,pu} + \tilde{I}_{c\,pu} + \tilde{I}_{m\,pu}$$

$$= 0.7696 - j0.7783 = 1.0946\underline{/-45.32°}$$

The primary current is

$$\tilde{I}_1 = I'_{2b}\tilde{I}_{1\,pu} = 2.7365\underline{/-45.32°}\ A$$

Power output on a per unit basis is

$$P_{o\,pu} = 1 \times 1 \cos(45) = 0.707$$

Power input on a per unit basis is

$$P_{in\,pu} = 1.0946 \times 1.0161 \cos(45.32 + 0.23)$$

$$= 0.7789$$

The efficiency is

$$\eta = \frac{0.707}{0.7789} = 0.9077 \quad \text{or} \quad 90.77\%$$

■ EXAMPLE 3.16

The per unit quantities of various elements of a 500-kVA, 10,000/2500-V, 60-Hz distribution transformer are $R_{1\,pu} = R_{2\,pu} = 0.003$, $X_{1\,pu} = X_{2\,pu} = 0.001$, $R_{c\,pu} = 40$, and $X_{m\,pu} = 50$. Calculate the exact values of these elements.

Solution

Base apparent power rating:

$$S_b = 500 \text{ kVA}$$

Base primary-voltage rating:

$$V_{1b} = 10,000 \text{ V}$$

Thus, base primary-current rating:

$$I_{1b} = 500 \times 100/10,000 = 50 \text{ A}$$

Base primary impedance:

$$Z_{1b} = 10,000/50 = 200 \text{ } \Omega$$

Base secondary-voltage rating:

$$V_{2b} = 2500 \text{ V}$$

Thus, base secondary current rating:

$$I_{2b} = 500 \times 1000/2500 = 200 \text{ A}$$

Base secondary impedance:

$$Z_{2b} = 2500/200 = 12.5$$

Thus the actual values of the transformer elements are

$$R_1 = 0.003 \times 200 = 0.6 \text{ } \Omega$$
$$X_1 = 0.001 \times 200 = 0.2 \text{ } \Omega$$
$$R_c = 40 \times 200 = 8000 \text{ } \Omega$$
$$X_m = 50 \times 200 = 10,000 \text{ } \Omega$$
$$R_2 = 0.003 \times 12.5 = 0.0375 \text{ } \Omega$$
$$X_2 = 0.001 \times 12.5 = 0.0125 \text{ } \Omega$$

3.6 OPEN- AND SHORT-CIRCUIT TESTS

The parameters of the approximate equivalent circuit of the transformer can be determined from open-circuit or no-load tests and short-circuit tests.

Open-circuit Test   As the name implies, one of the two windings of the transformer is left open while the other winding is excited by applying rated voltage. Under these conditions, the magnetic core of the transformer is carrying the magnetizing

FIGURE 3.23
Approximate
Equivalent
Circuit (Low-
Voltage Side)
for Open-Circuit
Test

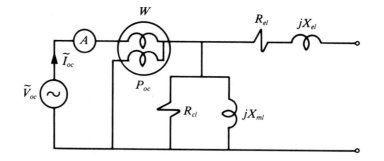

flux, and the excitation current in the winding is just sufficient to take care of the core losses and the related mmf drop. However, the question arises, which side of the transformer must be excited and which side must be left open? In any test facility, usually a low-voltage supply is readily available. Therefore, the rated voltage is applied to the low-voltage side, and the high-voltage side is left open. Since a step-up transformer has high voltage on the secondary side and a step-down transformer has the primary as the high-voltage side, let us use high and low instead of primary and secondary as the nomenclature to differentiate between the two windings.

The approximate equivalent circuit (Fig. 3.23) is drawn as referred to the low-voltage side because the test will be conducted on this side. Let V_{oc}, I_{oc}, and P_{oc} be the measured values of the applied voltage, current, and the power dissipated in the circuit, respectively. The apparent power at no-load is

$$S_{oc} = V_{oc} I_{oc}$$

at a lagging power factor of

$$\phi_{oc} = \cos^{-1} \frac{P_{oc}}{S_{oc}}$$

The core loss and the magnetizing current components at no load are $I_{oc} \cos \phi_{oc}$ and $I_{oc} \sin \phi_{oc}$, respectively. Thus, the equivalent core-loss resistance and magnetization reactance on the low-voltage side, respectively, are

$$R_{cl} = \frac{V_{oc}}{I_{oc} \cos \phi_{oc}} = \frac{V_{oc}^2}{P_{oc}} \qquad (3.49)$$

and

$$X_m = \frac{V_{oc}}{I_{oc} \sin \phi_{oc}} = \frac{V_{oc}^2}{Q_{oc}} \qquad (3.50)$$

where

$$Q_{oc} = \sqrt{S_{oc}^2 - P_{oc}^2}$$

Short-circuit Test   The second test that is performed to calculate the approximate parameters of the transformer is known as the short-circuit test. In this case, one of the two windings of the transformer is short-circuited while the other winding is excited by a low-voltage source such that it carries the rated current. To avoid the measurement of high currents in the winding, it is a normal practice to short-circuit the low-voltage side and excite the high-voltage side. The approximate equivalent circuit for the short-circuit test as referred to the high-voltage side is given in Fig. 3.24. Under short-circuit conditions, the approximate impedance of the windings, $\breve{Z}_{eh} = R_{eh} + jX_{eh}$, is in parallel with R_{ch} and jX_{mh}. In a well-designed transformer, $\breve{Z}_{eh}$ is very low in comparison with R_{ch} and jX_{mh}. Therefore, to circulate the full-load current in the windings, the applied voltage will only be a few percent of its rated value. Consequently, the mutual flux is very small and the core losses are almost negligible. Thus, the excitation current is nearly zero.

Let V_{sc}, I_{sc}, and P_{sc} be the measured values of the applied voltage, the short-circuit current, and the power input to the transformer, respectively. Since there is no output, all the applied power must be dissipated as heat by the equivalent resistance R_{eh}. Thus,

$$R_{eh} = \frac{P_{sc}}{I_{sc}^2} \tag{3.51}$$

The magnitude of the equivalent winding impedance, Z_{eh}, is

$$Z_{eh} = \frac{V_{sc}}{I_{sc}} \tag{3.52}$$

Thus,
$$X_{eh} = \sqrt{Z_{eh}^2 - R_{eh}^2} \tag{3.53}$$

In the open-circuit test, the parameters were obtained referred to the low-voltage side, and in the short-circuit test the parameters are obtained referred to the high-voltage side. To represent the equivalent circuit referred to any side, we

FIGURE 3.24
Approximate Equivalent Circuit (High-Voltage Side) for Short-Circuit Test

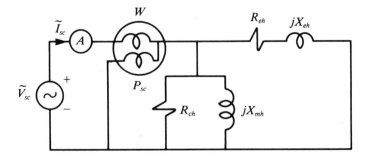

have to make a transformation of one set of parameters or the other. Let us define the transformation ratio as

$$a = \frac{N_h}{N_l} \tag{3.54}$$

In terms of the transformation ratio,

$$R_{eh} = R_h + a^2 R_l \tag{3.55}$$

and

$$X_{eh} = X_h + a^2 X_l \tag{3.56}$$

where
R_h = resistance of the high-voltage winding
R_l = resistance of the low-voltage winding
X_h = leakage reactance on the high-voltage side
X_l = leakage reactance on the low-voltage side

If the actual transformer is available, the winding resistances on both sides can be measured and compared with these values. However, in the case of leakage reactances there is no simple way to separate the two leakage reactances. If such a separation is desired, we will assume that the transformer has been designed in such a way that the power loss in its winding resistance on the high-voltage side is equal to the power loss in its winding resistance on the low-voltage side. That is,

$$I_h^2 R_h = I_l^2 R_l$$

But

$$a = \frac{N_h}{N_l}$$

Therefore,

$$R_h = a^2 R_l$$

In other words,

$$R_h = a^2 R_l = 0.5 R_{eh} \tag{3.57}$$

Similarly, it can be assumed that

$$X_h = a^2 X_l = 0.5 X_{eh} \tag{3.58}$$

■ **EXAMPLE 3.17**

A 10-kVA, 4000/125-V, 60-Hz transformer has been tested to obtain the following data:

Open-circuit test: voltage = 125 V, current = 0.25 A,
input power = 25 W

Short-circuit test: voltage = 100 V, current = 2.5 A,
power input = 125 W

Determine the approximate parameters of the transformer.

Solution

Open-circuit test: Since the applied voltage is equal to the rated voltage on the low-voltage side of the transformer, the test has been performed on the low-voltage side.

$$V_{oc} = 125 \text{ V}, \qquad I_{oc} = 0.2 \text{ A}, \qquad P_{oc} = 25 \text{ W}$$

$$R_{cl} = \frac{V_{oc}^2}{P_{oc}} = 125^2/25 = 625 \text{ } \Omega$$

The apparent power is

$$S_{oc} = 125 \times 0.25 = 31.25 \text{ VA}$$

The reactive power is

$$Q_{oc} = \sqrt{31.25^2 - 25^2} = 18.75 \text{ VAR}$$

The magnetizing reactance is

$$X_{ml} = \frac{V_{oc}^2}{Q_{oc}} = \frac{125^2}{18.75} = 833.33 \text{ } \Omega$$

Short-circuit Test   The measured current must be the rated current, and the test is usually performed by short-circuiting the low-voltage side and applying a voltage on the high-voltage side. Let us check if this is true. Assume the test is performed on the high-voltage side; the rated current would be $10 \times 1000/4000 = 2.5$ A, which agrees with the measured current.

$$V_{sc} = 100 \text{ V}, \qquad I_{sc} = 2.5 \text{ A}, \qquad P_{sc} = 125 \text{ W}$$

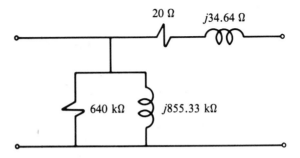

FIGURE 3.25
Approximate
Equivalent
Circuit of a
Transformer as
Referred to
High-Voltage
Side for
Example 3.17

Thus, the equivalent resistance is

$$R_{eh} = \frac{P_{sc}}{I_{sc}^2} = \frac{125}{2.5^2} = 20 \ \Omega$$

The equivalent impedance is

$$Z_{eh} = \frac{V_{sc}}{I_{sc}} = \frac{100}{2.5} = 40 \ \Omega$$

The leakage reactance is

$$X_{eh} = \sqrt{40^2 - 20^2} = 34.64 \ \Omega$$

The transformation ratio is $a = 4000/125 = 32$. For the well-designed transformer,

$$R_h = 0.5 \times 20 = 10 \ \Omega$$

$$R_l = \frac{10}{32^2} = 0.0098 \ \Omega$$

$$X_h = 0.5 \times 34.64 = 17.32 \ \Omega$$

and

$$X_l = \frac{17.32}{32^2} = 0.0169 \ \Omega$$

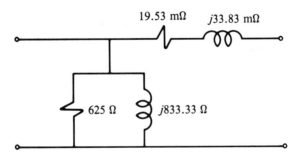

FIGURE 3.26
Approximate
Equivalent
Circuit of a
Transformer as
Referred to
Low-Voltage
Side for
Example 3.17

The approximate equivalent circuit referred to the high-voltage side is given in Fig. 3.25. Note that R_{cl} and X_{ml} have been multiplied by the square of the transformation ratio in order to represent them on the high-voltage side. A similar approximate equivalent circuit for the transformer referred to the low-voltage side is shown in Fig. 3.26.

3.7 AUTOTRANSFORMER

The two windings in an ordinary transformer, also referred to as a two-winding transformer, are electrically isolated from each other. Power is transferred inductively from one side to the other. However, an autotransformer consists of a single winding with part of the winding being common to both the primary and the secondary. Therefore, there is no electrical isolation between the input side and the output side. Hence, we can expect that the power transfer process will involve both induction and conduction. Autotransformers are also used to step up or step down transmission voltage levels. Often, different energy networks are interconnected by autotransformers. Also, they are utilized to start up various types of motors. Furthermore, they have an application in a number of testing and laboratory circuits as voltage dividers or variacs.

Let us consider a step-down autotransformer as shown in Fig. 3.27. The number of turns in the primary winding PG are $N_{PS} + N_{SG}$, while those in the secondary winding SG are N_{SG}. On impressing an alternating voltage $\tilde{V}_{1a}$ on the

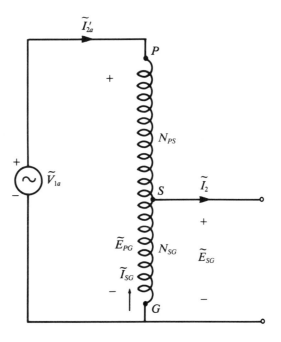

FIGURE 3.27
Step-Down
Autotransformer

primary, the secondary being left open, the resultant magnetizing current will set up a flux that induces a voltage E_{PG} in the primary and E_{SG} in the secondary. If the magnetizing current is assumed to be almost negligible, the induced voltage in the winding will be proportional to its number of turns. That is,

$$\frac{\tilde{E}_{PG}}{\tilde{E}_{SG}} = \frac{N_{PS} + N_{SG}}{N_{SG}} = a_T \tag{3.59}$$

where a_T is the transformation ratio for the autotransformer.

Application of KCl to node s under load conditions results in

$$I'_{2a} + \tilde{I}_{SG} = \tilde{I}_2 \tag{3.60}$$

To satisfy the mmf requirements, we have

$$N_{PS}\tilde{I}'_{2a} - N_{SG}\tilde{I}_{SG} = 0$$

or

$$\tilde{I}_{SG} = \frac{N_{PS}}{N_{SG}} \tilde{I}'_{2a} = (a_T - 1)\tilde{I}'_{2a} \tag{3.61}$$

Substituting for $\tilde{I}_{SG}$ in Eq. (3.60), we obtain

$$\frac{\tilde{I}_{2a}}{\tilde{I}'_{2a}} = a_T \tag{3.62}$$

Let R_{PS} and X_{PS} and R_{SG} and X_{SG} be the resistance and leakage reactance of PS and SG windings, respectively, as illustrated in Fig. 3.28. The impressed voltage can be expressed as

$$\tilde{V}_{1a} = (R_{PS} + jX_{PS})\tilde{I}'_{2a} + \tilde{E}_{PG} - (R_{SG} + jX_{SG})\tilde{I}_{SG} \tag{3.63}$$

where $\tilde{E}_{PG} = \tilde{E}_{PS} + \tilde{E}_{SG} \cong a_T\tilde{E}_{SG}$

However $\tilde{E}_{SG} = \tilde{I}_{2a}\hat{Z}_L + (R_{SG} + jX_{SG})\tilde{I}_{SG}$

or $\tilde{E}_{PG} = a_T\tilde{I}_{2a}\hat{Z}_L + a_T(R_{SG} + jX_{SG})\tilde{I}_{SG} \tag{3.64}$

With the help of Eqs. (3.61) and (3.64), Eq. (3.63) can be simplified as

$$\tilde{V}_{1a} = \tilde{I}'_{2a}[R_{PS} + jX_{PS} + (a_T - 1)^2(R_{SG} + jX_{SG}) + a_T^2\hat{Z}_L]$$
$$= \tilde{I}'_{2a}[R_e + jX_e + a_T^2 Z_l] \tag{3.65}$$

where $R_e = R_{PS} + (a_T - 1)^2 R_{SG} \tag{3.66}$

and $X_e = X_{PS} + (a_T - 1)^2 X_{SG} \tag{3.67}$

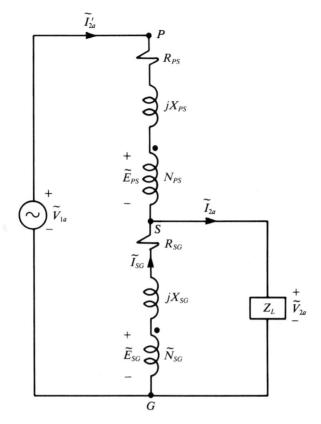

FIGURE 3.28
Step-Down
Autotransformer
with Winding
Impedances

An equivalent circuit that satisfies the voltage relation of Eq. (3.65) is shown in Fig. 3.29. This, therefore, is the equivalent circuit of an autotransformer referred to the primary side when the excitation current is neglected. We can, if we wish, approximately account for the excitation current by connecting an equivalent core-loss resistance R_c and magnetizing reactance X_m in parallel with the impressed voltage of the primary side as shown in Fig. 3.30.

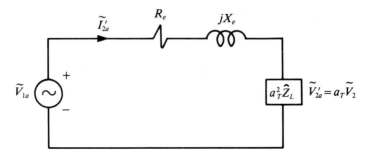

FIGURE 3.29
Equivalent
Circuit Based on
Equation 3.65

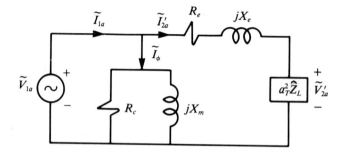

FIGURE 3.30
Equivalent
Circuit of an
Autotransformer
Including
Excitation
Impedance

The complex power supplied to the load is

$$\hat{S}_o = \tilde{V}'_{2a}\tilde{I}'^*_{2a} = a_T\tilde{V}_{2a}\tilde{I}'^*_{2a} \tag{3.68}$$

The average power output of the autotransformer is

$$P_o = R_e[a_T\tilde{V}_{2a}\tilde{I}'^*_{2a}] \tag{3.69}$$

and the power input is

$$P_{\text{in}} = R_e[\tilde{V}_{1a}\tilde{I}^*_{1a}] \tag{3.70}$$

where $\tilde{I}_{1a} = \tilde{I}'_{2a} + \tilde{I}_\phi$. The efficiency of the autotransformer can now be computed.

A two-winding transformer can be connected as an autotransformer by connecting its windings in series. In a step-down transformer the current $\tilde{I}_{SG}$, being the difference of $\tilde{I}_{2a}$ and $\tilde{I}'_{2a}$, is smaller than $\tilde{I}'_{2a}$ as long as the transformation ratio a_T is less than 2, which is usually true. Therefore, the high-voltage winding of a two-winding transformer can be used in place of the SG winding of the autotransformer. Consequently, the low-voltage winding can replace the PS winding. A two-winding transformer connected as an autotransformer for step-down application is shown in Fig. 3.31. Comparison of Fig. 3.31 with Fig. 3.28 shows that $\tilde{V}_{2a} = \tilde{V}_1$, $\tilde{I}_{SG} = \tilde{I}_1$, $\tilde{I}'_{2a} = \tilde{I}_2$, $\tilde{V}_{PS} = \tilde{V}_2$, $N_{SG} = N_1$, $N_{PS} = N_2$, and $a_T = (N_1 + N_2)/N_1$.

The complex power for the autotransformer, Eq. (3.68), can now be expressed in terms of the notations for the two-winding transformer as

$$\hat{S}_o = a_T\tilde{V}_1\tilde{I}^*_2$$

However, for the two-winding transformer, $\tilde{V}_1 \cong a\tilde{V}_2$. Thus,

$$\hat{S}_o = a_Ta\tilde{V}_2\tilde{I}^*_2$$

But

$$a_Ta = \frac{N_1 + N_2}{N_1}\frac{N_1}{N_2} = 1 + a$$

Hence
$$\hat{S}_o = \tilde{V}_2 \tilde{I}_2^*(1 + a) \tag{3.71}$$

Since $\tilde{V}_2 \tilde{I}_2^*$ is the complex power output of a two-winding transformer, it is evident from Eq. (3.71) that the power rating of a two-winding transformer increases by a factor of $1 + a$ when connected as an autotransformer. Consequently, for the same power rating an autotransformer will require less material, resulting in lower initial cost. On the other hand, for the same power output an autotransformer will have lower copper losses and better voltage regulation in comparison with a two-winding transformer. The main disadvantage of an autotransformer is that its secondary is not isolated from the primary. An open circuit in the common winding can result in dangerously high voltage across the load.

Equation (3.71) can be rewritten as

$$\hat{S}_o = \tilde{V}_2 \tilde{I}_2^* + a\tilde{V}_2 \tilde{I}_2^* \tag{3.72}$$
$$= \hat{S}_{\text{ind}} + \hat{S}_{\text{con}}$$

where $\hat{S}_{\text{ind}} = \tilde{V}_2 \tilde{I}_2^*$ is the complex power output due to induction, while $\hat{S}_{\text{con}} = a\tilde{V}_2 \tilde{I}_2^*$ is the power delivered to the load by conduction.

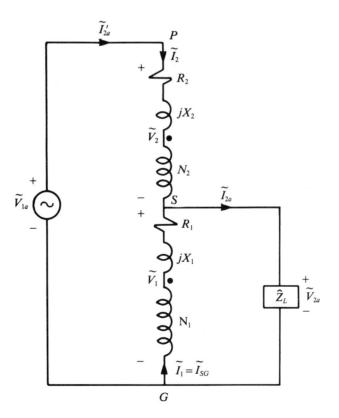

FIGURE 3.31
Two-Winding
Transformer
Connected as
an
Autotransformer

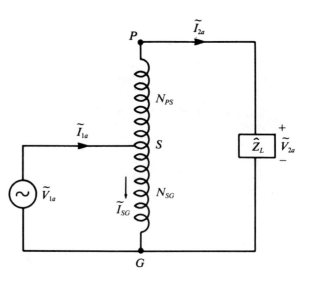

FIGURE 3.32
Step-Up
Autotransformer

An autotransformer can be used for step-up applications as shown in Fig. 3.32. It is left to the student to show that the complex power output satisfies the same relation as given in Eq. (3.71).

■ **EXAMPLE 3.18**

An 11-kVA, 220/440-V, 60-Hz, two-winding transformer has $R_1 = 0.08\ \Omega$ and $R_2 = 0.04\ \Omega$. (a) Calculate its efficiency at full load with unity power factor. (b) If the transformer is connected as a step-down 660/440-V autotransformer and supplies the same load as in part (a), determine its efficiency.

Solution

(a) As a two-winding transformer,

$$I_2 = \frac{11,000}{440} = 25\ \text{A}$$

and
$$I_1 = \frac{11,000}{220} = 50\ \text{A}$$

Total copper loss in the windings is

$$P_{\text{Cu}} = I_1^2 R_1 + I_2^2 R_2$$

$$= 50^2 \times 0.08 + 25^2 \times 0.04 = 225\ \text{W}$$

The efficiency is

$$\eta = \frac{11,000}{11,000 + 225} = 0.98 \quad \text{or} \quad 98\%$$

(b) As an autotransformer, the two windings and the load must be connected as shown in Fig. 3.33.

$$a_T = \frac{660}{440} \doteq 1.5$$

$$I_{2a} = 25 \text{ A} \quad [\text{same as in part(a)}]$$

and $$I'_{2a} = \frac{25}{1.5} = 16.667 \text{ A}$$

Thus, $$I_{SG} = 25 - 16.667 = 8.333 \text{ A}$$

Total copper loss is

$$P_{\text{Cu}} = 16.667^2 \times 0.08 + 8.333^2 \times 0.04 = 25 \text{ W}$$

The efficiency is, approximately,

$$\eta = \frac{11,000}{11,000 + 25} = 0.9977 \quad \text{or} \quad 99.77\%$$

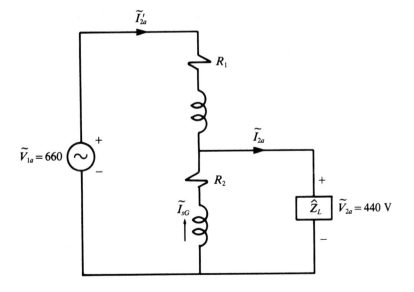

FIGURE 3.33
Connections of a Two-Winding Transformer as an Autotransformer for Example 3.18

■ EXAMPLE 3.19

A 5-kVA, 50/200-V, 60-Hz, two-winding transformer has a primary winding resistance of 0.0063 Ω and a secondary winding resistance of 0.088 Ω. If it is connected as a 50/250-V step-up autotransformer, determine (a) the maximum power that may be delivered to the resistive load without damaging the windings, (b) the efficiency of the transformer at that load, neglecting the magnetic losses, and (c) the power transferred conductively to the load.

Solution

(a) The two-winding transformer connected as a step-up autotransformer is shown in Fig. 3.34. It is apparent that the maximum load current $\tilde{I}_{2a}$ must be equal to the rated current for the 200-V winding of the two-winding transformer. Hence,

$$I_{2a} = I_2 = \frac{5000}{200} = 25 \text{ A}$$

The total power delivered to the resistive load will be

$$P_o = 250 \times 25 = 6250 \text{ W}$$

(b) The ratio of transformation for the autotransformer is

$$a_T = \frac{50}{250} = 0.2$$

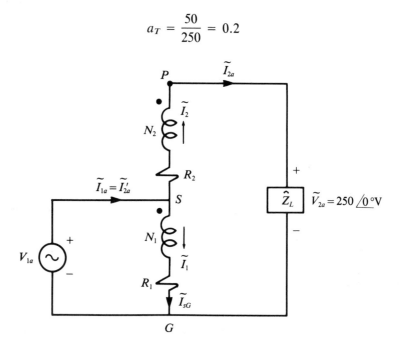

FIGURE 3.34
Autotransformer
Connections for
Example 3.19

The current in the 50-V winding is

$$I_{1a} = I'_{2a} = \frac{I_{2a}}{a_T} = \frac{25}{0.2} = 125 \text{ A}$$

Thus, the current in the common winding is

$$I_{SG} = I_1 = 125 - 25 = 100 \text{ A}$$

The total copper loss is

$$P_{Cu} = 25^2 \times 0.088 + 100^2 \times 0.0063 = 118 \text{ W}$$

The power input

$$P_{in} = 6250 + 118 = 6368 \text{ W}$$

Thus, the efficiency is, approximately,

$$\eta = \frac{6250}{6368} = 0.981 \quad \text{or} \quad 98.1\%$$

(c) The power transferred by conduction, P_{con}, is a times the power output of a two-winding transformer. For the two-winding transformer,

$$a = \frac{50}{200} = 0.25$$

$$P_o = 5000 \times 1 = 5000 \text{ W}$$

Thus, $$P_{con} = 0.25 \times 5000 = 1250 \text{ W}$$

The power transferred by induction will be $6250 - 1250 = 5000$ W and is equal to the power rating of the two-winding transformer.

■ EXAMPLE 3.20

An 11-kVA, 110/80-V, 50-Hz, step-down autotransformer is tested as follows: With low-voltage terminals open and the rated voltage applied on the high-voltage side, the current and power readings are 0.5 A and 30.25 W, respectively. When the rated current is maintained on the high-voltage side with the low-voltage terminals short-circuited, the voltage and power readings are 13.6 V and 400 W, respectively. Determine (a) the equivalent circuit parameters of the autotransformer, (b) the voltage regulation, and (c) the efficiency of the transformer at full load with 0.8 power factor lagging.

Solution

(a) *Open circuit test:* $V_{oc} = 110$ V, $I_{oc} = 0.5$ A, and $P_{oc} = 30.25$ W. For the approximate equivalent circuit as given in Fig. 3.30, the open-circuit current is equal to the excitation current for the transformer. Thus,

$$S_{oc} = 110 \times 0.5 = 55 \text{ VA}$$

$$\phi_{oc} = \cos^{-1} \frac{30.25}{55} = 56.63°$$

$$R_c = \frac{V_{oc}}{I_{oc} \cos \phi_{oc}} = \frac{110}{0.5 \cos (56.63°)} = 400 \ \Omega$$

$$X_m = \frac{V_{oc}}{I_{oc} \sin(\phi_{oc})} = \frac{110}{0.5 \sin(56.63°)} = 263.42 \ \Omega$$

Short-circuit test: $V_{sc} = 13.6$ V, $P_{sc} = 400$ W, and $I_{sc} = 11{,}000/110 = 100$ A.

Equivalent resistance: $R_e = \dfrac{P_{sc}}{I_{sc}^2} = \dfrac{400}{100^2} = 0.04 \ \Omega$

Equivalent impedance: $Z_e = \dfrac{V_{sc}}{I_{sc}} = \dfrac{13.6}{100} = 0.136 \ \Omega$

Equivalent reactance: $X_e = \sqrt{Z_e^2 - R_e^2} = \sqrt{0.136^2 - 0.04^2} = 0.13 \ \Omega$

(b) At full load with 0.8 pf lagging, the equivalent load current on the high-voltage side is

$$\tilde{I}_{2a}' = 100\underline{/-36.87°}$$

Thus, $\tilde{V}_{1a} = 110\underline{/0°} + (100\underline{/-36.87°})(0.04 + j0.13)$

$$= 121 + j8 = 120.74\underline{/3.78°} \text{ V}$$

The percent of voltage regulation is

$$VR\% = 100 \times \frac{120.74 - 110}{110} = 9.76\%$$

(c)
$$\tilde{I}_c = \frac{121 + j8}{400} = 0.303 + j0.02 \text{ A}$$

$$\tilde{I}_m = \frac{121 + j8}{263.42\underline{/90°}} = 0.03 - j0.459 \text{ A}$$

$$\tilde{I}_{1a} = \tilde{I}'_{2a} + \tilde{I}_c + \tilde{I}_m$$

$$= 80 - j60 + 0.303 + j0.02 + 0.03 - j0.459$$

$$= 80.333 - j60.439 = 100.53\underline{/-36.96°}$$

The power output is

$$P_o = 110 \times 100 \times 0.8 = 8800 \text{ W}$$

The power input is

$$P_{\text{in}} = R_e[(120.74\underline{/3.78°})(100.53\underline{/36.96°})]$$

$$= 9196.7 \text{ W}$$

The efficiency is

$$\eta = \frac{8800}{9196.7} = 0.957 \quad \text{or} \quad 95.7\%$$

3.8 INSTRUMENT TRANSFORMERS

In general, devices that detect electrical quantities in a system operate only at low voltage and current levels. For example, an ammeter can usually measure currents up to 100 A or so. However, current levels can be as high as thousands of amperes in a power-transmission system. Likewise, voltmeters generally have measuring capabilities of up to 1000 V. However, on a power-transmission line, the voltage levels are usually on the order of hundreds of kilovolts. On the other hand, no matter what the current or voltage levels are in an electrical system, these quantities must be measured by some means. For this purpose, instrument transformers, such as current and potential transformers, have been developed to reduce high currents or high voltages to some measurable level.

3.8.1 Current Transformers

Current transformers are classified into two categories. One is for measuring purposes and the other is to feed the protective devices in an electrical system.

The secondary winding of a current transformer is connected to an ammeter or the current coil of a wattmeter if it is used for measuring purposes. If it is used for the protection of a power system, the secondary winding feeds a protective device, such as a relay. Moreover, current transformers provide isolation between the detection and high-current circuits. Especially if the high-current circuit is a high-voltage transmission system, it is not safe or practical to measure the current in that circuit regardless of the current level. Generally, these transformers have one primary and one or two secondary windings. Usually, the primary winding of such a transformer consists of one or a small number of turns, with the conductors having a relatively large cross section. The secondary winding has a great number of turns, with smaller-sized conductors wound on a toroidal magnetic core as indicated in Fig. 3.35.

Usually, an ammeter or the current coil of a wattmeter is connected across the secondary winding of a current transformer, while its primary side is in series with the high-current circuit. Since the internal resistance of an ammeter is very small, it may be treated as a short circuit. This indicates that current transformers operate almost under short-circuit conditions.

If the magnetizing current, which is small under rated operating conditions for the current transformers, is neglected, the current in the secondary will be as given by Eq. (3.7). However, in practice the magnetizing current does exist and gives rise to an error in the measured current in the secondary of the transformer. This error is known as the *ratio error* and can be represented in percentage form as

$$\Delta I = \left(\frac{I_2 - aI_1}{I_2} \times 100 \right)\% \tag{3.73}$$

Furthermore, there is a phase shift in the magnetizing current with respect to the primary and secondary currents. This causes another source of error in measuring the current, known as the *phase-angle error*. Various standards specify the permissible levels of these error quantities depending on the accuracy class of the current transformers. For instance, if the accuracy class is 1.0, this indicates that the ratio error is $\pm 1\%$ and the phase-angle error is $\pm 10'$.

The current transformers used in high-voltage power systems exhibit a slight difference as compared to their low-voltage counterparts. Since these transformers are responsible for monitoring the current flow in a high-voltage circuit, the primary winding must be well isolated from the secondary winding, which is always at ground potential. As a result, the insulation design has to be developed as accurately as possible.

As mentioned earlier, current transformers operate under short-circuit conditions with a small magnetic flux density in their magnetic cores. If the meter across the secondary terminals is removed and left open, the transformer will be magnetized by the high current flowing through the primary winding. Consequently, the core will experience a very intense saturation, leading high-voltage

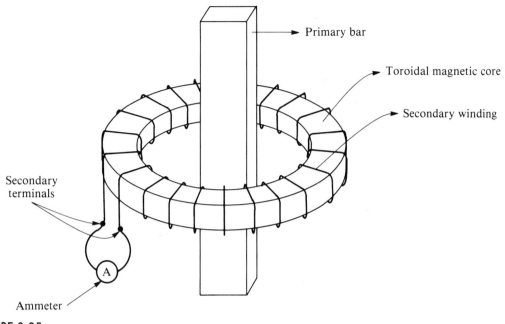

FIGURE 3.35
Toroidal Current Transformer

pulses across the secondary terminals, which may put the insulation of the transformer as well as operating personnel in danger. Furthermore, the primary current can cause excessive heat development in the transformer and may destroy the insulation. Hence, if the load is to be removed from the secondary terminals of a current transformer, the secondary should always be short-circuited and grounded.

3.8.2 Potential Transformers

Potential transformers are used to reduce high-voltage levels to measurable voltage levels. These transformers have a high-voltage primary winding and a low-voltage secondary winding wound concentrically. The secondary voltage is specified by the standards and can be $100/\sqrt{3}$ V, 100 V, or 100/3 V depending on whether the transformer is a phase-to-ground or phase-to-phase transformer. The magnetic core of a potential transformer usually has a shell-type construction in order to reduce the leakage flux and, consequently, have greater accuracy. Like current transformers, potential transformers are also permitted by the standards to have certain ratio and phase-angle errors.

Generally, the secondary winding of a potential transformer feeds a voltmeter, the voltage coil of a wattmeter, and/or a protective device. Therefore, a potential transformer operates under almost no-load conditions. If there are two secondary windings on the transformer, the one with greater accuracy is used for voltage measurements, while the other may be connected to a protective device.

The biggest problem with potential transformers is their insulation design. They have a high-voltage winding with thousands of turns using very thin conductors. Each coil layer has to be insulated so that no insulation breakdown will take place between the individual interlayers of the coil. Also, the high-voltage primary winding must be well isolated from the low-voltage secondary winding, as well as from the grounded parts of the transformer. The most common materials used in potential transformers are oil, oil-impregnated paper, some gaseous dielectrics such as sulfur hexafluoride, and epoxy resins.

3.9 THREE-PHASE TRANSFORMERS

In a three-phase electrical system, it is necessary to step-up or step-down the voltage levels at various points of the system. These transformations can be performed by means of transformer banks that consist of three identical single-phase transformers, one for each phase. Any type of three-phase connection can be carried out on both sides of the transformer bank, such as Y/Y, Y/Δ, Δ/Y and Δ/Δ. Since the individual transformers are separate, there is no common magnetic flux for the single-phase transformers in the three-phase bank. On the other hand,

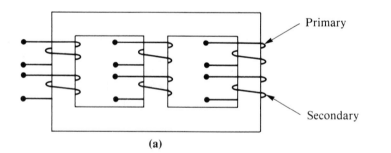

(a)

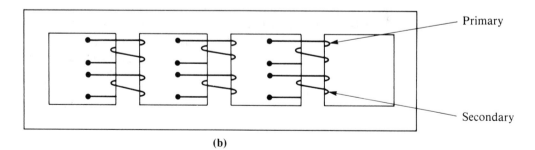

(b)

FIGURE 3.36
(a) Core and (b) Shell Construction of Three-Phase Transformers

it is not really practical to use three single-phase transformers since they require relatively larger space and their cost is relatively high. Consequently, three-phase transformers are designed with a common magnetic core as shown in Fig. 3.36, which helps to reduce core loss and thereby improves efficiency.

The major benefit of using a Y-connected winding in a transformer is that it provides a neutral point so that phase voltages are also available. Mainly, Δ-connected windings are used to suppress third harmonic currents in transformers, whereas third harmonic currents create a major problem in Y-windings. However, a major disadvantage of a Δ-winding in a transformer is that both ends of the individual windings are connected to the lines. Consequently, both terminals of the winding may experience overvoltages that occur in the system. Therefore, Δ-windings necessitate more expensive insulation as compared to Y-windings. To maintain the same flux density in the core, the mmf requirements for Y- and Δ-connected transformers must be equal. This requires the volt-per-turn ratio to be held constant. Therefore, a Δ-connected transformer requires $\sqrt{3}$ times as many turns for each of its phases as are needed for a Y-connected transformer.

3.9.1 Equivalent Circuit of Three-Phase Transformers

The equivalent circuit of a three-phase transformer is developed on a per phase basis for balanced loads. The representation of the equivalent circuit is the same as that for a single-phase transformer. However, the transformation ratio is defined in terms of the phase quantities as

$$a = \frac{V_{p1}}{V_{p2}}$$

The equivalent circuit parameters are obtained from the short-circuit and open-circuit tests on the transformer.

■ EXAMPLE 3.21

A 20-kVA, 220/110 V, 60-Hz, three-phase transformer yields the following data under short-circuit test:

$$V_{sc} = 16 \text{ V}, \qquad I_{sc} = 52.5 \text{ A}, \qquad P_{Cu} = 350 \text{ W}$$

If a Y-connected load with an impedance of $Z = 0.3 + j0.5 \; \Omega$ per phase is connected to the low-voltage winding of the transformer, determine the currents and power in both the high- and low-voltage windings. The rated voltage is maintained on the high-voltage side, and the core-loss resistance and the magnetizing reactance are negligible.

Solution

Since I_{sc} must be the rated current, we must determine on which side of the transformer the short-circuit test is performed. The current rating on the high-voltage side is

$$I_1 = \frac{20,000}{\sqrt{3} \times 220} = 52.5 \text{ A}$$

Hence, the test is performed on the high-voltage side, and the equivalent circuit constructed using the given test data will be referred to the high-voltage winding. On a per phase basis,

$$P_{Cu_p} = \frac{350}{3} = 116.67 \text{ W}$$

$$V_{sc_p} = \frac{16}{\sqrt{3}} = 9.24 \text{ V}$$

$$I_{sc_p} = 52.5 \text{ A}$$

Under load conditions,

$$V_{P1} = \frac{220}{\sqrt{3}} = 127 \text{ V}$$

$$V_{P2} = \frac{110}{\sqrt{3}} = 63.5 \text{ V}$$

and

$$a = \frac{127}{63.5} = 2$$

$$\phi_{sc} = \cos^{-1}\left(\frac{116.67}{9.24 \times 52.5}\right) = 76.08°$$

is the power-factor angle of the transformer during the short-circuit test. If $\tilde{V}_{sc} = 9.24\underline{/0°}$ V, then $\tilde{I}_{sc} = 52.5\underline{/-76.08°}$ A.

$$\hat{Z}_{e1} = \frac{\tilde{V}_{sc}}{\tilde{I}_{sc}} = \frac{9.24\underline{/0°}}{52.5\underline{/-76.08°}} = 0.176\underline{/76.08°}$$

$$= 0.042 + j0.170 \ \Omega$$

or

$$R_{e1} = 0.042 \ \Omega, \qquad X_{e1} = 0.170 \ \Omega$$

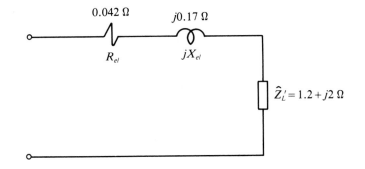

FIGURE 3.37
Approximate
Equivalent
Circuit of a
Three-Phase
Transformer on
a Per-Phase
Basis (for
Example 3.21)

Since the shunt elements are neglected, the per phase approximate equivalent circuit of the transformer is given in Fig. 3.37 referred to the high-voltage side, where $\hat{Z}'_L = a^2 \hat{Z}_L = 1.2 + j2.0 \ \Omega$.

The total impedance of the circuit is

$$\hat{Z} = 0.042 + j0.17 + 1.2 + j2.0 = 1.242 + j2.17$$

$$= 2.50\underline{/60.21°} \ \Omega$$

and

$$\tilde{I}_1 = \frac{\tilde{V}_1}{\hat{Z}} = \frac{127\underline{/0°}}{2.50\underline{/60.21°}} = 50.79\underline{/-60.21°} \ A$$

$$\hat{S}_1 = 3 \times \tilde{V}_{p1} \times \tilde{I}_1^*$$

$$= 3 \times 127\underline{/0°} \times 50.79\underline{/60.21°} = 19,352.53\underline{/60.21°} \ VA$$

$$S_1 = 19,352.53 \ VA$$

$$P_1 = 19,352.53 \times \cos 60.21 = 9614.77 \ W$$

$$Q_1 = 19,352.53 \times \sin 60.21 = 16,795.14 \ VAR$$

Since

$$\tilde{I}'_2 = \tilde{I}_1 = 50.79\underline{/-60.21°} \ A$$

$$\tilde{I}_2 = a\tilde{I}'_2 = 2 \times 50.79\underline{/-60.21°} = 101.58\underline{/-60.21°} \ A$$

$$\tilde{V}'_{P2} = 127\underline{/0°} - 50.79\underline{/-60.21°} \times 0.176\underline{/76.08°}$$

$$= 118.40 - j2.44 = 118.42\underline{/-1.18°} \ V$$

$$\tilde{V}_{P2} = \frac{\tilde{V}_2}{a} = \frac{118.42\underline{/-1.18}}{2} = 59.21\underline{/-1.18°} \ V$$

$$\tilde{V}_{2l} = \sqrt{3} \times 59.21\underline{/-1.18° + 30°} = 102.56\underline{/28.82°}$$

for positive phase sequence

$$\hat{S}_2 = 3 \times \bar{V}_2 \times \hat{I}_2^* = 3 \times 59.21\underline{/-1.18°} \times 101.58\underline{/60.21°}$$

$$= 18,043.63\underline{/59.03°} \text{ VA}$$

$$S_2 = 18,043.63 \text{ VA}$$

$$P_2 = 18,043.63 \cos 59.03° = 9285.06 \text{ W}$$

$$Q_2 = 18,043.63 \sin 59.03° = 15,471.27 \text{ VAR}$$

PROBLEMS

3.1. The magnetic core of a transformer is found to carry a flux of $0.075 \sin 314t$ Wb. Determine the rms value of the induced voltage in a 100-turn coil wound around the core using (a) Faraday's law and (b) the transformer equation. If the cross-sectional area of the core is 25 cm², what is the flux density in the core? Assume uniform flux distribution.

3.2. A coil wound around a magnetic material sets up a flux of 0.001 Wb (rms value) when 230 V, 60 Hz is applied to the coil. Under ideal conditions, determine the number of turns in the coil.

3.3. A transformer has 100 turns in the primary and 25 turns in the secondary. The secondary side is rated at 115 V. What is the nominal rating on the primary side?

3.4. A 10-kVA, 4000/200-V, single-phase ideal transformer is delivering the rated load (current) at unity power factor. The transformer is designed such that 2 V per turn are induced in its windings. Determine (a) the secondary winding current, (b) the primary winding current, (c) the number of turns in the secondary winding, and (d) the number of turns in the primary winding.

3.5. A 25-kVA transformer has 500 turns in the primary and 50 turns in the secondary. The primary is connected to a 3000-V, 50-Hz supply. Find the full-load currents on both sides, the induced voltage in the secondary winding, and the flux in the core. Assume ideal conditions.

3.6. If the transformer of Problem 3.5 is delivering the rated load at 0.8 power factor lagging, what is the power output of the transformer? Draw the phasor diagram.

3.7. A 300-kVA, 460/115-V ideal transformer is connected to a load impedance of $30 + j40 \Omega$. What is the primary winding current? What is the equivalent load impedance on the primary side? What is the power output? Draw the equivalent circuit as viewed from the primary side.

3.8. A 1-kVA, 480/120-V ideal transformer delivers the rated load at 0.6 power factor leading. Determine the load impedance.

3.9. A 200-turn coil wrapped around a magnetic material carries a current of 260 mA when connected to a 120-V, 60-Hz supply. The power input is 25 W. Calculate the equivalent circuit parameters of the coil. What is the inductance of the coil? What is the total reluctance of the magnetic path? Also determine the current in the core-loss and magnetizing branches.

3.10. In Problem 3.9 the mean length of the magnetic path is 100 cm and the cross-sectional area is 50 cm^2. Determine the relative permeability, flux density, and magnetic flux in the core using the transformer equation. Also calculate the induced voltage in the coil.

3.11. Express the transformer equation in terms of the self-inductance of the coil.

3.12. A 10-kVA, 3300/600-V step-down transformer draws no-load primary current of 0.8 A and dissipates 660 W as heat. Calculate the magnetizing and core-loss currents in the transformer. Also determine the equivalent core-loss resistance and magnetizing reactance of the core. Neglect the winding resistance and leakage reactance.

3.13. A 2200/220-V single-phase transformer is supplying a full load of 250 A at a lagging power factor of 0.707. The no-load current is 4 A at 0.2 pf lagging. Calculate the current in the primary winding.

3.14. A 100-kVA, 5000/500-V, 60-Hz transformer when tested gave the following data:

Open-circuit test: voltage = 500 V, current = 2.8 A, power = 1250 W

Short-circuit test: voltage = 400 V, current = 20 A, power = 4000 W

Determine the transformer parameters.

3.15. Using the approximate equivalent circuit referred to the low-voltage side of the transformer in Problem 3.14, determine the voltage that must be impressed on the high-voltage winding at full load with 0.8 pf lagging. Calculate the voltage regulation and efficiency of the transformer.

3.16. A 10-kVA, 330/220-V, 60-Hz transformer is tested with a 220-V winding open. With 330 V impressed, the input power is 87 W and the current is 1.1 A. If the test is performed by applying 220 V on the low-voltage side, determine the input power, no-load current, and power factor. What are the magnetizing reactance and equivalent core-loss resistance referred to the high-voltage side?

3.17. A 9.2-kVA, 230/115-V step-down transformer has the following parameters: $R_1 = 0.052$ Ω, $X_1 = 0.107$ Ω, $R_2 = 0.011$ Ω, $X_2 = 0.024$ Ω, $R_c = 434.78$ Ω, and $X_m = 243.9$ Ω. Determine the efficiency of the transformer under full-load condition with 0.6 pf lagging using an approximate equivalent circuit referred to the high-voltage side.

3.18. A 10-kVA, 2300/230-V transformer has the following parameters: $R_1 = 3$ Ω, $X_1 = 4 \Omega$, $R_2 = 0.03 \Omega$, $X_2 = 0.04 \Omega$, $R_c = 18 k\Omega$, and $X_m = 10 k\Omega$. Determine the efficiency and voltage regulation of the transformer at full load with unity power factor using an exact equivalent circuit.

3.19. The maximum flux density in a 6660/1665-V, 60-Hz transformer is 1.25 T with 1,000 turns in the primary winding. The relative permeability is 1500 at 1.25 T and the mean length of the transformer is 1.2 m. Assume the core loss is 4.2 W/kg and the density of the iron core is 7800 kg/m³. Determine (a) the net cross-sectional area of the core, (b) the no-load current in the primary, (c) the equivalent core-loss resistance, (d) the magnetizing reactance, and (e) no-load pf.

3.20. A 10-kVA, 3300/600-V step-down transformer has a core loss of 0.4 kW and a copper loss of 0.5 kW. Calculate the kVA loading at which the efficiency is maximum. At this loading, determine the efficiency if the transformer is operating at (a) 0.8 pf lagging, (b) unity pf, and (c) 0.8 pf leading.

3.21. Determine the per unit parameters for the transformer in Problem 3.18. Determine the efficiency and voltage regulation of the transformer at full load with unity power factor using per unit quantities and an exact equivalent circuit.

3.22. A 10-kVA, 100/900-V two-winding transformer is connected as a 1000/100-V step-down autotransformer with a load impedance of $3 + j4 \Omega$. Calculate load current, input current, and winding current in the autotransformer. What is the power delivered to the load? How much power is supplied by induction and by conduction?

3.23. A 1-kVA, 480:120-V ideal transformer is connected to form a 480:360-V autotransformer. What are the currents in the windings at full load with unity power factor? Determine the power transferred by induction and conduction.

3.24. A 12-kVA, 1732/200-V, Y-Δ connected three-phase transformer has 40 Ω and 1.0 Ω as the winding resistances measured between any two lines on the primary and secondary sides. The core loss is found to be 500 W. Determine the efficiency of the transformer on half-load at 0.8 pf lagging.

3.25. A 120-kVA, 10 kV/100 kV, Y-Y connected three-phase transformer has a total winding resistance of 10 Ω per phase and a leakage reactance of 30 Ω per phase referred to the primary side. Determine the voltage regulation and efficiency of the transformer on full load at unity power factor. Neglect the core loss.

CHAPTER 4

Direct-Current Generator

Commutator and Brush Arrangement (Courtesy
of Universal Electric Company)

The basic function of a direct-current (dc) generator is to provide a dc source. In all machines, either mechanical energy is converted into electrical energy or vice versa. In a dc generator, since the output is electrical energy, the input must be mechanical in nature. Therefore, a dc generator converts mechanical energy into electrical energy. In dc generators, the conductors or the coils are moved in a constant magnetic field so that the voltage is induced by flux-cutting action (motional emf). The constant magnetic field is provided by mounting permanent magnets or field windings on the inner surface of the outer part, known as a *stator*, of the generator. A rotating part, known as an *armature*, is mechanically supported within the stator by the shaft bearings. The armature is free to rotate in either direction and houses the coils or conductors.

4.1   MECHANICAL CONSTRUCTION

Physically, there is no difference between the construction of a dc generator and a dc motor. Some of the essential parts of a dc machine are shown in Fig. 4.1 and are described next.

Main Frame   The main frame or the field yoke of a dc machine serves the basic function of providing a highly permeable return path for the magnetic field. For small motors, it can be a rolled-ring structure welded at ends. For large machines,

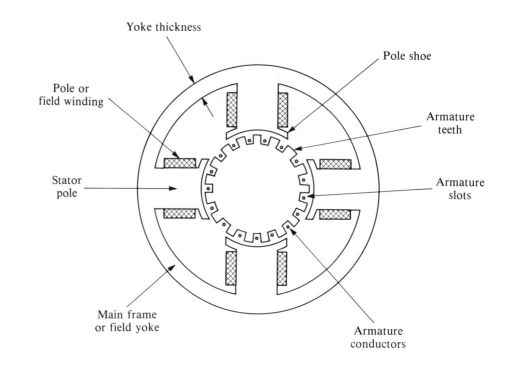

FIGURE 4.1
Parts of a dc
Machine

the yoke may be die cast in sections and finally bolted together. The maximum flux density requirements in the yoke stipulate the thickness of the yoke section.

Field Poles   Field poles are mounted inside the yoke and are properly designed to accommodate field windings. Usually, but not always, in dc machines, the field poles are made of thin laminations stacked together to form the required thickness of the pole to minimize the eddy-current losses due to the pole's proximity to the armature winding. For large machines, the field poles are built separately and then bolted to the main frame. A typical sketch of a field pole and the field winding is shown in Fig. 4.2. There may be more than one set of field windings on the same pole. The field poles must alternate in their polarity when placed inside the yoke. As the conductor moves under a pole, the voltage induced in that conductor builds up from zero to maximum and then back to zero. This corresponds to one-half of the induced voltage cycle (i.e., 180° electrical). Thus, theoretically, a pole must span a distance equivalent to 180° (electrical). Let us define the 180° (electrical) span of a pole as the *pole pitch*. However, from practical winding considerations, a pole usually covers only a part of the required span. Instead of the field windings, we can make use of permanent magnets to provide the required flux, as illustrated by Example 1.18.

Armature   The rotating part of the dc machine is known as the armature. The armature is made of thin, highly permeable steel laminations, as shown in Fig. 4.3, to reduce eddy-current losses. The armature coils or conductors are placed in slots punched in the armature laminations. Usually, copper wire is used for

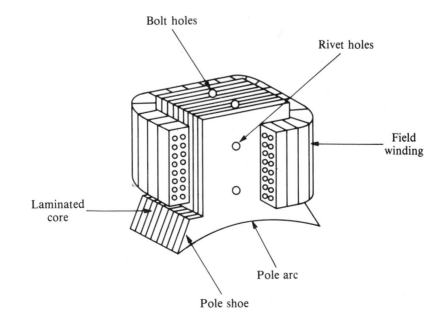

Bolt holes

Rivet holes

Field winding

Laminated core

Pole arc

Pole shoe

FIGURE 4.2
Sketch of a Field Pole and the Field Winding

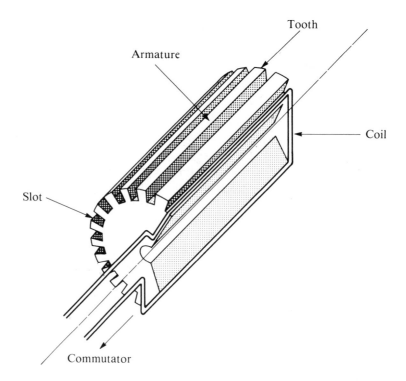

FIGURE 4.3
Sectional View
of an Armature

the armature winding due to its low resistivity. Since the magnetic flux is provided by the poles and the conductors are rotated at uniform speed for voltage generation, these types of machines are referred to as *revolving-armature types*.

Commutator   The commutator is made of hard-drawn copper segments, as shown in Fig. 4.4, which are insulated from one another by mica segments. Silver is often alloyed with copper to raise its conductivity and softening temperature in order to maintain its hardness and reduce wear. As we will see later, the commutator is a very well conceived device that serves the function of a rectifier. It converts alternating voltage induced in the armature coils into dc voltage.

Brushes   Brushes are used to provide a path for the current flow between the stationary and rotating parts of the generator. Brushes are made of carbon, carbon graphite, or a copper-filled carbon mixture. There are many different brush grades depending on their composition. These brushes make contact with the commutator segments. When properly seated under pressure, the brushes are normally selected to have a small voltage drop across them.

Armature Windings   Two types of widely used windings are the lap winding and the wave winding. The lap winding is the most common. The lap winding requires

Insulator (mica)

Commutator
segment
(copper)

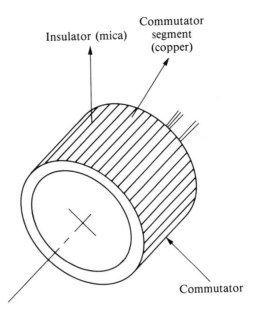

FIGURE 4.4
Commutator

Commutator

as many parallel paths for the armature winding currents to flow as there are poles. The wave winding is so arranged that it requires only two parallel paths. The winding may be simplex, duplex, and so on. Simplex windings are usually used. For simplex lap winding, the number of coils in the armature must be equal to the number of armature slots. In other words, each slot has two coil sides of two different coils. In all cases, the armature winding must close onto itself. Discussion of lap and wave windings is given in Appendix A.

4.2 INDUCED EMF IN A COIL

Consider the bipolar machine shown in Fig. 4.5 with a single-turn, full-pitch coil wound around the armature. The armature can be freely rotated between the north and south poles in either direction by coupling it to a prime mover. The mechanical power thus supplied to rotate the armature becomes the power input to the machine. Most of this power is converted into electrical power.

As the armature is made of thin laminations of highly permeable magnetic material, the lowest reluctance path for the magnetic field will be radial, as shown in Fig. 4.6a. Thus, the magnetic flux density distribution will theoretically be rectangular, as indicated by the developed diagram (Fig. 4.6b). However, the fringing of the flux at the pole tips rounds off the sharp corners, as shown in Fig. 4.6c. The area under each loop is proportional to the useful flux per pole and is independent of the shape of the curve, provided successive half-loops are symmetrical. The voltage induced in a conductor of length l, moving with a uniform

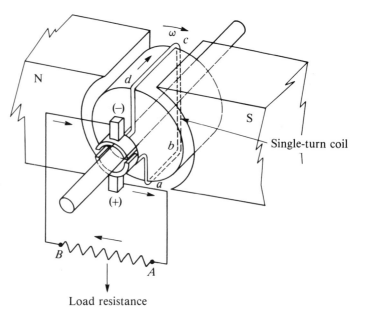

FIGURE 4.5
Elementary dc
Generator
Showing a
Single-Turn Coil
Connected to a
Two-Segment
Commutator

peripheral velocity v, will also have the same time variation as the space distribution of the flux density as illustrated in Figure 4.6(d). The induced voltage is definitely not sinusoidal.

The nonsinusoidal flux density distribution is periodic and single-valued and does not include any finite discontinuities. It can therefore be resolved into a fundamental sine wave and a series of higher-space harmonics in accordance with Fourier's theorem. Since the flux density distribution curve, as shown in Fig. 4.6c, is an odd function, there will not be any even-order harmonics. Figure 4.6e shows the fundamental-, third-, and fifth-harmonic flux density distribution for the bipolar machine. The voltage induced in each conductor will also show the presence of these harmonics. However, when the conductors are grouped together to form coils, the like frequency emfs will combine together to yield the resultant emfs. The manner in which they combine is not the same for all harmonics. Therefore, the final waveform of the induced voltage as a function of time will not be the same as the space waveform of flux density. Thus, to simplify the analysis we will concentrate only on the fundamental waveform, as shown in Fig. 4.6f. The voltage induced in each conductor will now follow this flux density distribution. However, the flux linking the coil would be as shown in Fig. 4.6g and can be expressed as

$$\phi = \phi_p \cos \theta \qquad (4.1)$$

where ϕ_p is the total flux per pole and θ represents the angular position of the

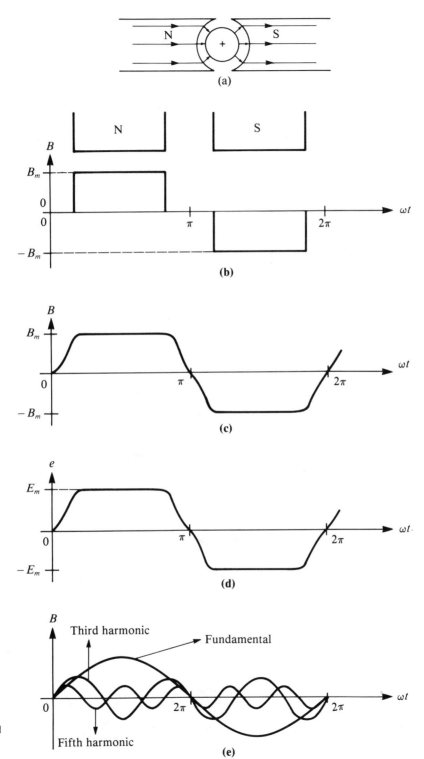

FIGURE 4.6
Flux Density and
Induced emf
Waveforms

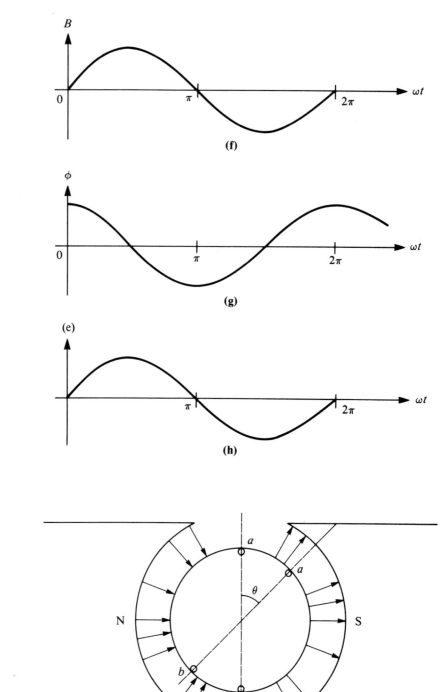

FIGURE 4.6
(continued)

FIGURE 4.7
Flux Linking a
Single-Turn Coil

coil at any time t, as shown in Fig. 4.7. The flux per pole is

$$\phi_p = \int_0^{\pi D/2} B_m \sin \theta \left(\frac{lD}{2}\right) d\theta = B_m Dl \qquad (4.2)$$

where D is the diameter of the armature, l is the length of the armature, and B_m is the maximum flux density as indicated in Fig. 4.6f. In accordance with Faraday's law, the induced voltage in the coil is

$$e = -\frac{d\phi}{dt} = \phi_p \sin \theta \frac{d\theta}{dt} \qquad (4.3)$$

where $d\theta/dt$ is the angular velocity of the coil, ω. Equation (4.3) can be rewritten as

$$e = \phi_p \omega \sin \omega t \qquad (4.4)$$

and its plot is given in Fig. 4.6h. It is evident that one complete revolution of the armature in a bipolar machine corresponds to one complete waveform of the induced voltage. In other words, it takes a pair of poles to generate a full cycle of induced voltage in a coil. If there are P poles, the frequency of the induced voltage in the armature conductors per revolution will be

$$f = \frac{P}{2} \text{ cycles/revolution} \qquad (4.5)$$

A sketch illustrating that the frequency of the induced voltage is proportional to the number of pair of poles is drawn in Fig. 4.8 for 2-, 4-, and 6-pole machines. If the armature is being rotated at a speed of N_m revolutions per minute (rpm), the frequency of the induced voltage will be

$$f = \frac{P}{2} N_m \text{ cycles/min}$$

or $$f = \frac{PN_m}{120} \text{ Hz (cycles/s)} \qquad (4.6)$$

If the speed of rotation of the armature is given in terms of its angular velocity ω_m (rad/s), the speed in rpm will be

$$N_m = \frac{60}{2\pi} \omega_m \text{ rpm} \qquad (4.7)$$

and $$f = \frac{P}{4\pi} \omega_m \text{ Hz} \qquad (4.8)$$

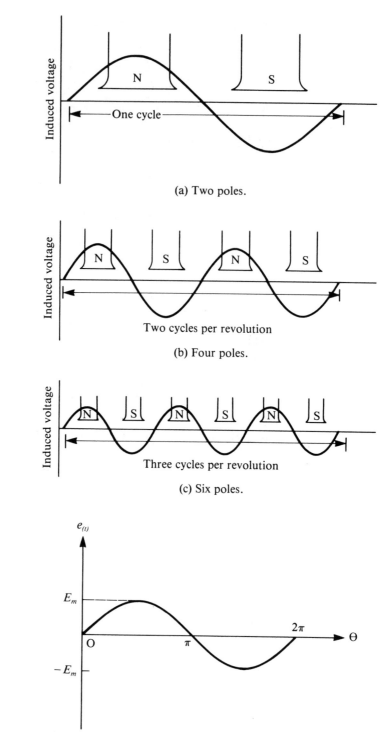

(a) Two poles.

(b) Four poles.

(c) Six poles.

FIGURE 4.8
Illustration Showing that the Cycles per Revolution Are Proportional to the Number of Poles: (a) Two Poles; (b) Four Poles; (c) Six Poles

FIGURE 4.9
Voltage Induced in a Coil

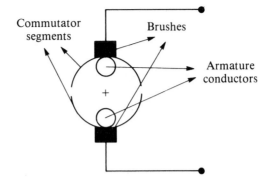

Commutator segments

Brushes

Armature conductors

+

FIGURE 4.10
Simplified Diagram of a Commutator and a Brush Arrangement

Since the angular frequency is

$$\omega = 2\pi f$$

then

$$\omega = \frac{P}{2} \omega_m \qquad (4.9)$$

or

$$\theta = \frac{P}{2} \theta_m \qquad (4.10)$$

where θ and θ_m are the electrical and mechanical angles of the coil at any instant of time t.

The induced voltage in a single-turn coil, Eq. (4.4), can now be expressed as

$$e = \frac{P}{2} \omega_m \phi_p \sin\left(\frac{P}{2} \omega_m t\right) \qquad (4.11)$$

If instead of one turn, there are in fact N_c turns in the coil, the total voltage induced will be

$$e_c = \frac{P}{2} \omega_m \phi_p N_c \sin\left(\frac{P}{2} \omega_m t\right) \qquad (4.12)$$

The induced voltage in the coil as a function of electrical angle θ is shown in Fig. 4.9. The maximum value of the induced voltage is

$$E_M = \frac{P}{2} \omega_m \phi_p N_c \qquad (4.13)$$

As is obvious from Eq. (4.12), the induced voltage in the armature winding is sinusoidal in nature and must be rectified for a direct voltage output. This can be achieved by the commutation action, which is related to the current reversal in the armature conductors when they pass through the zone where they are short-circuited by the brushes. A schematic diagram of a commutator and the brush arrangement is shown in Fig. 4.10 for a single-coil armature winding. As can be seen in this figure, each commutator segment is attached to one end of the coil, while the brushes providing connection to the external circuit touch the commutator segments. With this arrangement, a unipolar voltage is always obtained across the brushes, as illustrated in Fig. 4.11, regardless of the induced voltage in the armature winding.

The average value of the induced voltage is then

$$E = \frac{2}{\pi} E_M = \frac{P}{\pi} \omega_m \phi_p N_c \qquad (4.14)$$

Usually, more than one coil is wound around the armature and all of them are connected in series to form a closed armature winding. Let C be the total number of coils, which is also equal to the total number of slots on the armature. However, when the brushes are placed on the commutator, they help form the parallel paths for the current to flow in the armature winding. The number of parallel paths for the wave winding is always 2, whereas for the lap winding there are as many parallel paths as there are poles. In general, let us assume that there are a parallel paths. Total turns per parallel path are then equal to $N_c C/a$. Thus, the average voltage induced per parallel path is actually the generated voltage that is available at the brushes. The average value of this generated voltage is

$$E_g = \frac{P}{\pi} \omega_m \phi_p \frac{N_c C}{a} \qquad (4.15)$$

Since there are two conductors for each turn, the total number of conductors in the armature is

$$Z = 2N_c C \qquad (4.16)$$

or

$$N_c C = \frac{Z}{2}$$

Substituting for $N_c C$ in Eq. (4.15), we get

$$E_g = \frac{PZ}{2\pi a} \omega_m \phi_p \qquad (4.17)$$

This expression can be written as

$$E_g = K_a \omega_m \phi_p \qquad (4.18)$$

where

$$K_a = \frac{ZP}{2\pi a} \qquad (4.19)$$

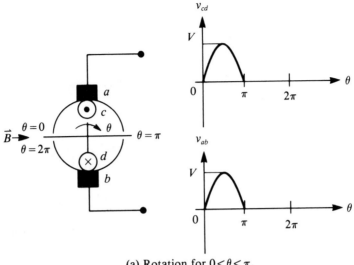

(a) Rotation for $0 < \theta < \pi$.

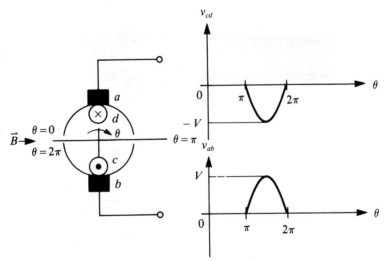

(b) Rotation for $\pi < \theta < 2\pi$.

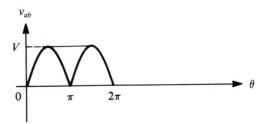

(c) Voltage across the brushes for one revolution of the armature.

FIGURE 4.11
Commutation in
a dc Generator.

K_a is a constant for a given machine and is usually referred to as the machine constant.

Equation (4.18) can also be written in terms of the armature speed in rpm, as follows:

$$E_g = K'_a \phi_p N_m \qquad (4.20)$$

where the new machine constant is

$$K'_a = \frac{ZP}{60a} \qquad (4.21)$$

For a given dc generator, the induced voltage is a function of armature speed and the flux per pole. In a wound machine with N_f turns per pole, this flux can be increased or decreased by controlling the field current I_f, whereas for a permanent-magnet machine the flux would be constant.

4.3 MAGNETIZATION CHARACTERISTIC

The magnetization or no-load characteristic for a dc generator is obtained by running the generator at its rated speed with no load and varying the field current. If the armature speed is held constant, the dc voltage generated will be proportional to the flux set up by the field winding. The field winding, therefore, is excited separately to control the flux. As the magnetizing flux per pole is increased by increasing the current in the field winding, the induced emf in the armature coils increases as well. In the beginning, the increase in the induced emf is almost linear with the increase in the field current. This is true because the magnetic circuit is far from its saturation. As the saturation of the magnetic circuit becomes a significant factor, the induced voltage deviates from the linear relation with the further increase in the field current. The curve that depicts the behavior of the induced emf as a function of field mmf or just the field current is known as the magnetization characteristic for the machine. Such a curve is shown in Fig. 4.12. Note that the curve does not start at zero. In other words, there is some induced voltage even when there is no current in the field winding. This is due to residual magnetism in the poles. Since this curve is produced by increasing the field current, it can be labeled the ascending curve. A descending curve can also be obtained by reducing the field current. If both curves are plotted on the same graph, the descending curve will be above the ascending curve due to hysteresis.

■ EXAMPLE 4.1

A 24-slot, 2-pole, dc generator has 18 turns per coil in its armature winding. The field excitation is adjusted to 0.05 Wb per pole and the armature angular velocity

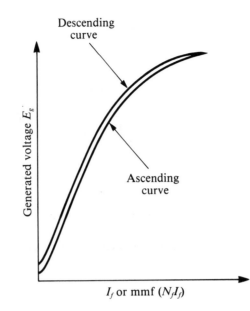

Descending
curve

Ascending
curve

Generated voltage E_g

I_f or mmf ($N_f I_f$)

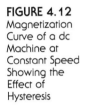

FIGURE 4.12
Magnetization
Curve of a dc
Machine at
Constant Speed
Showing the
Effect of
Hysteresis

is 183.2 rad/s. Determine (a) the total induced voltage, (b) induced voltage per turn, (c) induced voltage per coil, and (d) induced voltage per conductor.

Solution

$$\text{Number of coils} = \text{number of slots} = 24$$

$$\text{Total number of turns} = 18(\text{turns/coil}) \times 24(\text{coils}) = 432$$

$$\text{Total conductors} = Z = 2 \times 432 = 864$$

and

$$K_a = \frac{PZ}{2\pi a} = \frac{864 \times 2}{2 \times \pi \times 2} = 137.51$$

where $a = 2$.

(a) Thus, the total induced voltage is

$$E_g = K_a \phi_p \omega_m$$

$$= 137.51 \times 0.05 \times 183.2 = 1260 \text{ V}$$

(b) Since there are two parallel paths, the number of turns in each parallel path = total turns/a = 432/2 = 216. Therefore, E_g is the total voltage induced in 216 turns all connected in series. Thus, the induced voltage per turn is

$$E_g/\text{turn} = \frac{1260}{216} = 5.833 \text{ V}$$

(c) There are 18 turns per coil; the voltage induced per coil is

$$E_g/\text{coil} = 5.833 \times 18 = 105 \text{ V}$$

(d) Since a turn has two conductors, the voltage induced per conductor is

$$E_g/\text{conductor} = \frac{5.833}{2} = 2.92 \text{ V}$$

■ EXAMPLE 4.2

A simplex lap-wound, 6-pole dc generator has 360 conductors on the armature. The poles are 8 in. square and the flux density per pole is 75,000 lines/in.^{2}. Determine the generated voltage if the armature speed is 850 rpm.

Solution

The flux per pole is

$$\phi_p = BA = 75,000 \times 8 \times 8 = 4.8 \times 10^6 \text{ Lines}$$

$$= 0.048 \text{ Wb}$$

because 1 Wb = 10^8 lines of flux.

$$K'_a = \frac{ZP}{60a} = \frac{360 \times 6}{60 \times 6} = 6$$

as $a = P = 6$ (lap winding). The induced voltage is

$$E_g = K'_a \phi_p N_m$$

$$= 6 \times 0.048 \times 850 = 245 \text{ V}$$

4.4 DEVELOPMENT OF ELECTROMAGNETIC TORQUE

The load on the generator is defined in terms of the generator current. That is, a generator is said to be running under no load when the generator current is zero. On the other hand, the generator is operating on full load when it supplies the rated current.

Assume that the generator is supplying a load of I_a when the generated voltage is E_g. The power developed by the generator is

$$P_d = E_g I_a \tag{4.22}$$

As discussed in Chapter 2, the power developed can be expressed in terms of the torque developed by the machine as

$$T_d = \frac{P_d}{\omega_m} = \frac{E_g I_a}{\omega_m} \tag{4.23}$$

But

$$E_g = K_a \phi_p \omega_m$$

Thus,

$$T_d = K_a \phi_p I_a \text{ N-m} \tag{4.24}$$

4.4.1 Torque Developed from a Different Perspective

From electromagnetic field theory, the force experienced by a current-carrying conductor of length $\bar{l}$ immersed in a magnetic flux density, $\vec{B}$, is given as

$$\vec{F} = I(\bar{l} \times \vec{B}) \tag{4.25}$$

Let θ be the angle between the length of the conductor and the magnetic flux density; then the magnitude of the force is

$$F = lIB \sin \theta$$

For machines as shown in Fig. 4.13, $\theta = 90°$. Therefore,

$$F = lIB \tag{4.26}$$

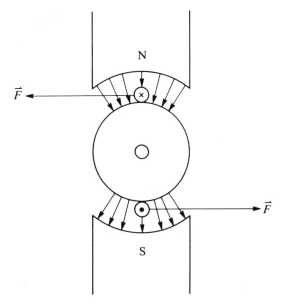

FIGURE 4.13
Force Acting on a Current-Carrying Conductor in a Magnetic Field

Since the armature conductors are immersed in a magnetic field and are also carrying the load current, a force is experienced by each of these conductors. Since the current in the conductors under the north pole is in opposite direction to that under the south pole and the same is true for the flux as it leaves the north pole and enters the south pole, the force acting on all the conductors will be in the same direction. The direction of the force acting on the two conductors of a single-turn generator discussed earlier is shown in Fig. 4.13. However, as the conductors are evenly distributed around the periphery of the armature, the flux density B linking all the conductors is not the same, as is evident from Fig. 4.6f.

Consider an arc of length dx around the armature circumference at a distance x from the center of the interpolar space, as shown in the developed diagram (Fig. 4.14), where the flux density is $B(x)$ and the number of conductors enclosed by this arc is $Z\,dx/\pi D$, where D is the diameter of the armature and Z is the total number of conductors. If I_a is the armature current and a is the number of parallel paths, then the current in each conductor will be I_a/a. The differential force experienced by these conductors is

$$df = B(x)l\,\frac{I_a}{a}\,\frac{Z\,dx}{\pi D} \tag{4.27}$$

where l is the axial length of the armature. The differential torque is

$$dT = \frac{1}{2}\,\frac{lZI_a}{\pi a}\,B(x)\,dx \tag{4.28}$$

The total torque is obtained by integrating this equation from $x = 0$ to $x = \pi D = P\tau$, where τ is arc length for each of the poles. Since the torque produced by each of the poles is the same, the integration can be carried out from $x = 0$ to $x = \tau$ and the result multiplied by P. Thus, the total electromagnetic torque

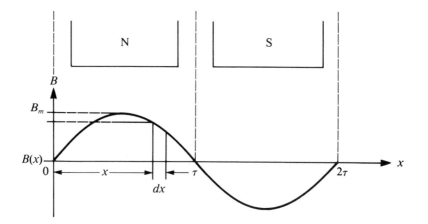

FIGURE 4.14
Approximate
Flux-Density
Distribution in a
Bipolar dc
Machine

developed is

$$T = \frac{ZPI_a}{2\pi a} \int_0^\tau lB(x)\, dx$$

However, as explained earlier,

$$\int_0^\tau lB(x)\, dx = \phi_p$$

Therefore, the torque equation becomes

$$T = K_a \phi_p I_a \tag{4.29}$$

where
$$K_a = \frac{ZP}{2\pi a}$$

This is the same result as obtained earlier from the power considerations.

4.5 TYPES OF DC GENERATORS

Direct-current generators may be divided into two classes based on the mode of excitation. They are (1) separately excited generators, and (2) self-excited generators.

4.5.1 Separately Excited Generators

Separately excited generators require a separate dc voltage source for the field winding and for that reason are not frequently used. However, separate excitation is the preferred technique used to determine the basic magnetization characteristics of a dc machine.

The equivalent-circuit representation of a separately excited generator is shown in Fig. 4.15 under steady-state conditions. Steady-state conditions imply that there are no appreciable changes in either the armature current or the rotational speed for any given load. In other words, there are no changes in the mechanical or magnetic energy of the system. Therefore, there is no need to include the inductance and the inertia of the system as part of the equivalent circuit.

In the equivalent circuit, E_g is the generated voltage in the armature windings, R_a is the effective armature winding resistance, which may include the resistance of the brushes, I_a is the armature current, which is also the load current, V_t is the terminal voltage, I_f is the field current, and V_f is the externally applied field voltage.

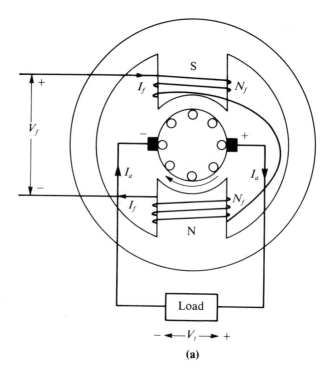

(a)

FIGURE 4.15
(a) Cross-
Sectional View
of a Separately
Excited
Generator; (b)
Equivalent-
Circuit
Representation

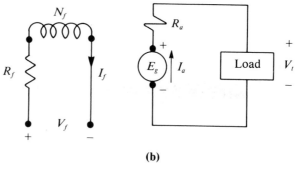

(b)

It is obvious from the circuit that the terminal voltage is

$$V_t = E_g - I_a R_a \tag{4.30}$$

Since the flux per pole is proportional to the field current (i.e., $\phi_p = k_f I_f$), the induced voltage, Eq. (4.18), can be rewritten as

$$E_g = K\omega_m I_f \tag{4.31}$$

where $K = K_a k_f$. Note that k_f depends on the permeability of the magnetic materials involved in the magnetic circuit. Under constant-speed operation, the induced voltage is proportional to the field current. On the other hand, if the field current is held constant, the induced voltage is proportional to the speed of rotation.

■ EXAMPLE 4.3

The rated speed of a 240-V, 40-A separately excited dc generator is 2000 rpm. The armature winding resistance is 0.4 Ω. Determine the no-load voltage at the rated speed. If the field current is held constant, what will the full-load voltage be at 1200 rpm?

Solution

The power output is $P_o = V_t I_a = 240 \times 40 = 9600$ W. Under no-load conditions, $I_a = 0$. From the equivalent circuit, the no-load voltage will be the same as the generated voltage. If the speed is held constant, the generated voltage will be the same under all loading conditions. Thus, at full load,

$$E_{g1} = V_t + I_a R_a = 240 + 40 \times 0.4 = 256 \text{ V}$$

For the constant field current, the voltage generated per unit speed is constant. That is,

$$\frac{E_{g1}}{\omega_{m1}} = \frac{E_{g2}}{\omega_{m2}}$$

or

$$E_{g2} = \frac{1200 \times 256}{2000} = 153.6 \text{ V}$$

The full-load current is still 40 A. Therefore, the full-load voltage at 1200 rpm will be

$$V_{t2} = E_{g2} - I_a R_a = 153.6 - 40 \times 0.4 = 137.6 \text{ V}$$

■ EXAMPLE 4.4

A 2000-kW, 2000-V, 2400-rpm separately excited dc generator has an armature circuit resistance of 0.02 Ω. The flux per pole is 0.4 Wb. Calculate (a) the induced voltage, (b) the machine constant, and (c) the torque developed at the rated conditions.

Solution

(a) The armature current is

$$I_a = \frac{P_o}{V_t} = \frac{2000 \times 10^3}{2000} = 1000 \text{ A}$$

The generated voltage is

$$E_g = V_t + I_a R_a = 2000 + 1000 \times 0.02 = 2020 \text{ V}$$

(b)
$$\omega_m = \frac{2\pi N_m}{60} = \frac{2 \times \pi \times 2400}{60} = 251.33 \text{ rad/s}$$

The machine constant is

$$K_a = \frac{E_g}{\phi_p \omega_m} = \frac{2020}{0.4 \times 251.33} = 20.09$$

(c) The torque developed is

$$T_d = K_a \phi_p I_a = 20.09 \times 0.4 \times 1000 = 8036 \text{ N-m}$$

4.6 SELF-EXCITED GENERATORS

Self-excited generators have field windings excited by their own generated voltages. Basically, there are three different types of self-excited generators.

1. Shunt generators
2. Series generators
3. Compound generators

4.6.1 Shunt Generator

The field winding of a shunt generator is connected in parallel with the armature. Since the generated voltage is usually very high, the resistance of the shunt field winding must be very high in order to limit the current in that winding. Therefore, the shunt winding has a comparatively high number of turns of relatively fine wire. The cross-sectional view of a shunt generator is shown in Fig. 4.16. Further reduction in field current can be obtained by adding external resistance in the field-winding circuit.

Voltage Buildup As long as there is some residual magnetism in the magnetic poles of the generator, the voltage buildup will take place if and only if the following conditions are satisfied:

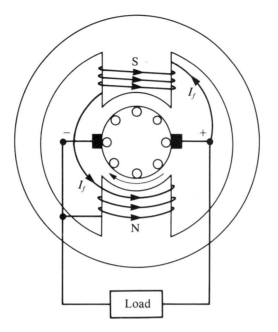

FIGURE 4.16
Cross-Sectional
View of a dc
Shunt Generator

1. The total shunt-winding resistance must be less than the *critical resistance*. Critical resistance is defined as the resistance in the shunt-field circuit below which the voltage buildup will take place. This value of the resistance can be determined from the magnetization curve. The slope of the magnetization curve in the linear region is the critical resistance. For a given field-winding resistance, a line known as the *resistance line* can be drawn on the magnetization curve of the generator. The point of intersection of the two curves corresponds to the voltage produced by the generator. Figure 4.17 shows the critical resistance line and a set of resistance lines for obtaining different voltages from the same generator.

2. Once the field resistance is known, there is a critical value of armature speed under which no voltage buildup will take place, as indicated by Fig. 4.18.

3. *Direction of rotation:* There is a definite relation between the direction of rotation and the field-winding connections to the armature terminals. Initially, the voltage being generated is due to the residual flux in the machine. This low voltage causes a current flow in the field windings, which in turn produces the magnetic flux. If this flux aids the residual flux, voltage *buildup* will take place as illustrated in Fig. 4.19. Otherwise, voltage *build down* will occur. The problem can be corrected either by reversing the direction of rotation or by interchanging the field terminals with respect to armature terminals, but not both.

When the generator is being run for the first time, there will be no residual

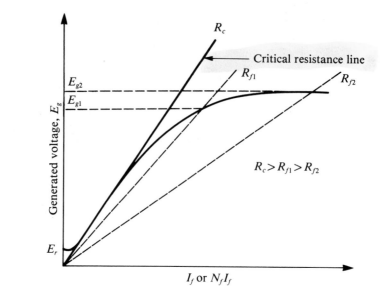

FIGURE 4.17
Voltage Buildup
for a Shunt
Generator for
Different Values
of Field
Resistance

flux in the machine. To establish the residual flux, the machine must be run as a separately excited generator.

Equivalent Circuit   An equivalent circuit of a shunt generator is shown in Fig. 4.20. To control the current in the shunt winding and thereby the flux per pole produced by it, an external resistance may be added as shown. Under no-load conditions, the armature current is equal to the field current. However, when the

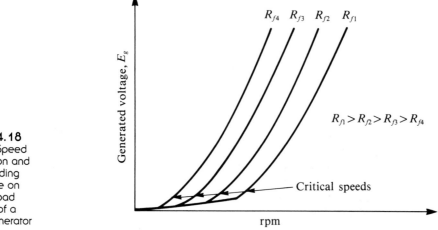

FIGURE 4.18
Effect of Speed
of Rotation and
Field-Winding
Resistance on
the No-Load
Voltage of a
Shunt Generator

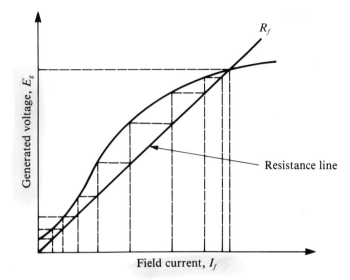

FIGURE 4.19
Voltage Buildup
in a Self-Excited
Shunt Generator

generator is loaded, the armature current supplies both the field current and the load current. Thus

$$I_a = I_L + I_f \qquad (4.32)$$

and the terminal voltage is

$$V_t = E_g - I_a R_a \qquad (4.33)$$

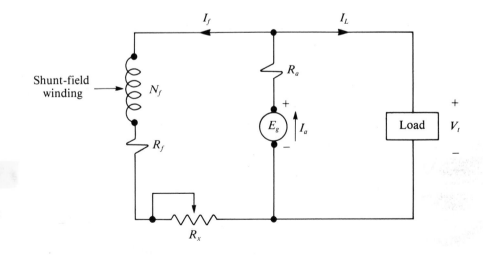

FIGURE 4.20
Equivalent-
Circuit
Representation
of a Shunt
Generator

4.6.2 Series Generator

As the name implies, the field winding is connected in series with the armature circuit, as shown in Fig. 4.21. Under no-load conditions the induced voltage is the voltage generated by the residual magnetism. If the direction of the current in the field winding under load is such that the flux produced by it aids the residual flux, voltage buildup will take place. Since the load current may be quite high, the series winding usually has few turns of relatively thicker wire. To control the slope of the generated voltage curve, a resistor known as a *series–field diverter* may be connected across the field winding.

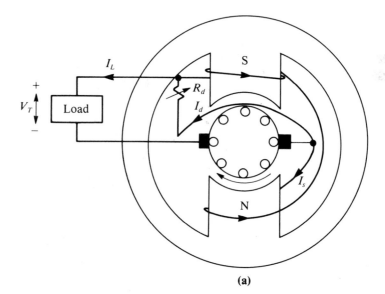

(a)

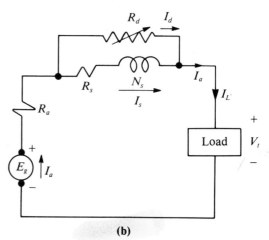

FIGURE 4.21
(a) Cross-Sectional View of a Series Generator; (b) Its Equivalent-Circuit Representation

(b)

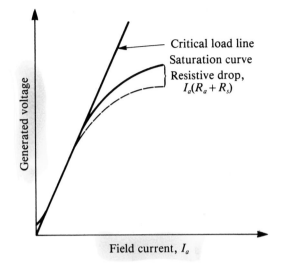

FIGURE 4.22
Voltage Buildup
in a Series
Generator

The voltage generated in a series generator is therefore an inverse function of the load resistance. A basic curve illustrating this principle is shown in Fig. 4.22.

If I_a, I_s, I_d, and I_L are the armature, series–field, diverter, and load currents, respectively, then

$$I_a = I_L = I_s + I_d \tag{4.34}$$

and

$$I_s R_s = I_d R_d \tag{4.35}$$

where R_s and R_d are the series–field and diverter resistances, respectively. The terminal voltage is then

$$V_t = I_L R_L = E_g - I_a R_a - I_s R_s \tag{4.36}$$

4.6.3 Compound Generator

A generator that has both series and shunt field windings is known as a compound generator. If the mmf of the series field is aiding the mmf of the shunt field, it is referred to as a *cumulative compound generator*. However, a generator is termed a *differential compound generator* if the mmfs of two fields oppose each other. Figures 4.23 and 4.24 show, respectively, the schematic representations for the cumulative and the differential compound generators.

Depending on the connections of the shunt-field windings with respect to armature terminals, a compound generator may be called a *short-shunt* (Fig. 4.25) or a *long-shunt* (Fig. 4.26) generator.

Three levels of compounding can be achieved by adjusting the series-field and shunt-field windings. Most often, the shunt-field winding is designed to pro-

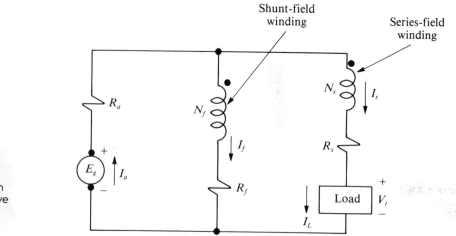

FIGURE 4.23
Schematic
Representation
of a Cumulative
Compound
Generator

vide almost all the flux in the generator. The series-field winding mainly provides control over the total flux. Therefore, the total flux in the generator may be higher at full load than at no load.

Normal or Flat Compound Generator A generator is said to be normally compounded if the effect of series winding is such that the generator has the same terminal voltage at both no load and full load.

Overcompound Generator In an overcompound generator, the full-load voltage is higher than the no-load voltage.

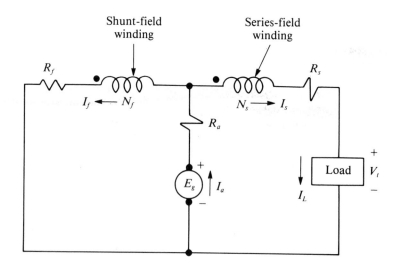

FIGURE 4.24
Schematic
Representation
of a Differential
Compound
Generator

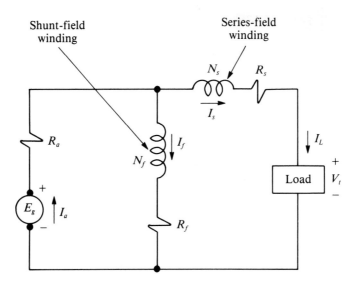

FIGURE 4.25
Short-Shunt
Compound
Generator

Undercompound Generator   In an undercompound generator, the no-load voltage is higher than the full-load voltage. Such a condition is often undesirable and is seldom incorporated in the design of a compound generator.

The normal practice is to design an overcompound generator. The adjustments can then be made in the series-field current by using the diverter resistance to satisfy the load requirements.

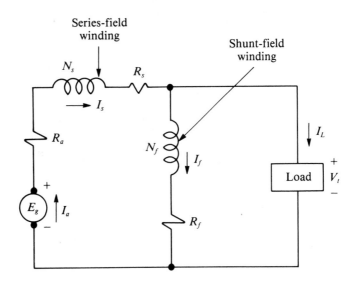

FIGURE 4.26
Long-Shunt
Compound
Generator

■ **EXAMPLE 4.5**

A 50-kW, 120-V shunt generator has an external resistance of 15 Ω in addition to its winding resistance of 30 Ω. The armature resistance is 0.09 Ω. Calculate (a) full-load current, (b) field current, (c) armature current, and (d) generated voltage.

Solution

(a) The load current is $I_L = 50,000/120 = 416.67$ A.

(b) In a shunt generator, $V_f = V_t = 120$ V. The total field resistance is $R_f = 30 + 15 = 45$ Ω. Therefore,

$$I_f = \frac{V_f}{R_f} = \frac{120}{45} = 2.667 \text{ A}$$

(c) The armature current is $I_a = I_L + I_f = 416.67 + 2.667 = 419.34$ A.

(d) From the equivalent circuit, the generated voltage is

$$E_g = V_t + I_a R_a = 120 + 419.34 \times 0.09 = 157.74 \text{ V}$$

4.7 ARMATURE REACTION

A bipolar generator with the current in the armature conductors is shown in Fig. 4.27. Its developed diagram is given in Fig. 4.28. Figure 4.28a illustrates the flux-density distribution for each pole. It also highlights the mechanical neutral position where the brushes must be positioned. This is also the magnetic neutral axis for the generator.

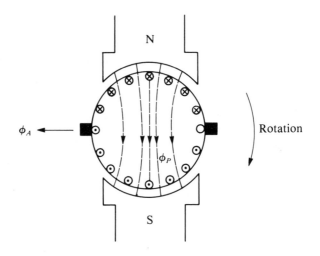

FIGURE 4.27
Direction of
Armature Flux
with Respect to
Main-Pole Flux
for a dc
Generator

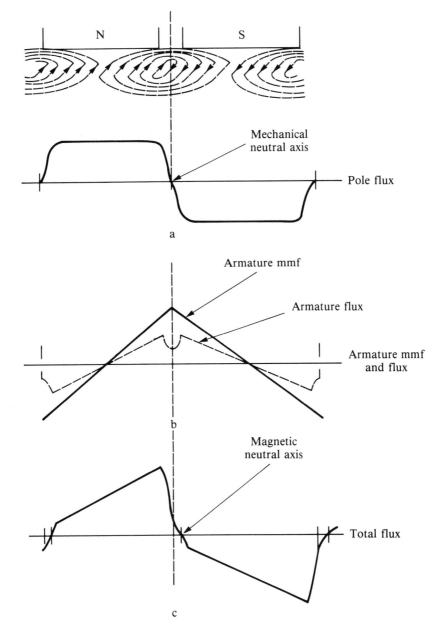

FIGURE 4.28
Effect of
Armature
Reaction in a
Generator

The mmf and the flux produced by the armature currents are shown in Fig. 4.28b. The drop in the armature flux in the interpolar region is due to the high reluctance of that region. It is apparent from Fig. 4.27 that the flux produced by the armature is orthogonal to the flux set up by the poles of the generator. Since both fields exist at the same time, the resultant field in the generator is the phasor

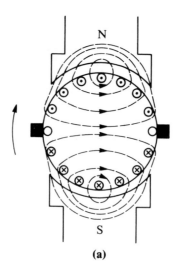

(a)

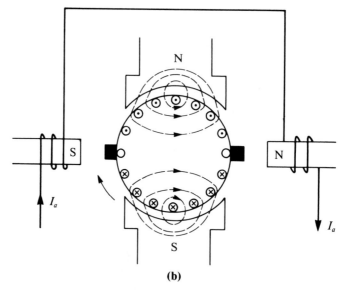

FIGURE 4.29
Armature Flux
Distribution: (a)
Without
Interpoles; (b)
With Interpoles

(b)

addition of the two fluxes, as shown in Fig. 4.28c. Note that the resultant flux is no longer symmetrical. In other words, the flux on one side of the pole has weakened, and the flux on the other side of the pole has been strengthened. If the magnetic poles are already close to the saturation point, the addition of the flux on one side will increase the saturation level, whereas on the other side the saturation level will decrease. Owing to the nonlinear behavior of the magnetic materials, the overall reduction in the flux on one side may be more than the overall increase in the flux on the other side. In other words, there may be a net loss in the total flux. This phenomenon is known as the *armature reaction*. Figure 4.28c reveals that the magnetic neutral axis has shifted to the right. Since the neutral zone must theoretically be the zone for the coils going under commutation to minimize any sparking, the brushes must be positioned accordingly. The armature flux varies with the load and so does its influence on the magnetic field set up by the poles. Thus, the shift in the magnetic neutral axis from the mechanical neutral axis will vary with the load.

Some of the measures being used to counteract armature reaction are summarized next.

1. The brushes may be advanced from their mechanical neutral axis to the magnetic neutral axis for a given load. This measure is the least expensive, but is only useful for constant-load generators.

2. Interpoles or commutating poles, as they are sometimes called, are narrow poles located between the main poles centered along the mechanical neutral axis of the generator and are used to minimize the armature reaction. The interpole windings are permanently connected in series with the armature winding to make them effective for varying loads. The direction of the flux produced by the interpoles be such that it opposes the direction of the field that would otherwise exist due to armature reaction. This makes the interpolar region free of any magnetic flux and hence favorable to sparkless commutation. Figure 4.29 shows the armature flux distributions for a generator with and without interpoles.

3. Another method to nullify the effect of armature reaction is to make use of compensating windings. These windings are placed in the slots cut in the poles near the armature region in such a way that the flux produced by these windings opposes the flux produced by the armature. These windings are

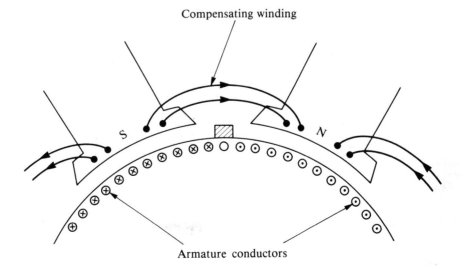

FIGURE 4.30
Compensating
Winding

connected in series with the armature so that the mmfs and thereby the fluxes are equal and opposite at all times. A typical arrangement of such windings is shown in Fig. 4.30.

4.8 EXTERNAL CHARACTERISTICS

The curve that shows the relationship between the terminal voltage and the load current under the condition that the speed of rotation is constant is known as the

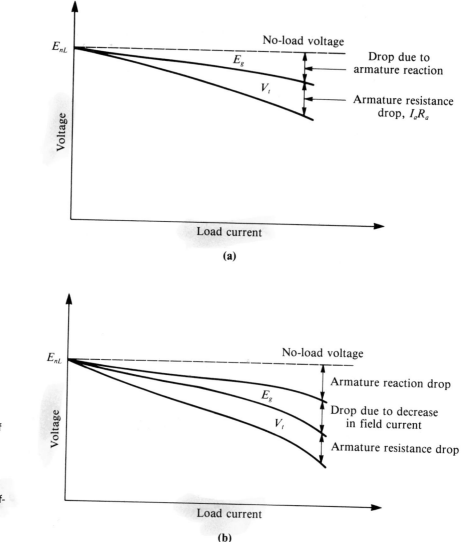

FIGURE 4.31
External
Characteristic of
a Shunt
Generator: (a)
For Separate
Excitation with
Constant Field
Current; (b) Self-
Excitation with
Constant Field
Resistance

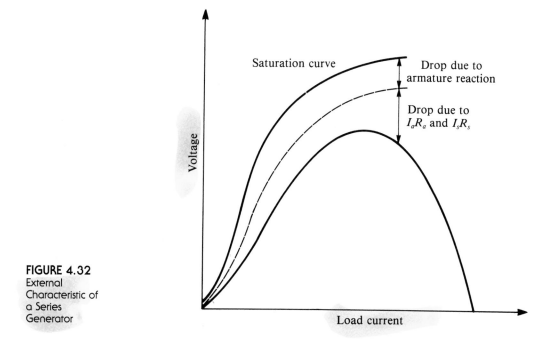

Saturation curve

Drop due to
armature reaction

Drop due to
I_aR_a and I_sR_s

Voltage

Load current

FIGURE 4.32
External
Characteristic of
a Series
Generator

external characteristic of the generator. As the load current increases, so does
the voltage drop across the armature resistance. Also, there is an increase in the
armature rection due to an increase in the load current. Consequently, the terminal
voltage decreases with the increase in load current. The external characteristics
of the separately excited and shunt generators are shown in Fig. 4.31. The terminal
voltage is lower in the shunt generator than in the separately excited generator
under load. This, in fact, is due to the decrease in the field voltage across the
field winding in a shunt generator. A reduction in field voltage reduces the field
current, because the field resistance is held constant. A reduction in the field
current decreases the field mmf and thereby decreases the flux per pole, which
in turn reduces the induced voltage. This process continues until the system becomes stable.

The external characteristic for the series generator is quite different from
the types just mentioned. In this case, the terminal voltage is very small under
no-load conditions. As the load current increases, the generated voltage increases
and so does the terminal voltage. Such a curve is shown in Fig. 4.32.

The external characteristics of compound generators are given in Fig. 4.33.
For comparison purposes, the external characteristics of other generators are also
included.

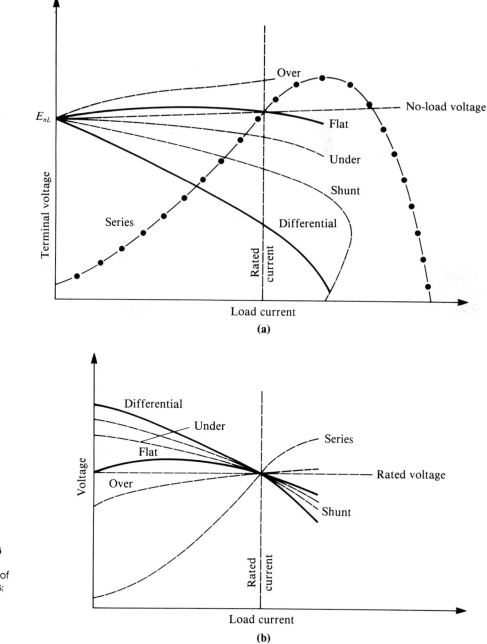

FIGURE 4.33
(a) External Characteristic of dc Generators; (b) Terminal Voltage Adjusted for Rated Value

4.9 VOLTAGE REGULATION

Voltage regulation is defined as follows:

$$\text{VR\%} = \frac{V_{nL} - V_{fL}}{V_{fL}} \, 100 \qquad (4.37)$$

where $\text{VR\%}$ = percent voltage regulation
V_{nL} = terminal voltage under no load
V_{fL} = terminal voltage under full load

■ EXAMPLE 4.6

A 100-kW, 1440-V shunt generator has a voltage regulation of 6.7%. Determine its no-load voltage.

Solution

$$6.7 = \frac{100(V_{nL} - 1440)}{1440}$$

or

$$V_{nL} - 1440 = \frac{6.7 \times 1440}{100} = 96.48$$

$$V_{nL} = 96.48 + 1440 = 1536.48 \text{ V}$$

4.10 LOSSES IN DC MACHINES

We are using the term "machines" in the discussion of the power losses due to the fact that no distinction need be made between the losses in the generator and the motor. The efficiency is the ratio of the output power to the input power. The input power must be equal to the output power plus all the losses in the machine. There are, in fact, three major classifications of losses: mechanical losses, electrical losses, and magnetic losses.

Mechanical Losses These losses are primarily due to the rotation of the armature and include bearing friction loss and the windage loss. Mechanical losses can be determined by rotating the armature of an unexcited machine (without its brushes) at its rated speed by coupling it to a calibrated motor. Since there is no electrical output, all the power supplied to the armature is converted into mechanical loss. If the machine is run again, but now with its brushes properly seated, the friction loss due to the brushes can be determined as the difference of two power readings.

Rotational losses can also be determined by running the machine as a motor at its rated speed under no load. Motor operation is discussed in the next chapter. The voltage applied to the armature circuit should be such that the motor has the same flux as it would when it operates under normal load. In the case of a generator, the applied voltage should be adjusted to

$$V_a = V_t + I_a R_a \tag{4.38}$$

However, for the motor, the applied voltage should be

$$V_a = V_t - I_a R_a \tag{4.39}$$

This is done to make sure that under no load the applied voltage, V_a, is nearly equal to the induced emf, E_g, in the armature. The speed can be adjusted by varying the external resistance in the field-winding circuit. When the preceding conditions are met, the total power input is

$$P_{\text{in}} = V_a I_L = V_a(I_a + I_f) = V_a I_a + V_a I_f \tag{4.40}$$

Since the power output is zero under no load, applied power is consumed by the losses in the machine. These losses are

$$V_a I_f + I_a^2 R_a + P_{\text{rot}} \tag{4.41}$$

Equating the power input to the losses, we get

$$P_{\text{rot}} = V_a I_a - I_a^2 R_a \tag{4.42}$$

However, under no load the armature current is very small and, therefore, the copper loss in the armature circuit, $I_a^2 R_a$, is so small that it can be ignored. Under such an assumption, the rotational loss is equal to the product of the applied voltage, V_a, and the armature current, I_a.

Magnetic Losses   Magnetic losses include eddy-current loss and hysteresis loss and are discussed in Chapter 1. These losses can be determined as follows:

1. Couple the dc machine to a motor and drive it at its rated speed under no load.

FIGURE 4.34
Power-Flow
Diagram of a
Self-Excited dc
Generator

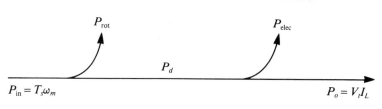

2. Separately excite the dc machine until the field current is equal to its rated value. This establishes the required flux in the machine.

3. The input mechanical power is equal to the sum of rotational and core losses.

Electrical Losses   In a dc machine, electrical or the copper losses can be segregated as follows:

1. Armature copper loss: $I_a^2 R_a$

2. Shunt-field copper loss: $I_f^2 R_f$

3. Series-field copper loss: $I_s^2 R_s$

4. Compensating-field winding loss: $I_a^2 R_c$

5. Interpole-field copper loss: $I_a^2 R_i$

Stray-load Loss   In a machine there are always some losses that cannot be easily accounted for; they are termed stray-load losses. As a rule of thumb, the stray-load loss is assumed to be nearly 1% of the output in large machines (above 100 horsepower) and can be neglected in smaller machines. Stray-load loss is caused by the distortions in the flux waveforms due to the armature reaction on one hand and the short-circuit currents in the coils undergoing commutation on the other.

Power-flow Diagram   In a dc generator, the mechanical energy supplied to the armature is converted into electrical energy. Right at the outset, some of the mechanical energy is lost as friction and windage loss. Additional energy, as stated earlier, is lost to take care of the core losses. Thus, the power that is actually available for conversion is the difference of total mechanical input and rotational losses. We will refer to this available power as developed power. From this developed power, which now exists in the electrical form, are subtracted all the copper losses in the dc generator to obtain the net power output. The basic power-flow diagram for the self-excited dc generator is shown in Fig. 4.34. The power supplied to the field winding may also be included as part of the input power.

■ EXAMPLE 4.7

A 440-V short-shunt compound generator is rated at 100 A. The shunt field current is 3.0 A. It has an armature resistance of 0.05 Ω, a series-field resistance of 0.01 Ω, and a diverter resistance of 0.04 Ω. Determine the generator voltage and the total power developed by the machine. What is the efficiency of the generator if the rotational loss is 1.2 kW? Draw the power diagram to show the power distribution.

Solution

A schematic circuit of a short-shunt generator is shown in Fig. 4.35a with the diverter resistance in parallel with the series field. The load current flows through

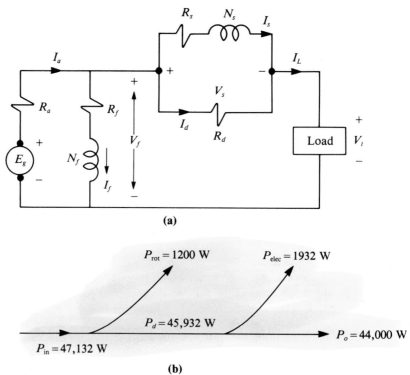

FIGURE 4.35
(a) Equivalent
Circuit of a
Short-Shunt
Compound
Generator; (b)
Power-Flow
Diagram

the parallel combination of series-field resistance and diverter resistance. Therefore,

$$I_s = \frac{I_L R_d}{R_s + R_d} = \frac{100 \times 0.04}{0.01 + 0.04} = 80 \text{ A}$$

Thus $I_d = 100 - 80 = 20$ A

The voltage drop across the series-field or diverter resistance is

$$V_s = I_s R_s = 80 \times 0.01 = 0.8 \text{ V}$$

Thus, $V_f = V_t + V_s = 440 + 0.8 = 440.8$ V

and the armature current is

$$I_a = I_L + I_f = 100 + 3 = 103 \text{ A}$$

The generated voltage is then

$$E_g = V_f + I_a R_a = 440.8 + 103 \times 0.05 = 445.95 \text{ V}$$

The power developed is

$$P_d = E_g I_a = 445.95 \times 103 = 45,933 \text{ W}$$

Since rotational loss is known, total power input is

$$P_m = P_d + P_{\text{rot}} = 45,933 + 1200 = 47,133 \text{ W}$$

Copper losses are:

1. Armature: $I_a^2 R_a = 103^2 \times 0.05 = 530 \text{ W}$
2. Series field: $I_s^2 R_s = 80^2 \times 0.01 = 64 \text{ W}$
3. Shunt field: $V_f I_f = 440.8 \times 3 = 1322 \text{ W}$
4. Diverter: $I_d^2 R_d = 20^2 \times 0.04 = 16 \text{ W}$

$$P_{\text{elec}} = \text{total copper loss} = 530 + 64 + 1322 + 16 = 1932 \text{ W}$$

The power output is

$$V_t I_L = 440 \times 100 = 44,000 \text{ W}$$

However, the power developed must be equal to the sum of the power output and the total copper loss. That is,

$$P_d = 44,000 + 1932 = 45,932 \text{ W}$$

which is the same as calculated previously, within the accuracy of our calculations. Efficiency is

$$\eta = \frac{P_o}{P_{\text{in}}} = 44,000/47,133 = 0.934 \quad \text{or} \quad 93.4\%$$

The power-flow diagram for this generator is given in Fig. 4.35b.

4.11 MAXIMUM EFFICIENCY CRITERION

Either by performing an actual load test on a given dc machine or just by calculating its performance at different load levels, we will observe that its efficiency increases with the increasing values of load current at first, reaching a maximum value, and then decreases with further increases in the load current. A typical efficiency versus load current curve is shown in Fig. 4.36.

The losses can be grouped into two categories: fixed losses and variable losses. Fixed losses are losses that do not change with the load when the machine

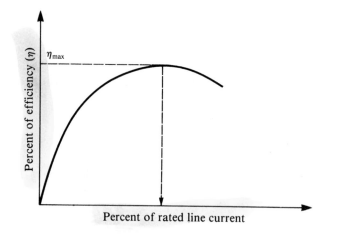

FIGURE 4.36
Typical Efficiency
versus Full-Load
Current
Characteristic of
a dc Machine

is operating at a known speed. Therefore, rotational losses are one part of fixed losses. In the case of a wound machine, where the field current is required to set up the required flux in the machine, the power supplied to the field winding is also considered a fixed loss. Therefore, the sum of rotational loss and field loss is the total fixed loss in the machine. On the other hand, variable losses are losses that vary with the load current. All the copper losses are included in this category.

The efficiency of a separately excited dc generator can be expressed as

$$\eta = \frac{P_o}{P_{in}} = \frac{P_o}{P_o + P_{loss}} = \frac{V_t I_a}{V_t I_a + P_c + I_a^2 R_a} \tag{4.43}$$

where $\qquad P_c = P_{rot} + V_f I_f = \text{fixed losses}$

For maximum efficiency,

$$\frac{d\eta}{dI_a} = \frac{(V_t I_a + P_c + I_a^2 R_a)V_t - V_t I_a(V_t + 2I_a R_a)}{(V_t I_a + P_c + I_a^2 R_a)^2} = 0 \tag{4.44}$$

from which $\qquad\qquad P_c = I_a^2 R_a \tag{4.45}$

Equation (4.45) states that the efficiency of a separately excited generator is maximum when the variable losses are equal to the fixed losses. Therefore, the maximum efficiency of a separately excited generator is

$$\eta_{max} = \frac{V_t I_a}{V_t I_a + 2I_a^2 R_a} \tag{4.46}$$

Expression (4.46) is true for the shunt as well as other dc machines if it is assumed that the load current is very nearly the same as the armature current.

■ EXAMPLE 4.8

A 115-hp, 440-V separately excited dc generator has an armature resistance of 0.2 Ω, a field resistance of 120 Ω, and a maximum efficiency of 85%. Calculate (a) the load current at the maximum efficiency and (b) the rotational losses.

Solution

(a) Let I_L be the load current at which the maximum efficiency occurs; then from the maximum efficiency, Eq. (4.46),

$$0.85 = \frac{440I_L}{440I_L + 2I_L^2(0.2)} = \frac{440}{440 + 0.4I_L}$$

from which $I_L = 194.1$ A

(b) The power output for maximum efficiency is

$$P_o = 440 \times 194.1 = 85,404 \text{ W}$$

$$= \frac{85,404}{746} = 114.5 \text{ hp}$$

From Eq. (4.45), $P_c = 194.1^2 \times 0.2 = 7535$ W.

The field loss is

$$\frac{V_f^2}{R_f} = \frac{440^2}{120} = 1613 \text{ W}$$

The rotational loss is $P_{\text{rot}} = 7535 - 1613 = 5922$ W.

■ EXAMPLE 4.9

A 50-kW, 115-V long-shunt compound generator is supplying a load of 400 A at rated voltage. The shunt-field current is 10 A. Armature resistance is 0.05 Ω, and series-field resistance is 0.02 Ω. Rotational power loss is 2000 W. Calculate the efficiency of the generator. Determine the load current at which the efficiency is approximately a maximum.

Solution

The equivalent circuit of a long-shunt generator is shown in Fig. 4.26.

Armature current: $I_a = I_f + I_L = 10 + 400 = 410 \text{ A}$

Power output: $P_o = V_t I_L = 115 \times 400 = 46{,}000 \text{ W}$

Copper losses:

Shunt field: $V_f I_f = 115 \times 10 = 1150 \text{ W}$

Armature: $I_a^2 R_a = 410^2 \times 0.05 = 8405 \text{ W}$

Series field: $I_s^2 R_s = 410^2 \times 0.02 = 3362 \text{ W}$

Thus,

$$\text{Fixed losses} = P_{\text{rot}} + V_f I_f = 2000 + 1150 = 3150 \text{ W}$$

$$\text{Variable losses} = 8405 + 3362 = 11{,}767 \text{ W}$$

$$\text{Total losses} = \text{fixed} + \text{variable} = 11{,}767 + 3150 = 14{,}917 \text{ W}$$

Power input: $P_{\text{in}} = P_o + \text{total losses} = 46{,}000 + 14{,}917 = 60{,}917 \text{ W}$

Efficiency: $\eta = \dfrac{P_o}{P_{\text{in}}} = \dfrac{46{,}000}{60{,}917} = 0.755 \quad \text{or} \quad 75.5\%$

For efficiency approximately to be maximum,

$$I_a^2 (R_a + R_s) = P_{\text{rot}} + V_f I_f$$

$$I_a^2 (0.05 + 0.02) = 3150$$

or $I_a = 212 \text{ A}$

Load current at maximum efficiency: $I_L = I_a - I_f = 212 - 10 = 202 \text{ A}$

Power output at maximum efficiency: $P_o = 115 \times 202 = 23{,}230 \text{ W}$

Total losses at maximum efficiency: $2 \times 3150 = 6300 \text{ W}$

The maximum efficiency is

$$\eta_{\text{max}} = \frac{23{,}230}{23{,}230 + 6300} = 0.787 \quad \text{or} \quad 78.7\%$$

The generator must be derated to 23.23 kW if it is to be operated at its maximum efficiency.

PROBLEMS _____

4.1. In a dc machine the air-gap flux density is 6 kG and the pole face area is 3 cm by 5 cm. Determine the flux per pole in that machine.

4.2. A series generator requires 620 A-t per pole to deliver a load of 10 A. How many turns must be there on each of its poles?

4.3. Calculate the armature current per path of a 6-pole, 60-kW, 240-V separately excited generator if wound using (a) a simplex wave, and (b) a simplex lap winding.

4.4. In a 4-pole simplex lap-wound generator the induced emf is 135 V and armature current is 200 A. Calculate (a) the number of parallel paths, (b) the current in each path, (c) the voltage induced in each path, and (d) the power developed by the armature.

4.5. A 100-hp, 10-pole, 500-V generator has simplex lap winding. Determine the voltage, current, and power rating of the generator if it were wave wound.

4.6. What is the frequency of the alternating emf in the armature winding of an 8-pole generator that operates at a speed of 600 rpm?

4.7. Determine the mmf requirements to set up a flux density of 69,000 lines per square inch in the 0.16-in. air gap of a machine. Assume there is no other mmf loss.

4.8. A separately excited generator develops 440 V at 1600 rpm. What would be the induced emf if the speed changes to 2000 rpm? Assume no change in the flux.

4.9. A 1200-rpm, 4-pole, lap-wound generator has 240 conductors on the armature. It has 9-in. square poles with an average flux density of 80,000 lines per square inch. What is the induced emf in the armature?

4.10. The armature of a 4-pole lap-wound generator has 28 slots with 10 conductors in each slot. The flux per pole is 0.03 Wb and the armature speed is 1200 rpm. Calculate (a) the frequency of the alternating voltage induced in the conductor, and (b) the induced emf in the armature.

4.11. When the generator of Problem 4.10 was required to deliver a load current, each of its conductors had to carry a current of 2 A. What is the armature current? What are the power and torque developed by the machine?

4.12. If the armature of the machine in Problem 4.10 is wave wound, what is the induced voltage? What is the armature current if the current in each conductor is still 2 A? Calculate the power and torque developed by the machine.

4.13. A 6-pole lap-wound armature has 126 slots and is wound with 5 turns per coil using double-layer winding. The induced voltage is 440 V at 120 rad/s. Determine the flux per pole.

4.14. A 600-rpm, 10-pole generator has a simplex lap-wound armature with 1600 conductors. Each pole is 36 in.^{2} and the average flux density is 100 kilolines per square inch (kL/in.^{2}). What is the induced emf in the armature?

4.15. A simplex, double-layer, wave-wound generator has 8 poles, 100 coils of 5 turns each, and a flux per pole of 0.5 Wb. Determine the induced emf at 300 rpm.

4.16. A 1200-A short-shunt compound generator has a series resistance of 0.02 Ω and diverter resistance of 0.1 Ω. There are 10 turns per pole in the series-field winding. Calculate the mmf produced by the series-field winding with and without the diverter in the circuit.

4.17. A 125-V 10-A shunt generator has an armature resistance of 0.8 Ω, and the shunt-field current is 1.5 A. Determine the voltage regulation of the generator assuming no change in shunt-field current.

4.18. The induced emf in a separately excited generator is 135 V while its rated terminal voltage is 115 V. What is the percent voltage regulation?

4.19. A 230-V shunt generator has a voltage regulation of 12%. What is its terminal voltage when the load is reduced to zero?

4.20. A 50-kW, 250-V separately excited generator has an armature resistance of 0.1 Ω. Determine the terminal voltage, power output, and power developed by the armature when the load current is 150 A.

4.21. A 40-kW, 400-V shunt generator has an armature resistance of 0.1 Ω and a shunt-field resistance of 200 Ω. Determine the power developed at its rated load. What is the electrical power loss in the generator?

4.22. In a 25-kW, 125-V short-shunt compound generator, the current in the series winding must be 160 A for flat compounding under full-load conditions. Calculate the value of the diverter resistance when the series-field resistance is 0.1 Ω.

4.23. A 230-V shunt generator has a total armature circuit resistance of 0.2 Ω and a shunt-field resistance of 65 Ω. The induced emf due to residual magnetism is 10 V. Assume the generator is short-circuited prior to its running. Will there be a voltage buildup? What will be the short-circuit current?

4.24. A 2-pole, 460-kW, 4.6-kV shunt generator has armature and field resistances of 1.0 and 230 Ω, respectively. If the generator is operating at its rated speed of 3000 rpm, what is the no-load voltage of the generator? What is the terminal voltage at half-load?

4.25. A 200-kW, 125-V separately excited generator has a no-load voltage of 170 V. Assuming a straight-line relationship between the armature current and the terminal voltage, determine the power output of the machine when the terminal voltage is 150 V.

4.26. A 10-kW, 125-V short-shunt compound generator has a series-field resistance of 0.01 Ω and a diverter resistance of 0.04 Ω. The series-field winding has 10 turns per pole. Calculate the mmf per pole set up by the series field at full load.

4.27. A 100-kW, 250-V compound generator has an armature resistance of 0.05 Ω, series-field resistance of 0.02 Ω, and shunt-field resistance of 100 Ω. Calculate the induced emf at rated load when the generator is connected as (a) short shunt, and (b) long shunt.

4.28. A 10-kW, 125-V long-shunt compound generator has an armature resistance of 0.01 Ω, a series-field resistance of 0.02 Ω, and a shunt-field resistance of 50 Ω. Under no-load conditions the terminal voltage reduces to 100 V. The series-field winding has 5 turns per pole, while the shunt field has 500 turns per pole. To make the generator flat compounded, the series field must produce 300 A-t. Calculate (a) the diverter resistance to accomplish the change, (b) the total mmf per pole at full load and at no load, and (c) the efficiency if the rotational losses are 740 W.

4.29. A short-shunt compound generator is designed to take a load of 75 A at 440 V. The shunt-field current is 2.5 A, the series-field resistance is 0.5 Ω, the diverter resistance is 1.5 Ω, and the armature resistance is 0.2 Ω. Calculate (a) the power developed by the armature, (b) all the electrical losses, and (c) the efficiency of the generator if the rotational loss is 1 kW.

4.30. A dc machine has a core loss of 960 W of which three-fourths is the hysteresis loss when operated at 1800 rpm. Under constant flux density conditions, what is the core loss when the speed is (a) 2400 rpm, and (b) 1200 rpm?

4.31. A 25-kW, 125-V generator is running at half-load at 1725 rpm with an efficiency of 85%. What are the total losses and power input to the machine? What are the torque requirements of the prime mover driving the armature?

4.32. A 10-kW, 125-V flat compound generator has an armature resistance of 0.1 Ω, total shunt-field resistance of 50 Ω, a series-field resistance of 0.02 Ω, a rotational loss of 800 W, and a brush voltage drop of 2 V. Calculate its efficiency at full load.

CHAPTER 5

Direct-Current Motors

Series Motors (Courtesy of Bodine Electric
Company)

In Chapter 4 we devoted our attention to the operation and characteristics of different types of direct-current generators. With the help of dc generators we were able to convert mechanical energy into electrical energy. On the other hand, electric motors are employed to transform electrical energy into mechanical energy. In this chapter we will study the electromechanical energy-conversion process by concentrating on direct-current or simply dc motors.

The mechanical construction of a dc motor is no different than that of the dc generator. The induced voltage is referred to as the generated voltage when the machine is operating as a generator and the back or counter emf when the machine is used as a motor.

5.1   TYPES OF DC MOTORS

The factors that are of prime importance in dc motor selection are the speed and the torque. The curve that defines the relation between these two factors is known as the speed–torque curve. Thus, from the performance point of view, a dc motor can be classified as follows:

1.   Constant-speed motor
2.   Variable-speed motor
3.   Multiple-speed motor

Just like dc generators, the dc motors can also be classified on the basis of their winding connections as follows:

1.   Series motor
2.   Shunt motor
3.   Compound motor

A shunt motor can be operated as a separately excited motor. A large number of dc motors are used for constant-speed operations and shunt motors are selected for that purpose. However, the major applications where the dc motors are successfully employed are:

1.   In automobiles as starter motor, heater motor, and the like
2.   Traction motors for railways
3.   Subway trains and trams
4.   Electric shavers
5.   Audio and video cassette recorders
6.   Computers

5.2   CLASSIFICATIONS OF DC MOTORS

As pointed out earlier, there are basically three types of motors if classified on the basis of winding connections.

5.2.1 Series Motor

In a series motor the field winding is connected in series with the armature winding as shown schematically in Fig. 5.1. In this motor the field excitation is a function of load current. As the load current increases or decreases, so does the flux produced by the field winding. It can therefore be said that in a series motor the flux is proportional to the load current. However, the load current is same as the armature current. Thus, the torque developed and the back emf expressions can be rewritten as

$$T_d \propto I_a^2 \tag{5.1}$$

and

$$E_g \propto I_a \omega_m \tag{5.2}$$

When the motor is operating under no load, the torque developed by the motor is just sufficient to overcome friction and windage losses in the machine. Since these losses are generally a very small fraction of the full-load torque, the torque developed by the motor is very small. The armature current must also be very small, as the torque is proportional to the square of the armature current. If the armature current is very small, the back emf must be nearly equal to the applied voltage. For a small value of armature current, the speed of the motor must be very high. It is therefore possible for such motors to self-destruct under no-load conditions.

On the other hand, under full-load conditions the torque developed by the motor is high enough to result in large current flows in both the field and armature windings. Due to this fact, the back emf is now smaller and the speed of the motor is lower. Thus, a series motor is a variable-speed motor, and its speed is a function of the applied load. The speed–torque characteristic for such a motor is given in Fig. 5.2.

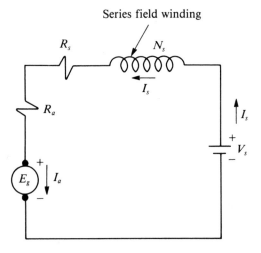

FIGURE 5.1
Schematic
Representation
of a Series
Motor

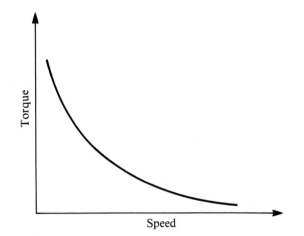

FIGURE 5.2
Speed–Torque
Curve of a
Typical Series
Motor

The torque is proportional to the square of the current only when the motor is operating in the linear region of the magnetization characteristic. As the armature current increases, so does the flux produced by the field winding, which enhances the level of saturation in the motor. When the motor is saturated, there is only a very gradual increase in the flux, and the developed torque is no longer proportional to the square of the current. The torque versus armature current relation is given in Fig. 5.3.

From the equivalent circuit of the motor, it is evident that

$$E_g = V_s - I_a(R_a + R_s) \tag{5.3}$$

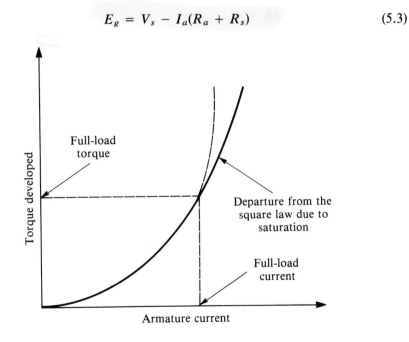

FIGURE 5.3
Typical
Armature
Current versus
Torque
Characteristic of
a Series Motor

where V_s is the supply voltage. But

$$E_g = K_a \phi_p \omega_m$$

and
$$\phi_p \propto I_a \quad \text{or} \quad \phi_p = k_f I_a$$

From these equations we can obtain an expression for the angular velocity:

$$\omega_m = \frac{V_s - I_a(R_a + R_s)}{K_a k_f I_a} \tag{5.4}$$

It is evident that as the armature current, I_a, becomes zero under ideal no-load conditions, there will be no flux in the motor and the speed of the motor will approach infinity.

It is obvious from the speed–torque curve that the torque developed by a series motor is quite high when the speed of the motor is low. For this reason, the series motor is suitable for hoists, cranes, electric trains, and other such applications where the starting torque requirements are quite high.

The torque developed by the motor can be controlled by controlling the applied voltage to the motor. If the applied voltage is increased, the torque developed by the motor will also increase. Another method to obtain higher torque is to reduce the total armature and field-winding resistance.

■ EXAMPLE 5.1

A 10-hp, 115-V dc series motor takes 40 A when operating at a full-load speed of 1800 rpm. If the line current changes to 50 A, what torque is developed by the motor when it operates within its linear operating region?

Solution

Within the linear region of operation the flux per pole is directly proportional to the current in the field winding. That is,

$$\phi_p = k_f I_a$$

Full-load speed: $N_m = 1800$ rpm   or

$$\omega_m = \frac{2\pi N_m}{60} = 60\,\pi \text{ rad/s}$$

Rated output power: $P_o = 10 \times 746 = 7460$ W

Full-load torque: $T_{40} = \dfrac{7460}{60\pi} = 39.58$ N-m

But
$$T_d = K_a k_f I_a^2 = K I_a^2.$$

Thus, at full-load,
$$K = \frac{39.58}{40^2} = 0.0247$$

The torque developed when the line current changes to 50 A is

$$T_{50} = 0.0247 \times 50^2 = 61.84 \text{ N-m}$$

■ EXAMPLE 5.2

What torque is developed by the motor in Example 5.1 if a 25% increase in full-load current results in a 16% increase in the flux due to the demagnetizing effect of armature reaction?

Solution

Since $T_d = K_a I_a \phi_p$ at 40 A, $39.58 = 40 K_a \phi_{p40}$. And at 50 A, $T_{50} = 50 K_a \phi_{p50}$. Thus

$$T_{50} = 39.58 \frac{\phi_{p50}}{\phi_{p40}} \frac{50}{40}$$

With a 25% increase in full-load current, the increase in the motor flux is 16%. That is,

$$\frac{\phi_{p50}}{\phi_{p40}} = 1.16$$

Therefore
$$T_{50} = 39.58 \times 1.16 \times 1.25 = 57.39 \text{ N-m}$$

■ EXAMPLE 5.3

The magnetization curve for the motor in Example 5.1 is given in Fig. 5.4 at 1800 rpm. Calculate the torque, speed, and power output of the motor when the line current is 60 A. The armature winding resistance is 0.075 Ω and the field winding resistance is 0.05 Ω. Neglect the effect of armature reaction.

Solution

From the magnetization curve, the no-load voltages are

$$\text{At } I_a = 40 \text{ A:} \quad E_{g40} = 110 \text{ V}$$
$$\text{At } I_a = 60 \text{ A:} \quad E_{g60} = 126 \text{ V}$$

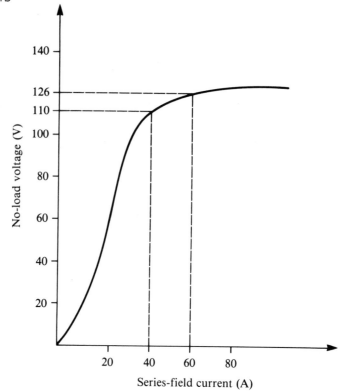

FIGURE 5.4
Magnetization
Curve for a
Series Motor
(Example 5.3)

Since $E_g = K_a\phi_p\omega_m$ at 1800 rpm,

$$\frac{E_{g60}}{E_{g40}} = \frac{\phi_{p60}}{\phi_{p40}} = \frac{126}{110} = 1.145$$

$$\frac{I_{60}}{I_{40}} = \frac{60}{40} = 1.5$$

Since $T_d = K_a\phi_pI_a$,

$$\frac{T_{60}}{T_{40}} = \frac{\phi_{p60}}{\phi_{p40}}\frac{I_{60}}{I_{40}} = 1.145 \times 1.5 = 1.7175$$

Thus $T_{60} = 39.58 \times 1.7175 = 67.98$ N-m

However, $E_g = K_a\phi_p\omega_m = V_s - I_a(R_a + R_s)$

Thus $K_a\phi_{p40}\omega_{m40} = 115 - 40(0.075 + 0.05) = 110$ V

and $K_a\phi_{p60}\omega_{m60} = 115 - 60(0.075 + 0.05) = 107.5$ V

Thus
$$\frac{\phi_{p60}}{\phi_{p40}}\frac{\omega_{m60}}{\omega_{m40}} = \frac{107.5}{110}$$

or
$$\omega_{m60} = \omega_{m40}\frac{107.5}{110}\frac{\phi_{p40}}{\phi_{p60}}$$

$$= 60\,\pi\,\frac{107.5}{110}\frac{1}{1.145}$$

$$= 51.21\pi \text{ rad/s}$$

Speed in rpm is

$$N_{m60} = \frac{60\omega_{m60}}{2\pi} = \frac{60 \times 51.21\pi}{2\pi} = 1536 \text{ rpm}$$

The power output is $P_{o60} = 67.98 \times 51.21\pi = 10,936.7$ W. The power output can also be expressed in terms of the commonly used unit of horsepower as $10,936.7/746 = 14.66$ hp.

5.2.2 Shunt Motor

In a shunt motor, the field winding is connected in parallel with the armature circuit as shown in Fig. 5.5. The speed–voltage relationship for the shunt motor can be derived from its equivalent circuit. For the armature circuit,

$$E_g = V_s - I_aR_a$$

or
$$\omega_m = \frac{V_s - I_aR_a}{K_a\phi_p} \tag{5.5}$$

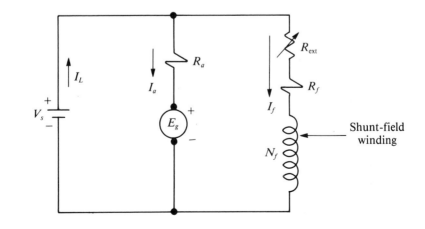

FIGURE 5.5
Schematic Representation of a Shunt Motor. An External Resistance, R_{ext}, is Shown to Limit the Field-Winding Current

where I_aR_a represents the effective voltage drop in the armature circuit. Any increase in the torque will cause the armature current to rise and thereby result in further reduction in the speed of the motor. With the increase in armature current, the armature reaction will become more significant. The effective flux in the air-gap region of the motor goes down, which in turn results in the increase in speed. If the motor is operated above the knee of the saturation curve, the increase in speed due to increased armature reaction is lower than the drop in speed due to the increase in voltage drop in the armature. The net effect is a decrease in speed with the increase in armature current. The speed versus armature current characteristic for a shunt motor is shown in Fig. 5.6.

The armature current can be written in terms of the developed torque as

$$I_a = \frac{T_d}{K_a\phi_p} \tag{5.6}$$

If we substitute this relation for the armature current in the speed–voltage relation obtained earlier, we get

$$\omega_m = \frac{V_s}{K_a\phi_p} - \frac{T_dR_a}{(K_a\phi_p)^2} \tag{5.7}$$

This equation describes the speed–torque characteristics of a shunt motor. It is a straight line with a negative slope. The speed is maximum under no-load conditions but has a finite value. The speed–torque characteristics for the shunt motor are given in Fig. 5.7 with and without armature reaction. Since the effect of armature reaction is to decrease the overall flux in the motor, the reduction in flux increases the speed of the motor.

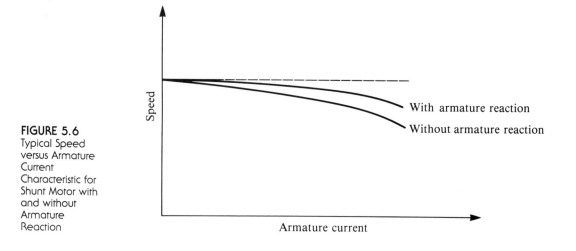

FIGURE 5.6
Typical Speed versus Armature Current Characteristic for Shunt Motor with and without Armature Reaction

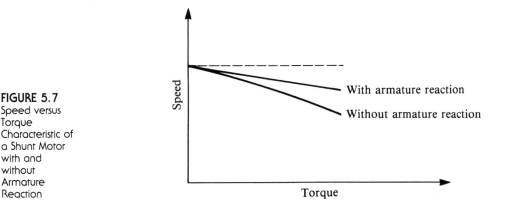

FIGURE 5.7
Speed versus Torque Characteristic of a Shunt Motor with and without Armature Reaction

From the speed–voltage relationship it is evident that the speed of the shunt motor can be increased or decreased by increasing or decreasing the effective resistance in the armature circuit. However, the resistance of the armature winding has an impact on the efficiency of the motor. It is, however, apparent that the change in the resistance of the armature circuit has no effect on the no-load speed of the motor. It only changes the slope with which the speed of the motor falls under load. The speed–torque relation for various values of total armature circuit resistance is shown in Fig. 5.8.

The speed control for shunt motors can also be accomplished by increasing or decreasing the applied voltage. It is apparent that the changes in the applied voltage directly affect the no-load speed of the motor. However, the rate with which the speed decreases with the load remains the same. Thus, the speed torque curves are a series of parallel lines for various applied voltages as given in Fig. 5.9.

The speed of a shunt motor can also be controlled by controlling the current in the field winding, which can be accomplished by adding an external resistance in series with it. Basically, we are changing the overall flux in the motor, which

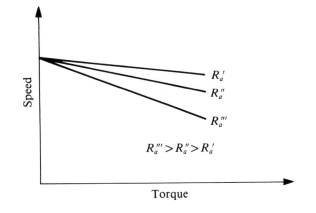

FIGURE 5.8
Effect of Armature Circuit Resistance upon Speed versus Torque Characteristic of a Shunt Motor

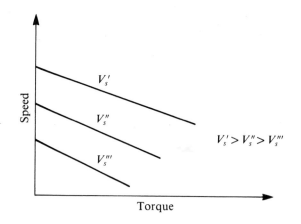

FIGURE 5.9
Effect of Supply Voltage upon the Speed–Torque Curve of a Shunt Motor

directly affects the back emf. Therefore, the decrease in the field-winding current lowers the back emf in the motor. Since the armature current is

$$I_a = \frac{V_s - E_g}{R_a}$$ (5.8)

and the armature resistance, R_a, is usually kept as low as possible, a small decrease in the induced emf results in a high percentage increase in the armature current. Thus, the torque developed by the shunt motor increases. If the load on the motor is kept the same, it now operates at a higher speed. However, an increase in motor speed increases the induced emf, causing the armature current and thereby the developed torque to decrease. This continues until the developed torque is equal to the load torque and the speed of the motor settles at a higher value than before. Figure 5.10 shows the speed–torque curves for various values of the field resistance.

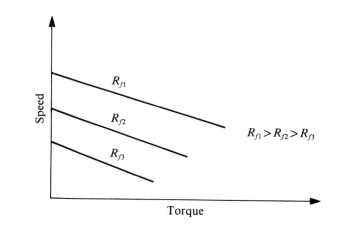

FIGURE 5.10
Speed–Torque Curve of a Shunt Motor for Various Values of the Field Resistance

■ **EXAMPLE 5.4**

A 6-pole, 3-hp, 120-V, dc lap-wound shunt motor has 960 conductors in the armature. It takes 25 A from the supply at full load. The armature resistance if 0.75 Ω and the flux per pole is 10 mWb. The field winding current is 1.2 A. Calculate the speed and torque available at the shaft.

Solution

Power output at full load: $P_o = 3 \times 746 = 2238$ W

Armature current: $I_a = I_L - I_f = 25 - 1.2 = 23.8$ A

No-load voltage: $E_g = V_s - I_a R_a = 120 - 23.8 \times 0.75$

$$= 102.15 \text{ V}$$

Motor constant: $K_a = \dfrac{Zp}{2\pi a} = \dfrac{960 \times 6}{2\pi \times 6} = 152.79$

Since $E_g = K_a \phi_p \omega_m$,

$$\omega_m = \frac{102.15}{152.79 \times 10 \times 10^{-3}} = 66.856 \text{ rad/s}$$

or

$$N_m = \frac{66.856 \times 60}{2\pi} = 638 \text{ rpm}$$

The torque available at the shaft is

$$T = \frac{P_o}{\omega_m} = \frac{2238}{66.856} = 33.48 \text{ N-m}$$

■ **EXAMPLE 5.5**

The data for the no-load characteristics of a 4-pole, 6-hp, 120-V dc shunt motor at 2400 rpm are

E_g (V)	60	88	115	134	150	164	174	185	192
I_f (A)	0.2	0.4	0.6	0.8	1.0	1.2	1.4	1.6	1.8

The combined resistance of the armature circuit is 0.1 Ω. The shunt field winding has 1600 turns per pole with a total resistance of 200 Ω. The full-load current is 40 A and the demagnetizing effect of the full-load armature current is 0.025 A expressed in terms of shunt-field current. The no-load losses are 230 W. Determine (a) the speed of the motor at full load, (b) the torque developed by the motor, and (c) the electrical losses in the motor.

Solution

(a) The plot for the no-load characteristics of a dc shunt motor is shown in Fig. 5.11. The field current is

$$I_f = \frac{120}{200} = 0.6 \text{ A}$$

At the full-load current, the effective field current is

$$I_{f\text{eff}} = 0.6 - 0.025 = 0.575 \text{ A}$$

The back emf in the motor at its rated speed of 2400 rpm at a field current of 0.575 A is 111.6 V from the magnetization characteristic. The full-load armature current is

$$I_a = I_L - I_f = 40 - 0.6 = 39.4 \text{ A}$$

The back emf is

$$E_{g2} = V_s - I_a R_a = 120 - 39.4 \times 0.1 = 116.06 \text{ V}$$

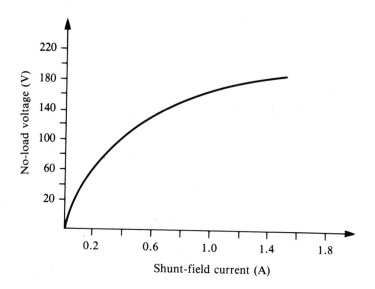

FIGURE 5.11
No-Load
Characteristic of
a Shunt Motor
(Example 5.5)

The motor speed, therefore, is

$$N_{m2} = N_{m1} \frac{E_{g2}}{E_{g1}} = 2400 \frac{116.06}{111.6} = 2496 \text{ rpm}$$

or $\qquad \omega_{m2} = 261.38 \text{ rad/s}$

It is evident that the full-load speed is higher than the no-load value. This is because of the reduction in flux due to the armature reaction. Such an effect is undesirable and is usually overcome by adding a few series turns. Such a winding is usually referred to as the *stabilizing* winding.

(b) The power developed is

$$P_d = E_{g2} I_a = 116.06 \times 39.4 = 4572.76 \text{ W}$$

The power available at the shaft is

$$P_o = P_d - P_{\text{rot}} = 4572.76 - 230 = 4342.76 \text{ W}$$

$$= \frac{4342.76}{746} = 5.82 \text{ hp}$$

Thus, the torque available at the shaft is

$$T_s = \frac{4342.76}{261.38} = 16.61 \text{ N-m}$$

(c) The electrical losses are:

Armature-winding loss: $\quad I_a^2 R_a = 39.4^2 \times 0.1 = 155.24 \text{ W}$

Field-winding loss: $\qquad V_f I_f = 120 \times 0.6 \quad = 72.0 \text{ W}$

The total electrical losses are $230 + 155.24 + 72 = 457.24$ W.

■ **EXAMPLE 5.6**

A 220-V dc shunt motor draws 10 A at 1800 rpm. The armature circuit resistance is 0.2 Ω and the field-winding resistance is 440 Ω. What is the torque developed by the motor? If the field current is unchanged, determine the speed and line current when the motor develops a torque of 20 N-m.

Solution

When the motor is operating at 10 A and 1800 rpm

Field current:

$$I_f = \frac{V_f}{R_f} = \frac{220}{440} = 0.5 \text{ A}$$

Armature current: $I_a = I_L - I_f = 10 - 0.5 = 9.5 \text{ A}$

Back emf: $E_{ga} = V_s - I_a R_a = 220 - 9.5 \times 0.2 = 218.1 \text{ V}$

Power developed: $P_d = E_{ga} I_a = 218.1 \times 9.5 = 2071.95 \text{ W}$

Angular velocity $\omega_{ma} = \dfrac{2\pi N_m}{60} = \dfrac{2\pi \times 1800}{60} = 188.5 \text{ rad/s}$

Torque developed: $T = \dfrac{P_d}{\omega_{ma}} = \dfrac{2071.95}{188.5} = 11.0 \text{ N-m}$

When the motor develops a torque of 20 N-m.   Since the current in the field winding is held constant, the flux produced by the motor stays the same under different loads. From the preceding information we can determine the product of flux per pole and the motor constant as follows:

$$K_a \phi_p = \frac{E_{ga}}{\omega_{ma}} = \frac{218.1}{188.5} = 1.157$$

Since $T_d = K_a \phi_p I_a$,

$$I_a = \frac{T_d}{K_a \phi_p} = \frac{20}{1.157} = 17.286 \text{ A}$$

and the load current is

$$I_L = I_a + I_f = 17.286 + 0.5 = 17.786 \text{ A}$$

The back emf is

$$E_{gb} = 220 - 17.286 \times 0.2 = 216.54 \text{ V}$$

Thus, the angular velocity is

$$\omega_{mb} = \frac{E_{gb}}{K_a \phi_p} = \frac{216.54}{1.157} = 187.16 \text{ rad/s}$$

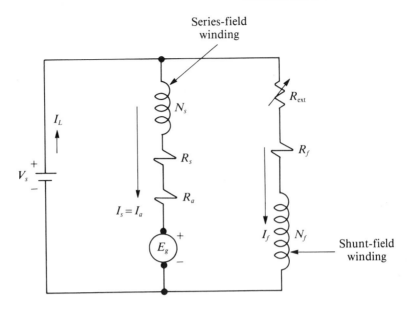

FIGURE 5.12
Long-Shunt
Compound dc
Motor

or the speed is

$$N_{mb} = \frac{60\omega_{mb}}{2\pi} = \frac{60 \times 187.16}{2\pi} = 1787 \text{ rpm}$$

There is only a change of 13 rpm even when the torque developed by the motor has almost doubled. This is why a shunt motor is considered as a fairly constant speed motor.

5.2.3 Compound Motor

Similar to compound generators, a compound motor has both series- and shunt-field windings. A compound motor may be connected short shunt or long shunt. In a long-shunt motor, the shunt-field winding is directly connected across the supply and therefore provides a constant flux. The field excitation varies with the load for the series winding. The schematic representations for long- and short-shunt motors are given in Figs. 5.12 and 5.13, respectively. A compound motor may be cumulative compound or differential compound. For direct comparison purposes, the speed–torque characteristics for all types of dc motors are shown in Fig. 5.14.

From the equivalent circuit of a long-shunt compound motor (Fig. 5.12), we can obtain the expression for its speed as

$$\omega_m = \frac{V_s - I_a(R_a + R_s)}{K_a(\phi_{sh} + k_f I_a)} \tag{5.9}$$

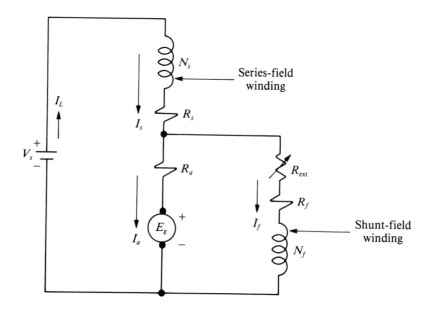

FIGURE 5.13
Short-Shunt
Compound
Motor

The numerator is very similar to that obtained for the series motor, Eq. (5.4), while the denominator is the sum of both series and shunt cases. In Eq. (5.9), ϕ_{sh} is the flux set up by the shunt-field winding and is assumed to be constant. The flux for the series-field winding has been replaced by $k_f I_a$ because it is proportional to the armature current. The torque developed by a long-shunt (cumulative) compound motor is

$$T_d = K_a I_a \phi_{sh} + K_a k_f I_a^2 \tag{5.10}$$

In the case of a long-shunt (differentially) compound motor, the speed–voltage relationship becomes

$$\omega_m = \frac{V_s - I_a(R_a + R_s)}{K_a(\phi_{sh} - k_f I_a)} \tag{5.11}$$

In this equation the numerator is the same as that for the long-shunt (cumulative) compound motor, and the denominator differs by the minus sign because the flux produced by the shunt-field winding is being opposed by the flux set up by the series-field winding. When the fluxes set up by the two windings are equal in magnitude, the denominator becomes zero and the motor accelerates to dangerously high speed. Under this condition the armature current is also very high as it is only limited by the total resistance in the armature circuit.

The torque relationship for the long-shunt (differentially) compound motor can be expressed as

$$T_d = K_a I_a \phi_{sh} - K_a k_f I_a^2 \qquad (5.12)$$

In a differentially compound motor the speed increases with the load due to the decrease in the overall flux in the motor, but the speed decreases because of the voltage drop across the total armature circuit resistance. The overall change in the speed is quite small, and for that reason a differentially compound motor can be used for constant speed applications. However, this motor is rarely used because a shunt motor provides subtantially constant speed at a lower initial cost.

The field of application of a cumulative compound motor lies principally in driving machines that are subjected to sudden changes in load, such as rolling mills, shears, and punches. It can also be used to replace a series motor in elevators and cranes.

■ EXAMPLE 5.7

To stabilize the shunt motor of Example 5.5, let us assume that a series-field winding with 4 turns per pole is added with a winding resistance of 0.025 Ω. Determine the new speed of the motor. What are the developed power and the torque in this case?

Solution

Since the series-field winding is added to stabilize the shunt motor, the overall flux in the motor must increase. Therefore, the motor must be a cumulative com-

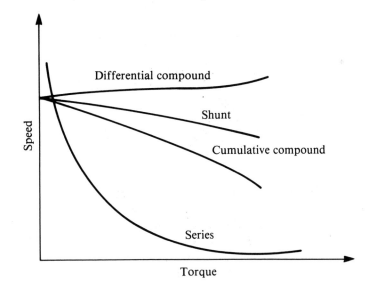

FIGURE 5.14
Direct Comparison of Speed–Torque Curves for dc Motors

pound type. At full load, the armature current is 39.4 A. The mmf provided by the series-field winding is $39.4 \times 4 = 157.6$ A-t/pole. Therefore, the equivalent shunt-field current would be

$$\frac{157.6}{1600} = 0.0985 \text{ A}$$

Thus the effective shunt-field current is

$$0.6 + 0.0985 - 0.025 = 0.6735 \text{ A}$$

From the magnetization curve, the no-load voltage at 2400 rpm or 251.33 rad/sec is

$$E_{g1} = 122 \text{ V}$$

Since

$$E_g = K_a \phi_p \omega_m,$$

$$K_a \phi_p = \frac{122}{251.33} = 0.4854$$

The back emf at full-load is

$$E_{g2} = 120 - 39.4(0.1 + 0.025) = 115.08 \text{ V}$$

The speed is

$$\omega_{m2} = \frac{E_{g2}}{K_a \phi_p} = \frac{115.08}{0.4854} = 237.08 \text{ rad/s} \quad \text{or} \quad 2264 \text{ rpm}$$

The power developed by the motor is

$$P_d = E_{g2}I_a = 115.08 \times 39.4 = 4534.15 \text{ W}$$

The torque developed by the motor is

$$T_d = \frac{P_d}{\omega_{m2}} = \frac{4534.15}{237.08} = 19.13 \text{ N-m}$$

5.3 POWER-FLOW DIAGRAM

In a dc motor, the electrical power supplied to the motor is converted into mechanical power. Some of the electrical power is dissipated as heat due to the electrical losses in the windings of the machine. The electrical power that can be converted into mechanical power is therefore the difference between the total electrical power input and all the electrical power lost in the windings of the motor.

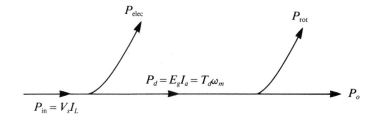

FIGURE 5.15
Power-Flow
Diagram for a
dc Motor

This power is known as the *available* power or the *developed* power. From this developed power we must now subtract all the rotational losses to obtain the net mechanical power available at the shaft of the motor. Figure 5.15 shows the basic power-flow diagram for a dc motor.

The efficiency of the motor can be expressed as

$$\eta = \frac{P_{in} - P_{loss}}{P_{in}} \tag{5.13}$$

where P_{in} is the power input and P_{loss} is the sum of all the losses in the motor.

■ EXAMPLE 5.8

A 120-V dc motor develops 10 hp at its rated speed of 1800 rpm. What is the torque developed by the motor? If the efficiency of the motor is 85%, determine the current taken by the motor and cost of energy absorbed if the motor runs continuously at full load for a period of 10 hours. Assume the cost of electrical energy is $0.05 per kilowatt-hour.

Solution

Power output at full load: $P_o = 10 \times 746 = 7460$ W

Angular velocity of the motor: $\omega_m = \dfrac{2\pi \times 1800}{60} = 188.5$ rad/s

Torque developed: $T_s = \dfrac{P_o}{\omega_m} = \dfrac{7460}{188.5} = 39.58$ N-m

Power input: $P_{in} = \dfrac{P_o}{\eta} = \dfrac{7460}{0.85} = 8776.5$ W

Line current: $I_L = \dfrac{P_{in}}{V_s} = \dfrac{8776.5}{120} = 73.14$ A

Energy absorbed in 10 hours: $W_e = 8776.5 \times 10 = 87765$ W-h
$= 87.765$ kW-h

Cost of energy: $C = 87.765 \times 0.05 = \$4.39$

■ **EXAMPLE 5.9**

A 120-V, 0.75-hp dc shunt motor takes 2 A at no load and 7 A at full load. The armature-circuit resistance is 0.8 Ω and the shunt-field resistance is 240 Ω. Determine (a) the rotational power loss in the motor, and (b) its efficiency at full load.

Solution

(a) From no-load data:

Field current:
$$I_f = \frac{V_f}{R_f} = \frac{120}{240} = 0.5 \text{ A}$$

Armature current:
$$I_a = I_L - I_f = 2 - 0.5 = 1.5 \text{ A}$$

Armature winding loss:
$$I_a^2 R_a = 1.5^2 \times 0.8 = 1.8 \text{ W}$$

Shunt-field winding loss:
$$V_f I_f = 120 \times 0.5 = 60.0 \text{ W}$$

Thus, the total electric power loss at no load is

$$P_{\text{elec}} = 1.8 + 60 = 61.8 \text{ W}$$

The power input at no load is

$$P_{\text{in}} = V_s I_L = 120 \times 2 = 240 \text{ W}$$

Therefore, the rotational loss is

$$P_{\text{rot}} = P_{\text{in}} - P_{\text{elec}} = 240 - 61.8 = 178.2 \text{ W}$$

(b) At full-load:

Power input:
$$P_{\text{in}} = 120 \times 7 = 840 \text{ W}$$

Shunt-field loss:
$$120 \times 0.5 = 60 \text{ W}$$

Armature winding loss:
$$6.5^2 \times 0.8 = 33.8 \text{ W}$$

Full-load electrical loss:
$$60 + 33.8 = 93.8 \text{ W}$$

Power developed:
$$P_d = 840 - 93.8 = 746.2 \text{ W}$$

Power output:
$$P_o = P_d - P_{\text{rot}} = 746.2 - 178.2 = 568 \text{ W}$$

Efficiency of the motor:
$$\eta = \frac{P_o}{P_{\text{in}}} = \frac{568}{840} = 0.676 \text{ or } 67.6\%$$

5.4 SPEED REGULATION OF DC MOTORS

Both shunt and compound motors are essentially constant-speed motors and are useful for fixed-speed operations. On the other hand, the speed of a series motor is very high at no load and decreases with the increase in the load. Since the no-load speed of a series motor is limited only by residual magnetism and friction, a compound motor is quite often used in place of a series motor to obtain a fixed no-load speed.

The term *speed regulation* is used to define the change in speed that can be expected when the load on the motor is gradually reduced from full load to zero. It is usually expressed as a percent of the speed at the rated load. Thus,

$$SR\% = \frac{\omega_{nL} - \omega_{fL}}{\omega_{fL}} \times 100 \qquad (5.14)$$

or

$$SR\% = \frac{N_{nL} - N_{fL}}{N_{fL}} \times 100 \qquad (5.15)$$

where

$SR\%$ = percent of speed regulation
N_{nL} = no-load speed in rpm
ω_{nL} = no-load speed in rad/s
N_{fL} = full-load speed or rated speed in rpm
ω_{fL} = full-load speed or rated speed in rad/s

The speed regulation is a measure of a motor's ability to maintain its speed with varying load. A speed regulation of zero implies that the motor speed is independent of the load and is equal to its rated value. On the other hand, a speed regulation of 100% corresponds to no-load speed, which is twice as much as the full-load speed.

■ EXAMPLE 5.10

The speed of a shunt motor falls from 1800 rpm at no load to 1750 rpm at rated load. What is the percentage of speed regulation?

Solution

The no-load speed is N_{nL} = 1800 rpm. The full-load speed is N_{fL} = 1750 rpm. Thus,

$$SR\% = \frac{1800 - 1750}{1750} \times 100 = 2.875\%$$

5.5   STARTING OF DC MOTORS

The impressed voltage on a dc motor is balanced by the voltage drop due to the effective resistance of the armature circuit and the back emf. To minimize the power losses in the motor and thereby improve its efficiency, the effective resistance in the armature circuit is held as low as possible. However, at the time of starting the armature is at a standstill and the back emf is zero. In this case, the applied voltage is balanced only by the voltage drop across the armature-circuit resistance only. The application of rated voltage at standstill can therefore result in excessive current flow and may cause damage to the motor. On the other hand, excessive current may also result in large voltage fluctuation in the power lines. Since the torque developed by the motor is proportional to the current in the armature circuit, the developed torque is higher than the rated. Such a high torque may result in too rapid acceleration and may damage the load. Thus, all the effects of starting a dc motor at its rated voltage are harmful. Therefore, we need a starting device that can keep the starting current in a motor within the acceptable limits. The basic purpose of the starting device is to reduce the applied voltage at the terminals of the motor during the starting period and thereby control the time used for acceleration. It can either be a manual or an automatic operation.

5.6   CONVENTIONAL METHODS OF SPEED CONTROL

Direct-current motors are suitable for applications that require considerable adjustment in speed due to the fact that the speed of these motors can be varied over a wide range. The back emf in a dc motor is directly proportional to the flux per pole in the motor. The flux per pole can be changed by changing the field current. On the other hand, the back emf can be controlled by changing the effective resistance in the armature circuit. These are, in fact, the conventional methods of controlling the speed of a dc motor.

5.6.1   Armature Resistance Control

In this method a variable resistor is inserted in series with the armature. The field winding is directly connected across the full line voltage in order to maintain a constant flux for shunt and compound motors. The variable-resistor connections for the purpose of adjusting the speeds of a series, shunt, and compound motors are illustrated in Fig. 5.16. Since the full armature current flows through this variable resistor, it must be large enough to handle the large values of armature current. The back emf equation can be written, in general, as

$$E_g = V_s - I_a(R_a + R_c + R_s) \qquad (5.16)$$

where V_s = full-line voltage

R_c = resistance of external variable resistor

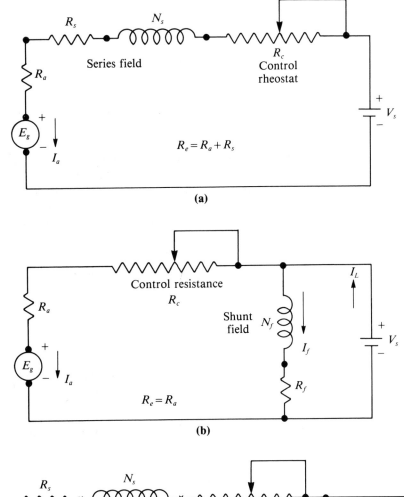

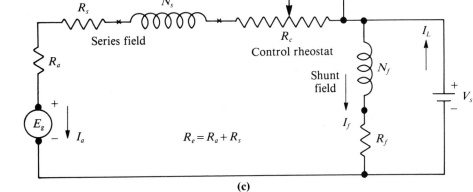

FIGURE 5.16
Armature
Resistance
Speed Control
for (a) Series
Motor; (b) Shunt
Motor; and (c)
Compound
Motor

The operating speed of the motor is

$$\omega_m = \frac{E_g}{K_a\phi_p}$$

which can be written as

$$\omega_m = \frac{V_s - I_a(R_a + R_c + R_s)}{K_a\phi_p} \tag{5.17}$$

Any increase or decrease in R_c will decrease or increase the speed of the motor in accordance with Eq. (5.17). Speed–torque characteristics of series, shunt, and compound motors for various values of the control resistor are given in Fig. 5.17, which shows that a wide range of speed can be obtained by this method. The motor will, however, deliver any desired torque over its working range because the torque depends on the flux and the armature current.

As pointed out earlier, the full armature current flows through the control resistance and, therefore, there is a considerable power loss in that resistance, which lowers the efficiency of the motor. This, in fact, is the main disadvantage of such a speed-control system.

■ EXAMPLE 5.11

A 2-hp, 120-V, 2400-rpm dc shunt motor takes 2 A at no load and 14.75 A at full load. The armature-circuit resistance is 0.4 Ω and the field-winding resistance is 160 Ω. If an external resistance of 3.6 Ω is connected in series with the armature circuit, calculate the motor speed, the power loss in the external resistance as a percent of total power input, and the efficiency of the motor.

Solution

From no-load data

Field-winding current: $\quad I_f = V_f/R_f = 120/160 = 0.75$ A

Armature current: $\quad I_a = I_L - I_f = 2 - 0.75 = 1.25$ A

Back emf at no load: $\quad E_{gnL} = V_s - I_aR_a = 120 - 1.25 \times 0.4 = 119.5$ V

Since the power supplied to the motor at no load accounts for all the losses, the rotational losses are

$$P_{\text{rot}} = P_{\text{in}} - P_{\text{elec}} = V_sI_L - V_fI_f - I_a^2R_a$$

$$= 120 \times 2 - 120 \times 0.75 - 1.75^2 \times 0.4$$

$$P_{\text{rot}} = 148.78 \text{ W}$$

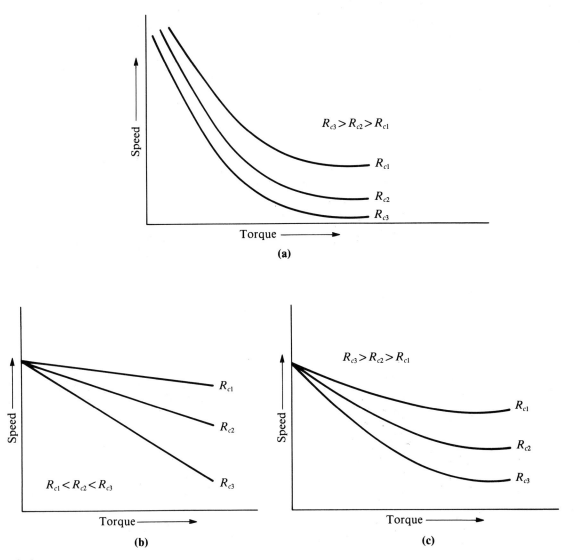

FIGURE 5.17
Speed–Torque Characteristics for (a) Series; (b) Shunt; and (c) Compound Motors Depicting the Effect of Adding a Variable Resistance in the Armature Circuit to Control the Motor Speed

Full-load data

Full-load armature current:	$I_a = 14.75 - 0.75 = 14$ A
Back emf at full-load:	$E_{gfL} = 120 - 14 \times 0.4 = 114.4$ V
Full-load speed:	$N_{fL} = 2400$ rpm

The no-load speed can be calculated from the following equation:

$$\frac{N_{nL}}{N_{fL}} = \frac{E_{gnL}}{E_{gfL}}$$

$$\frac{N_{nL}}{2400} = \frac{119.5}{114.4}$$

Thus

$$N_{nL} = 2507 \text{ rpm}$$

The power output at full-load is

$$P_o = P_{\text{in}} - P_{\text{elec}} - P_{\text{rot}} = V_s I_L - V_f I_f - I_a^2 R_a - P_{\text{rot}}$$

$$= 120 \times 14.75 - 120 \times 0.75 - 14^2 \times 0.4 - 148.78$$

$$= 1452.82 \text{ W}$$

The motor efficiency is

$$\eta = \frac{P_o}{P_{\text{in}}} = \frac{1452.82}{120 \times 14.75} = 0.82 \quad \text{or} \quad 82\%$$

With external control resistance in the armature circuit: Let us assume that the motor still develops the full-load torque. Thus, the full-load current in the armature circuit remains the same.

New full-load back emf:

$$E_{gn} = 120 - 14.75 \times (0.4 + 3.6) = 61 \text{ V}$$

New full load speed:

$$N_n = 2400 \times \frac{61}{114.4} = 1280 \text{ rpm}$$

Total power input:

$$P_{\text{in}} = 120 \times 14.75 = 1770 \text{ W}$$

Power loss in the control resistance:

$$P_{cr} = 14.75^2 \times 3.6 = 783.23 \text{ W}$$

Percent of power loss in control resistance:

$$\frac{783.23}{1770} \times 100 = 44.25\%$$

Power output:

$$P_o = P_{\text{in}} - V_f I_f - I_a^2 R_a - P_{cr} - P_{\text{rot}}$$

$$= 1770 - 120 \times 0.75 - 14.75^2 \times 0.4 - 783.23 - 148.78$$

$$= 660.97 \text{ W}$$

Motor efficiency:

$$\eta = P_o/P_{\text{in}} = 660.97/1770 = 0.373 \quad \text{or} \quad 37.3\%$$

As is evident in this example, almost half the power is lost in the external control resistance. At the rated load, the motor speed is almost half as much with the control resistance than without it. There is, of course, no change in the torque developed by the motor but the efficiency of the motor has reduced considerably.

5.6.2 Shunt-field Current Control

As highlighted earlier, the second approach to control the speed of the motor involves the control of field current, which thereby regulates the flux in the motor. The current in the field winding can be controlled by directly connecting a variable resistor in series with the field winding. Since the current in the field winding of a shunt or a compound motor is comparatively very small, the external resistance does not have to dissipate a lot of power. This is therefore an economical way to control the speed of the motor. Furthermore, the efficiency of the motor will be higher as compared to the series resistance control. This would be the preferred way of speed control for some motors. Adding an external resistance in the field circuit, will cause the field current and thereby the flux in the motor to decrease. Since the speed is inversely proportional to the flux, the motor speed will increase. Therefore, this method of speed control provides adjustments in speed that are above the rated values only.

As the torque developed by the motor is directly proportional to the product of the armature current and the flux in the motor, the armature current must increase by the same proportion that the flux in the motor decreases in order to obtain the required torque from the motor. This method of speed control is therefore not satisfactory for compound motors, because any decrease in the flux produced by the shunt-field winding is offset by an increase in the flux produced by the series-field winding owing to the increase in armature current. Figure 5.18 illustrates how an external resistance is added directly in the field-winding circuit of a shunt motor for the purpose of adjusting its speed at a value higher than its rated speed.

■ EXAMPLE 5.12

If in the motor discussed in Example 5.11 an external resistance of 80 Ω is inserted in series with the shunt-field winding instead of the armature circuit, determine

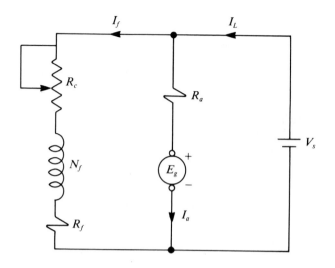

FIGURE 5.18
Shunt Motor with
an External
Resistance in its
Field Circuit to
Control its
Speed of
Operation

the new motor speed, the power loss in the external resistance, and the motor efficiency. Assume that the flux is proportional to the square root of the field winding current.

Solution

The new field current is

$$I_{fn} = \frac{120}{160 + 80} = 0.5 \text{ A}$$

Let ϕ be the flux at full load when the field current is 0.75 A and ϕ_n be the flux when the field current is 0.5 A. The new flux in the motor is then

$$\phi_n = \phi \sqrt{\frac{I_{fn}}{I_f}} = 0.816\phi \text{ Wb}$$

Since the torque developed by the motor must be the same, the new armature current must be

$$I_{an} = I_a \frac{\phi}{\phi_n} = 14 \times \frac{1}{0.816} = 17.15 \text{ A}$$

The new counter emf is

$$E_{gn} = 120 - 17.15 \times 0.4 = 113.14 \text{ V}$$

Since $E_g = K'_a \phi N$, then

$$114.4 = K'_a \phi N$$

and
$$113.14 = K'_a \phi_n N$$

Thus
$$N_n = N \frac{113.14}{114.4} \left(\frac{\phi}{\phi_n} \right)$$

$$= 2400 \times \frac{113.14}{114.4} \times \frac{1}{0.816}$$

$$= 2909 \text{ rpm}$$

Power loss in the control resistance:

$$I_f^2 R_c = 0.5^2 \times 80 = 20 \text{ W}$$

Power loss in the field winding:

$$I_f^2 R_f = 0.5^2 \times 160 = 40 \text{ W}$$

Power loss in the armature circuit:

$$I_a^2 R_a = 17.15^2 \times 0.4 = 117.65 \text{ W}$$

Total electrical loss:

$$P_{\text{elec}} = 117.65 + 40 + 20 = 177.65 \text{ W}$$

Power input:

$$P_{\text{in}} = 120 \times (17.15 + 0.5) = 2118 \text{ W}$$

Percent of power loss in control resistance:

$$\frac{20}{2118} \times 100 = 0.94\%$$

To determine the efficiency of the motor, we need the information on the rotational losses. Both the flux and the rated speed of the motor have changed. The motor should have low core loss due to the reduction in the flux and higher friction and windage loss due to the increase in speed. It is possible that one may offset the other. On that possibility, let us assume that the rotational losses in the motor stay the same. The motor power output is then

$$P_o = P_{\text{in}} - P_{\text{elec}} - P_{\text{rot}}$$

$$= 2118 - 177.65 - 148.78 = 1791.57 \text{ W}$$

The motor efficiency is

$$\eta = \frac{P_o}{P_{\text{in}}} = \frac{1791.57}{2118} = 0.845 \quad \text{or} \quad 84.5\%$$

5.7 WARD–LEONARD CONNECTION

The chief drawback of an armature resistance control method to adjust the speed of the motor is the power loss in that resistance. The aim of such a control is to apply a variable voltage to the armature circuit by adjusting the voltage drop across the control resistance. If we can have an independent variable voltage that can be directly applied to the armature of the motor, we can effectively control the speed of the motor without any additional power loss. A system where a variable voltage is applied to the armature circuit and the field voltage is held constant is known as *armature-voltage control*. On the other hand, we can maintain the armature voltage constant and vary the motor's field current in order to change its operating speed. This is called as a *field-controlled system*. A system that incorporates both the control schemes is referred to as the *Ward–Leonard system*, as shown in Fig. 5.19. This method of speed control requires a constant voltage that can be applied to the field winding of a separately excited dc generator and a separately excited dc motor. This constant voltage is obtained from another generator, which is known as the *exciter*. A three-phase motor is used as the prime mover to rotate the armatures of the two generators, one of which has been labeled as the exciter. The armature of the other generator is directly connected to the armature of the separately excited dc motor. By varying the field current in the generator, the desired voltage across the motor armature terminals is obtained. The speed is controlled in accordance with the basic equation

$$\omega_m = \frac{V_{sv} - I_a R_a}{K_a \phi_p} \tag{5.18}$$

where V_{sv} is the variable voltage provided by the dc generator.

If the shunt-field excitation is held constant, then the flux in the motor will be constant. For the constant torque requirements, the armature current must remain the same. Equation (5.18) can then be rewritten as

$$\omega_m = k_1 V_{sv} - k_2 \tag{5.19}$$

which is the equation of a straight line, where

$$k_1 = \frac{1}{K_a \phi_p} \quad \text{and} \quad k_2 = \frac{I_a R_a}{K_a \phi_p}$$

In the Ward–Leonard system, by maintaining the load and field currents of the dc motor at their rated values, we can vary the speed by adjusting the armature terminal voltage. The typical torque and power characteristics of a dc motor in a Ward–Leonard system are illustrated in Fig. 5.20. The two regions of importance are clearly highlighted. These are constant-power and constant-torque regions. The constant-power region corresponds to field current control while the constant-torque region corresponds to the armature resistance control. The Ward–Leonard system of speed control offers excellent possibilities for the adjustment of the speed of the motor above or below its base value. Base speed is the speed at the rated field current and armature voltage on full load. It is, however, an expensive system of control because it requires a set of two generators and a three-phase motor.

■ EXAMPLE 5.13

A 39-hp, 440-V, dc motor operates at 1000 rpm on full load. The motor efficiency is 86.72% and armature resistance is 0.377 Ω. The speed of the motor will be

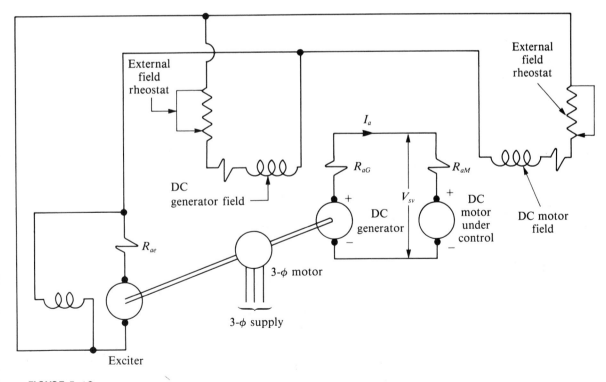

FIGURE 5.19
Schematic Representation of a Ward–Leonard Speed-Control System for a Shunt Motor

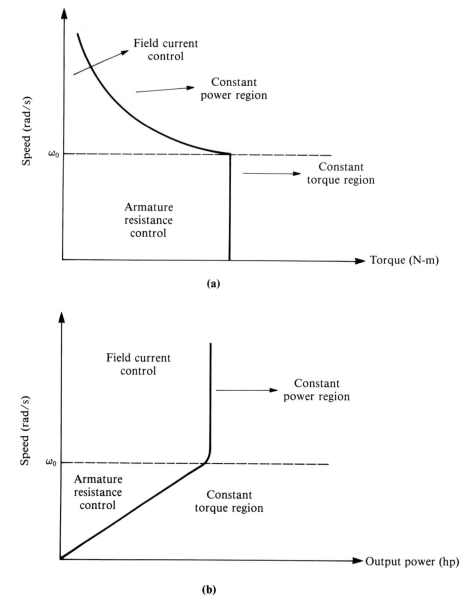

FIGURE 5.20
Typical (a)
Torque and (b)
Power
Characteristics
for a Ward–
Leonard Speed-
Control System

adjusted to 750 rpm at the rated load using suitable Ward–Leonard system. The armature resistance of the dc generator is 0.336 Ω. Determine (a) the speed versus current characteristics of this motor for the given operating conditions, and (b) the speed adjustment range of this motor such that the armature current does not exceed twice its rated value in the constant-torque region.

Solution

(a) A simple circuit diagram of a dc motor in a Ward–Leonard system is shown in Fig. 5.21a. Since the speed–current characteristic of a separately excited dc

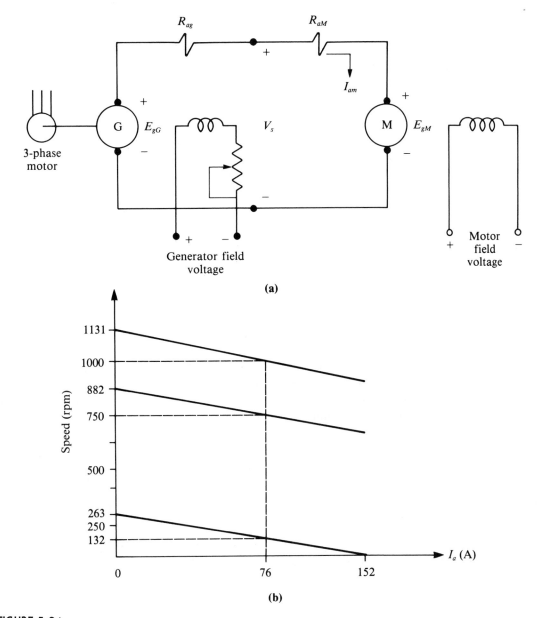

(a)

(b)

FIGURE 5.21
(a) Simplified Diagram of a Ward–Leonard System; (b) Speed–Current Characteristics for a dc Motor (Example 5.13)

motor is basically a straight line, knowledge of two operating points is enough to determine it. The rated operating conditions are given, and we can easily determine the no-load operating conditions as follows: The input power to the motor is

$$P_{in} = \frac{P_o}{\eta} = \frac{39 \times 746}{0.8672} = 33,549 \text{ W}$$

The armature current at full-load is

$$I_a = \frac{33,549}{440} \cong 76 \text{ A}$$

The back emf is

$$E_{gM} = V_s - I_a R_{aM} = 440 - 0.377 \times 76 = 411.35 \text{ V}$$

$$\omega_m = \frac{2\pi N_m}{60} = \frac{2\pi \times 1000}{60} = 104.72 \text{ rad/s}$$

$$K_a\phi_p = \frac{E_{gM}}{\omega_m} = \frac{411.35}{104.72} = 3.93$$

To meet these requirements, the emf generated by the dc generator must be

$$E_{gG} = V_s + I_a R_{aG} = 440 + 0.336 \times 76 = 465.53 \text{ V}$$

Under no-load conditions, the armature current is just enough to supply the rotational losses. Since the rotational losses are not given, we will assume them to be zero. In that case, the no-load current is also zero. Consequently, the generator's emf is equal and opposite to the back emf of the motor. That is, $E_{gG} = E_{gM}$. Thus, the no-load speed is

$$\omega_o = \frac{E_{gM}}{K_a\phi_p} = \frac{465.53}{3.93} = 118.46 \text{ rad/s} \quad \text{or} \quad 1131 \text{ rpm}$$

The speed–current characteristic is shown in Fig. 5.21b.

Operation at 750 rpm on full load

$$\omega_m = \frac{2\pi \times 750}{60} = 78.54 \text{ rad/s}$$

Since the motor field current is kept constant at its rated value, the flux per pole is the same at all operating conditions. Thus, the no-load voltage at 750 rpm is

$$E_{gM} = 3.93 \times 78.54 = 308.66 \text{ V}$$

The emf of the generator must be

$$E_{gG} = 308.66 + (0.377 + 0.336) \times 76 = 362.85 \text{ V}$$

For proper operation, the generator's emf is lower when the motor runs at 750 rpm on full load than that at 1000 rpm. We must decrease the current in the field winding of the generator to lower its emf. Once again, under no-load conditions, $E_{gG} = E_{gM}$. Thus the no-load speed is

$$\omega_o = \frac{362.85}{3.93} = 92.33 \text{ rad/s} \quad \text{or} \quad 882 \text{ rpm}$$

and the operating characteristic is given in Fig. 5.21b.

(b) In this case we are required to determine the lowest speed at which the motor can operate on full load. The maximum allowable current is $I_M = 2 \times 76 = 152$ A, and it will be maximum under a blocked rotor condition. Thus, $\omega_m = 0$, which implies that $E_{gM} = 0$. The generator's emf must be

$$E_{gG} = I_M(R_{aG} + R_{aM}) = 152 \times (0.336 + 0.377) = 108.376 \text{ V}$$

The no-load speed is now

$$\omega_o = \frac{E_{gG}}{K_a \phi_p} = \frac{108.376}{3.93} = 27.58 \text{ rad/s} \quad \text{or} \quad 263 \text{ rpm}$$

On full load the back emf of the motor is

$$E_{gM} = E_{gG} - (R_{aG} + R_{aM})I_a$$
$$= 108.376 - (0.336 + 0.377) \times 76 = 54.188 \text{ V}$$

The full-load speed is

$$\omega_m = \frac{54.188}{3.93} = 13.79 \text{ rad/s} \quad \text{or} \quad 132 \text{ rpm}$$

To satisfy the given constraints, the full-load speed of the motor must be between 132 rpm and 1000 rpm.

5.8   TORQUE MEASUREMENTS

Two well-known methods are commonly used to measure the torque developed by a motor: the *Prony brake* test and a *dynamometer.* Let us discuss briefly how torque measurements are made using these methods.

5.8.1   Prony Brake Test

The basic arrangement of this test is shown in Fig. 5.22. In this method a pulley is mounted on the shaft extension of the motor, which acts as a brake drum. A brake band with or without brake blocks is wrapped around the brake drum as shown. For subfractional-horsepower motors, a string can be used instead of a brake band. One end of the brake band may be permanently fastened to the torque arm while the other end can be adjusted by a thumbscrew for tightening against the surface of the brake drum. For high-horsepower motors, cooling of the brake drum might be necessary. The brake band is slowly tightened until the motor runs at the desired speed. The tendency of the torque arm to move with the drum is resisted by the force of the spring (scale) attached at the far end of the torque arm. The deflection of the spring can be calibrated either in force units or in torque units. If the calibration is done in terms of force units, the torque is simply the product of the force times the effective length of the torque arm. The effective length is the distance between the center of the pulley and the place where the spring is attached.

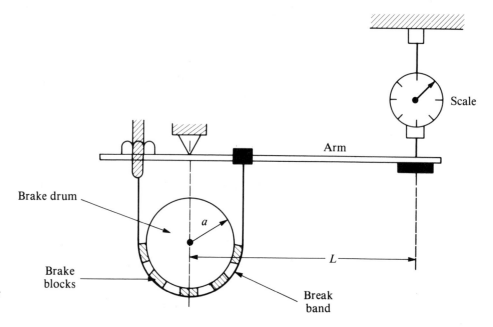

FIGURE 5.22
Typical
Arrangement for
Prony Brake Test

If F is the force in newtons acting on the spring and L is the effective length of the torque arm, the torque developed by the motor is

$$T_s = FL \text{ N-m} \qquad (5.20)$$

In our discussion we have assumed that the torque arm is perfectly horizontal before the beginning of the test and the dead (tare) weight of the lever arm is zero. In actual practice, the dead weight of the torque arm must be taken into consideration, as illustrated by the following example.

■ EXAMPLE 5.14

A 120-V, 0.75-hp motor is tested at 2400 rpm using the Prony brake test. The input current is 7 A and the deflection force on the spring is 4.57 N. The effective length of the torque arm is 50 cm and its dead weight is 0.03 N. Determine the torque developed and the efficiency of the motor.

Solution

Net force exerted on the spring:

$$F = 4.57 - 0.03 = 4.54 \text{ N}$$

Effective length of the torque arm:

$$L = 0.5 \text{ m}$$

Torque developed by the motor:

$$T_s = FL = 4.54 \times 0.5 = 2.27 \text{ N-m}$$

Angular velocity of the shaft:

$$\omega_m = \frac{2\pi N_m}{60} = \frac{2\pi \times 2400}{60} = 251.33 \text{ rad/s}$$

Power output:

$$P_o = T_s \omega_m = 2.27 \times 251.33 = 570.52 \text{ W}$$

Power input:

$$P_{in} = 120 \times 7 = 840 \text{ W}$$

Efficiency of the motor:

$$\eta = \frac{P_o}{P_{in}} = \frac{570.52}{840} = 0.679 \quad \text{or} \quad 67.9\%$$

5.8.2 Dynamometer Method

The chief disadvantages of the Prony brake test are the vibrations caused by the brake blocks and the necessity of constantly cooling the brake drum. These drawbacks are overcome by the electrodynamometer. The general construction details are shown in Fig. 5.23. The dynamometer is a dc machine whose field winding is usually excited by a separate voltage source. The flux can, however, be controlled by adding an external variable rheostat in series with the field winding. The only difference between the dynamometer and the dc motor is that the stator of the dynamometer is also free to rotate, whereas it is rigid in the dc motor. The machine is mounted on low-friction ball bearings. On the outside of the stator yoke is fastened a torque arm, the other end of which is attached to the spring (scale) in the same fashion as for the Prony brake.

The armature of the dynamometer is rotated by coupling it to the shaft of the motor under test. Due to the flux-cutting action, a voltage is induced in the armature winding of the dynamometer (generator action). If the armature circuit is completed by connecting its terminal to a resistive load, a current will flow in the armature winding. The amount of current, however, depends on the flux and the total resistance in the armature circuit of the dynamometer. As soon as there is a current flow in the armature winding of the dynamometer, a force is produced by the interaction of stator flux and armature current (motor action). The direction of force is such that it tends to resist the rotation of the armature winding. Since the stator, which houses the field winding producing the flux, is free to rotate, it will be pulled around equally by the motor action that exists within the dynamometer. The only restraining force acting on the stator is provided by the spring (scale) as in the Prony brake method.

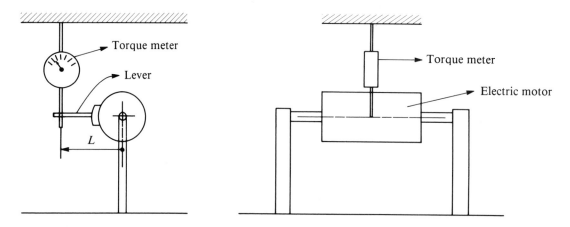

FIGURE 5.23
Typical Electrodynamometer Test Setup

■ EXAMPLE 5.15

A 5-hp motor rated at 1200 rpm is to be tested on a dynamometer whose torque arm is 40 cm in length. If the scale is calibrated in terms of force, what must be the reading on the scale?

Solution

Power output: $P_o = 5 \times 746 = 3730$ W

Shaft speed: $\omega_m = \dfrac{2\pi N_m}{60} = \dfrac{2\pi \times 1200}{60} = 125.66$ rad/s

Torque developed: $T_s = \dfrac{P_o}{\omega_m} = \dfrac{3730}{125.66} = 29.68$ N-m

Effective torque arm length: $L = 0.4$ m

Force reading: $F = \dfrac{T_s}{L} = \dfrac{29.68}{0.4} = 74.2$ N

If the scale is calibrated in load units such as kilograms, the load on the scale is $74.2/9.81 = 7.564$ kg.

5.9 BRAKING OR REVERSING DC MOTORS

In certain applications, it becomes necessary to quickly stop the motor or even to reverse its direction of rotation. The motor can be stopped with the help of mechanically operated brakes, which are blocks held against the brake drum by springs or weights. The operation is far from being smooth and is difficult to control. Therefore, it is desirable to use other means of stopping and reversing the motor. The three commonly used methods to either stop or reverse the direction of rotation of the motor are:

1. Plugging
2. Dynamic braking
3. Regenerative breaking

5.9.1 Plugging

Reversing the direction of rotation of a motor by merely reversing the power connections to the armature terminals is known as plugging. The field-winding connections are left undisturbed. Just prior to plugging, the back emf is opposing the applied voltage. Since the armature-circuit resistance is usually very small, the back emf is almost equal and opposite to the applied voltage. At the moment the motor is plugged, the back emf and the applied voltage are in the same direction. Therefore, the total voltage in the armature circuit is almost twice as much as the applied voltage. To protect the motor from the sudden jump in armature current, an external resistance must be added in the circuit. The current in the armature reverses its direction and produces a force that tends to rotate the armature in a direction opposite to its initial rotation, thereby causing the motor to stop. This technique can be used to brake the motor if the power to the motor is disconnected at the instant the armature is passing through zero speed in its attempt to reverse itself as is obvious from the derivation given next. The circuit connections, in their simplest forms, for shunt and series motors are shown in Figs. 5.24a and 5.24b, respectively.

At any time during the plugging action, let E_g be the back emf. The armature current is given by

$$I_a = \frac{V_s + E_g}{R + R_a} = \frac{V_s}{R + R_a} + \frac{E_g}{R + R_a}$$

$$= \frac{V_s}{R + R_a} + \frac{K_a \phi_p \omega_m}{R + R_a} \tag{5.21}$$

We can now determine the breaking torque as

$$T_b = K_a I_a \phi_p = \frac{K_a \phi_p V_s}{R + R_a} + \frac{K_a^2 \phi_p^2 \omega_m}{R + R_a}$$

or

$$T_b = K_1 \phi_p + K_2 \phi_p^2 \omega_m \tag{5.22}$$

where

$$K_1 = \frac{K_a V_s}{R + R_a} \quad \text{and} \quad K_2 = \frac{K_a^2}{R + R_a}$$

For the series motor, the flux also depends on the armature current, which in turn depends on the speed of rotation. However, the flux in the shunt motor is constant, and the expression for the braking torque can be rewritten as

$$T_b = K_3 + K_4 \omega_m \tag{5.23}$$

where

$$K_3 = K_1 \phi_p \quad \text{and} \quad K_4 = K_2 \phi_p^2$$

Even when the speed of rotation is zero there is still some braking torque. Thus, if the supply voltage is not disconnected at the instant the motor reaches zero speed, it will accelerate in the reverse direction.

5.9.2 Dynamic Braking

To bring a motor to rest quickly, the dynamic braking technique makes use of the generator action in a motor. If the armature winding of a motor is suddenly

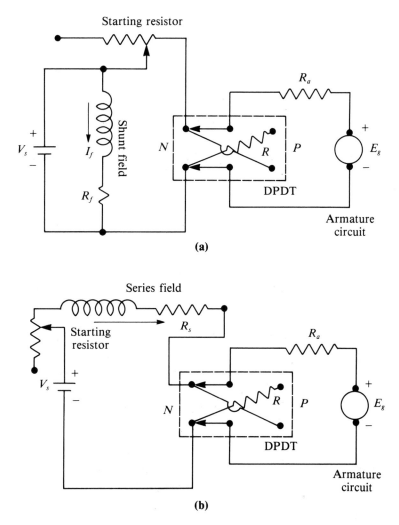

FIGURE 5.24
(a) Plugging of
a Shunt Motor;
(b) Plugging of
a Series Motor.
N, Normal
Operation, P,
Plugged
Operation

disconnected from the source while the field is left energized, the motor will coast to a stop. The time taken for the motor to come to rest depends on the kinetic energy stored in the rotating system. If, on the other hand, the armature winding immediately after being disconnected from the source is connected across a resistance, the back emf will produce a current whose direction is opposite to the direction of current when the machine was operating as a motor. Since the motor is now acting as a generator and is producing power that is being dissipated in the resistor, torque is required to turn the armature. The amount of power being dissipated in the resistor therefore determines how fast the kinetic (mechanical) energy is being expended. The resistor is selected to limit the inrush of current typically to 150% of the rated current of the motor. As the speed of the motor falls, so does the back emf and the current in the external resistance. Therefore, the dynamic braking action is maximum at first and then diminishes to zero as the motor comes to a stop. A simple circuit illustrating the dynamic braking action for a shunt motor is given in Fig. 5.25a. A sketch illustrating the principle of dynamic braking for a series motor is shown in Fig. 5.25b.

The armature current in this case is

$$I_a = \frac{E_b}{R + R_a} = \frac{K_a \phi_p \omega_m}{R + R_a} \tag{5.24}$$

and the braking torque is given by

$$T_b = K_a \phi_p I_a = \frac{K_a^2 \phi_p^2 \omega_m}{R + R_a} = K_2 \phi_p^2 \omega_m \tag{5.25}$$

Therefore, for a series motor the braking torque is $T_b \propto \phi_p^2 \omega_m$ or $T_b \propto I_a^2 \omega_m$. For shunt motors, the braking torque is $T_b \propto \omega_m$. It is evident that T_b decreases as the motor slows down and disappears altogether when the motor comes to a stop.

5.9.3 Regenerative Braking

Regenerative braking is used in applications where the motor speed is likely to increase from its rated value. Such applications include trains, elevators, cranes, and hoists. Let us consider the shunt motor used in trains. Under normal operation the back emf is slightly less than the applied voltage to the armature terminals of the motor. If the flux in the shunt motor is held constant, the back emf is directly proportional to the speed of the motor. Now assume the train is going downhill. As its speed increases, so does the back emf in the motor. If the back emf becomes higher than the applied voltage, the current in the armature winding reverses its direction, and the motor now operates as a generator and therefore sends the power back to the source or other devices operating from the same source. The generator current is directly related to the speed. Since the direction of current in the armature has reversed, the direction of electromagnetic torque developed will also be reversed. Thus, the regenerative braking action will tend (1) to slow down the motor and thereby the train and (2) to produce power.

■ EXAMPLE 5.16

A 1200-V dc shunt motor draws 30 A while supplying full load at 100 rad/s. The armature resistance is 1.0 Ω and the field-winding resistance is 600 Ω. Determine the external resistance that must be added in series with the armature circuit so that the armature current does not exceed 150% of its full-load value when

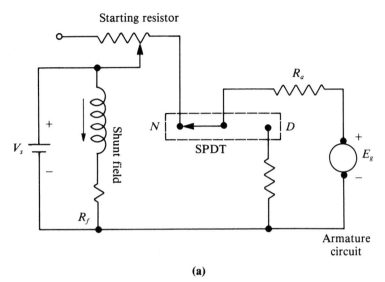

(a)

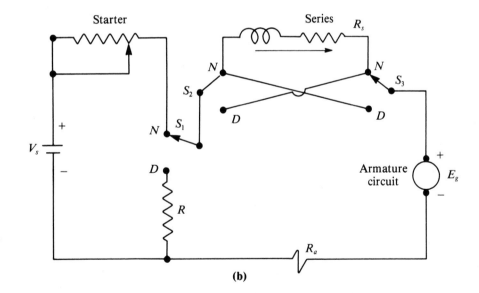

FIGURE 5.25
(a) Dynamic
Braking of a
Shunt Motor; (b)
Dynamic
Braking of a
Series Motor

(b)

plugged. Determine the braking torque (a) at the instant of plugging and (b) when the motor attains zero speed.

Solution

The field current is $I_f = 1200/600 = 2.0$ A and the armature current is $I_a = I_L - I_f = 30 - 2 = 28$ A. The back emf in the motor is

$$E_g = V_s - I_a R_a$$

$$= 1200 - 28 \times 1 = 1172 \text{ V}$$

The total voltage at the instant of plugging is $1200 + 1172 = 2372$ V. The maximum allowable current in the armature is $1.5 \times 28 = 42$ A. The total resistance must be $2372/42 = 56.48$ Ω. The external resistance that must be added is $R = 56.48 - 1 = 55.48$ Ω.

The braking torque at any speed for the shunt motor is given as

$$T_b = K_3 + K_4 \omega_m \quad \text{and} \quad K_a \phi_p = \frac{1172}{100} = 11.72$$

where

$$K_3 = \frac{K_a \phi_p V_s}{R + R_a} = \frac{11.72 \times 1200}{56.48} = 249.01$$

and

$$K = \frac{(K_a \phi_p)^2}{R + R_a} = \frac{11.72^2}{56.48} = 2.43$$

The braking torque becomes

$$T_b = 249.01 + 2.43 \omega_m$$

(a) At the instant of plugging, $\omega_m = 100$ rad/s, and

$$T_b = 249.01 + 2.43 \times 100 = 492.01 \text{ N-m}$$

(b) When the motor attains zero speed, the braking torque is 249.01 N-m.

PROBLEMS

5.1. A 4-pole dc motor has 28 slots with two conductors per slot. The induced voltage in its simplex lap-wound armature is 320 V at 1800 rpm. What is the flux per pole in the motor?

5.2. A simplex lap-wound, 6-pole dc motor has 180 conductors uniformly distributed on the surface of its armature. If the armature current is 25 A and the flux per pole is 0.5 Wb, determine the torque developed by the motor. What is the induced voltage in the armature if the motor is running at 900 rpm? What is the current in each conductor?

5.3. A 2-pole wave-wound motor has 1600 conductors in its armature. The flux per pole is 50 mWb and the armature current is 80 A. Determine (a) the torque developed neglecting armature reaction, and (b) the torque developed if the equivalent effect of armature reaction is to reduce the air-gap flux by 12%.

5.4. A 4-pole dc motor draws 25 A when it is wave wound. Determine the armature current for a lap-wound motor delivering the same torque.

5.5. The torque developed by a dc motor is 20 N-m when the armature current is 25 A. If the field intensity is reduced by 20% of its original value while the armature current is increased by 40%, what is the new torque developed?

5.6. A 5-hp, separately excited, motor is taking 37.9 A from a 120-V dc source at full load. The armature circuit resistance is 0.1 Ω. What are the counter emf and mechanical power developed by the motor?

5.7. The armature resistance of a 120-V dc motor is 0.25 Ω. When operating at full load, the armature current is 16 A at 1400 rpm. Determine the speed of the motor if the flux is increased by 20% and the armature current is 26 A.

5.8. The speed of a dc motor falls from 1200 rpm at no load to 1000 rpm at full load. What is the percent of speed regulation?

5.9. A dc shunt motor has a percent of speed regulation of 2.5%. If its full-load speed is 1800 rpm, what is its no-load speed?

5.10. A 120-V, 2-hp dc motor has an armature resistance of 0.1 Ω. At no load, the motor is taking 1 A at 1600 rpm. (a) What must be the external resistance connected in series with the armature circuit to reduce its speed to 1200 rpm at full-load current of 15 A? (b) How much power is being converted into mechanical power? (c) How much power is lost in the external resistor? (d) What is the percent of speed regulation? Neglect armature reaction.

5.11. A 120-V dc shunt motor takes 10 A at full load. Total electrical losses are 125 W. Friction and windage loss is 75 W. The core loss is 100 W. Determine the efficiency of the motor.

5.12. A 230-V dc shunt motor draws 40 A at full load. The shunt field current is 2.5 A. Armature circuit resistance is 0.5 Ω. Total rotational loss is 200 W. Determine (a) the power loss in the armature circuit, (b) the power loss in the shunt-field circuit, (c) the total power loss in the motor, (d) the net (shaft) power output, (e) the efficiency of the motor, and (f) the maximum efficiency of the motor.

5.13. A 230-V, separately excited, dc motor draws 15 A when operating at full-load speed of 900 rpm. The armature circuit resistance is 0.2 Ω. Determine the torque developed by the motor. If the field excitation is held constant, determine the armature current and speed of rotation when the motor develops 20 N-m.

5.14. A 10-hp, 220-V, 3000-rpm dc shunt motor has to be designed at its maximum efficiency of 90% with 0.5 A in the shunt-field winding. Determine the (a) armature current, (b) shunt-field resistance, (c) torque developed by the motor, and (d) armature circuit resistance.

5.15. A 120-V dc shunt motor develops 1 hp at the shaft when rotating at 1200 rpm. The field resistance is 160 Ω and the armature resistance is 0.5 Ω. The various losses are core loss = 10 W, friction and windage loss = 8 W, brush contact loss = 2 W, and stray load loss = 3 W. Determine (a) the power developed by the motor and (b) the efficiency of the motor.

5.16. A separately excited motor takes a line current of 20 A at 125 V while running at 1725 rpm. The armature resistance is 0.5 Ω. When the load on the machine is changed, the line current reduces to 10 A. Calculate the speed of the motor under new loading conditions assuming the field current is held constant.

5.17. A 230-V shunt motor draws 5 A at no-load speed of 2400 rpm. The shunt-field resistance is 184 Ω and the armature resistance is 0.1 Ω. At full load the motor draws 25 A. Calculate (a) rotational losses, (b) full-load speed, (c) motor efficiency, and (d) speed regulation.

5.18. A 125-V shunt motor has an armature resistance of 0.4 Ω and a shunt-field resistance of 50 Ω. When the motor is running steady, the line current is 17.5 A. What is the back emf of the motor? If the motor is started from rest at rated voltage, what would be the current intake at starting? What is the ratio of starting to running line currents? What is the power developed by the armature?

5.19. A 125-V, 10-hp series motor delivers the rated load at 2750 rpm. What is the torque developed by the motor? If the motor is derated to operate at 125 V and delivers 8 hp at 3200 rpm, what is the percent of flux reduction in the motor? Neglect the rotational losses.

5.20. A 230-V, 1800-rpm, long-shunt compound motor has shunt-field resistance of 460 Ω, series-field resistance of 0.2 Ω, and armature resistance of 1.3 Ω. The full-load line current is 10.5 A. Calculate the speed of the motor when the line current rises to 15.5 A under increased load. Under this condition, assume that there is a 20% increase in overall flux in the motor.

5.21. A 230-V, 15-A dc series motor is rated at 1800 rpm. It has an armature resistance of 2.1 Ω and a series-field resistance of 1.25 Ω. Under light load the machine draws 5 A from the source. What is the speed of the motor? Assume the flux produced is proportional to the current.

5.22. A 120-V dc shunt motor is tested using the Prony brake test and the following data are recorded: effective length of torque arm = 0.75 m, initial tare = 1.25 N, and test reading = 9.5 N at 1000 rpm. The line current is 7.5 A. Calculate the available torque at the shaft and the efficiency of the motor.

5.23. An electric motor develops 0.75 hp at 1725 rpm. If the pulley has a radius of 10 cm, determine the force acting on the pulley.

CHAPTER 6

Synchronous Generators

Laboratory Model of a Synchronous Machine
(Courtesy of Lab-Volt)

A synchronous machine, similar to a dc machine, is also a doubly fed machine. In a dc machine, the stator houses the field winding and the voltage is induced in the armature winding on the rotor. A synchronous machine, on the other hand, is designed to house the field winding on its rotating member, and the voltages are induced in the armature windings wound on the inside of the stationary member. Such a construction has been nicknamed an "inside-out" construction when compared to its dc counterpart. The field winding of a synchronous machine also carries the dc current and therefore produces a pole with a constant magnetic field very similar to the field produced by the dc machine. When the synchronous machine is acting as a generator, it is also referred to as an *alternator*. The word synchronous means that the rotating magnetic field inside the machine has the same angular velocity as that of the rotor. The frequency of the induced voltage is directly proportional to the number of poles and the speed of rotation of the rotor. From our study of dc machines, the frequency of the induced voltage is

$$f = \frac{PN_s}{120} \tag{6.1}$$

where
f = frequency of the induced voltage in Hz
P = number of poles
N_s = (synchronous) speed of the rotor in rpm

This relationship is true for all synchronous machines. Thus a 4-pole, 60-Hz synchronous motor rotates at a speed of 1800 rpm. On the other hand, the induced voltage in a 4-pole alternator when the rotor is being rotated at 1500 rpm has a frequency of 50 Hz.

Basic Reasons for Inside-out Construction

1. In a dc generator, the induced voltage and thereby the current are brought outside the armature with the help of brushes sliding over the commutator. Such an arrrangement is not economical due to brush contact losses and arcing in high-power applications. Furthermore, it becomes difficult to cool the machine due to high power losses, resulting in heat generation, and to provide proper isolation between electrical connections inside the machine. An inside-out generator does not require any brushes to distribute the generated voltage to the external load and is easier to cool from the outside.

2. In a generator, our aim is to produce an alternating voltage and current and, therefore, there is no need for the commutator.

3. The field current in the generator is quite low and there is no requirement to switch its polarities. Thus, the slip rings are normally used to connect the dc power source to the rotating field winding on the rotor. If the field is supplied by permanent magnets, the slip rings can also be dispensed with.

4. The stationary member, or stator, can be designed with deep slots to accommodate comparatively thicker copper conductors in order to reduce the copper losses in its power windings.

6.1 CONSTRUCTION

Stator The stator of a synchronous machine is made of thin laminations of highly permeable steel in order to reduce the core losses. The inside of the stator has a plurality of slots with normally parallel sides for easy winding. Figure 6.1 shows a typical stator lamination. The total number of slots depends, among many other factors, on the size of the machine, the number of poles, and the number of coils per phase per pole or the phase group. The larger the number of coils in a phase group is, the closer will be the induced voltage to its sinusoidal waveform.

For machines of larger size, the stator winding is usually a three-phase winding. That is, each pole has a coil or a group of coils connected in series for each of its three phases, and there are as many such groups as there are poles. Most of the windings are of a double-layer type (i.e., each slot has two coil sides that belong to two different coils). Depending on the requirements, the windings may be either wye (Y) connected or delta (Δ) connected. In a wye-connected machine, the common or the neutral point may also be brought outside.

Rotor The rotor is also made of punched laminations that are stacked together for the same reasons as for the stator. For high-speed operation, the cylindrical or round rotor is commonly used with slots on its periphery to house the field windings. Figure 6.2 shows the construction and winding arrangement for a 4-pole cylindrical rotor. Salient pole construction may be used for low-speed applications. Salient rotor construction for 2 and 4 poles is shown in Fig. 6.3. It is evident that a cylindrical rotor has more strength and less windage loss. The rotor

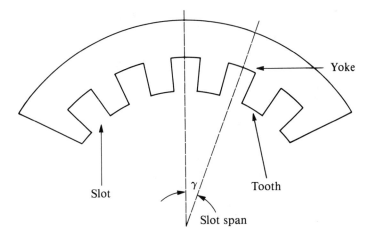

FIGURE 6.1
Stator
Lamination for
an Induction
Motor

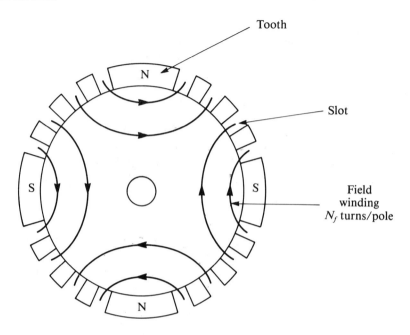

FIGURE 6.2
Four-Pole
Round-Rotor
Construction and
Field Winding

must be wound for the same number of poles as the stator. The rotor winding carries dc current so as to produce constant flux per pole. It is therefore the rotation of the rotor with respect to the stator that causes the induction of voltage in the stator windings. However, whenever there is a current flow in the three-phase windings of the stator, they produce the revolving field as explained in Appendix B. This is the field that revolves with synchronous speed. As this field is produced by the field windings on the rotor, the synchronous speed is the speed of the rotor.

Stator Windings Most high-power generators have three-phase windings for the generation of three-phase power. To produce induced voltages that are 120° apart in phase with respect to each other, the phase windings must be placed 120° apart in space. As far as the type of windings is concerned for three-phase machines, we can use either a lap winding or a wave winding as explained in Appendix C. The lap winding is most commonly used. Figure 6.4 shows the winding arrangement for one of the poles for a 4-pole, 24-slot stator for a synchronous machine. A single coil that spans the entire length of the pole is known as a *full-pitch* coil. Thus, a full-pitch coil also spans 180°. Both the pole pitch and coil pitch can also be expressed in terms of the number of stator slots. Usually, a coil spans less than 180°, and there is more than one coil under a pole in each of its three phases. A coil with a span of less than 180° is known as a *fractional-pitch* coil. If we know the number of slots spanned by a pole, we can determine the electrical angle from one slot to the next. This is known as the *slot span*.

Since a coil usually spans a fraction of the pole and there is more than one coil in a given phase under the pole, all these coils are connected in series to form a group that is known as the *phase belt* or simply the *phase group*. The advantages of a fractional pitch winding are:

1. It reduces the presence of harmonics by producing a truer sinusoidal waveform.

2. It shortens the ends of the windings and thereby results in the use of less copper.

3. The shorter coils can be conveniently managed, which reduces the buildup of end turns on both sides of the stack. This reduces the overall length of the machine and minimizes flux leakages.

The main disadvantage of the fractional-pitch winding is that the magnitude of the voltage induced in the armature is smaller as compared to the full-pitch

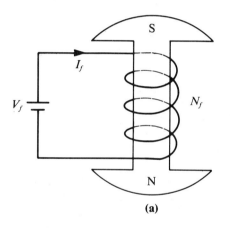

(a)

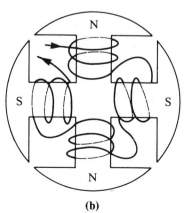

FIGURE 6.3
(a) Two-Pole
Salient Rotor;
(b) Four-Pole
Salient Rotor

(b)

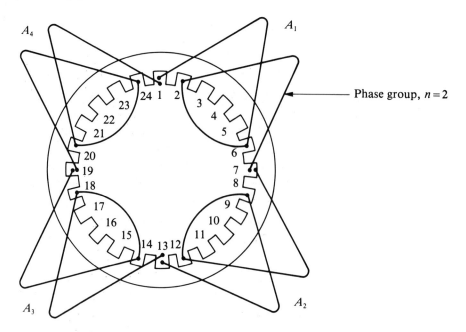

FIGURE 6.4
Phase A Coils for
the Four Poles of
a Synchronous
Machine

coil. Thus, either a larger number of turns is needed or the flux produced by the field winding must be increased to compensate for the reduction of the induced voltage due to fractional-pitch coils. For double-layer windings, the total number of coils, C, is equal to the number of stator slots, S_1. The number of coils in a phase group, n, is the same as the number of coils per pole per phase. That is,

$$n = \frac{C}{Pq_1} \tag{6.2}$$

where q_1 is the number of phases in the machine.

■ EXAMPLE 6.1

A 54-slot stator of a synchronous machine must be wound for three-phase, 6-pole applications. Determine the coil span and the number of coils in a phase group.

Solution

Since there are 54 total slots in the stator, for a double-layer winding, the total number of coils is $C = 54$. For 6-pole applications, the number of slots per pole is $54/6 = 9$. A full-pitch coil for each of the phases will also span 9 slots.

Since a pole spans 180° (electrical) and there are 9 slots per pole, the slot span is $\gamma = 180°/9 = 20°$. The number of coils per phase is $C_\phi = 54/3 = 18$. The number of coils per pole is $C_p = 54/6 = 9$. Since there are 9 coils per pole, the number of coils in a phase group is $n = 9/3 = 3$. Thus each phase group has three coils that must be connected in series to form one of the phases under a pole. The coil group must also span 180°.

Thus the coils must be placed in such a way that the beginning end of the first coil goes in that slot which is under the beginning of the pole and the finishing end of the third coil must be placed in a slot that is under the trailing end of the pole. Each coil must span an equal number of slots.

Let us illustrate this placement of the coils with the help of the developed diagram of Fig. 6.5. The slots are numbered for convenience from 1 to 10. The pole span includes only half of the first slot and half of the tenth slot. The starting end of coil 1, s_1, is placed in slot 1. Similarly, the starting ends s_2 and s_3 of coils 2 and 3 are placed in slots 2 and 3, respectively. The finishing end of coil 3 must be placed in slot 10 because the pole span is 9 slots. Note that the first slot is not counted when determining the pole span. Another way to determine the pole span is to count the number of teeth. As the number of teeth spanned is the same as the number of slots, a pole must embrace 9 teeth. There are 9 teeth between slot 1 and slot 10. Once we have established where the finishing end of coil 3 goes, the finishing ends for coils 2 and 1 must go in slots 9 and 8, respectively. Note that each coil now spans 7 slots. Since the slot span is 20° (electrical), the coil span is $20 \times 7 = 140°$ (electrical).

6.1.1 Typical Winding Layouts and Connections

As pointed out earlier, the phase windings must be displaced from one another by 120° in space to obtain the proper three-phase relationships between the induced voltages. This type of winding displacement is known as a 120° phase belt. In this case, all the coils for each of the phases are wound in the same direction, and their magnetic axes are 120° apart. Such placement of windings for a 2-pole machine is shown in Fig. 6.6. Note that only a single coil is shown for each of the phase groups just to highlight the winding arrangements.

The schematic arrangement of the windings for a 6-pole generator is shown in Fig. 6.7 with a permanent magnet rotor. Let us consider one of its phases for the purpose of connections. The other two phases would be connected the same way. Terminals 1, 2, and 3 are the external terminals for A, B, and C phases, respectively. All the phase groups under similar poles are connected in series. That is, the phase groups under all north poles are connected in series as shown for phase A, with terminal 4 as the finishing end of phase group A under pole $N3$. Similarly, all the phase groups under all the south poles are connected in series, as shown by the dashed line connections Fig. 6.7 for phase A with terminals 7 and 10. For each phase we now have two winding groups, one for all the north poles and other for all the south poles. The total induced voltage in each winding

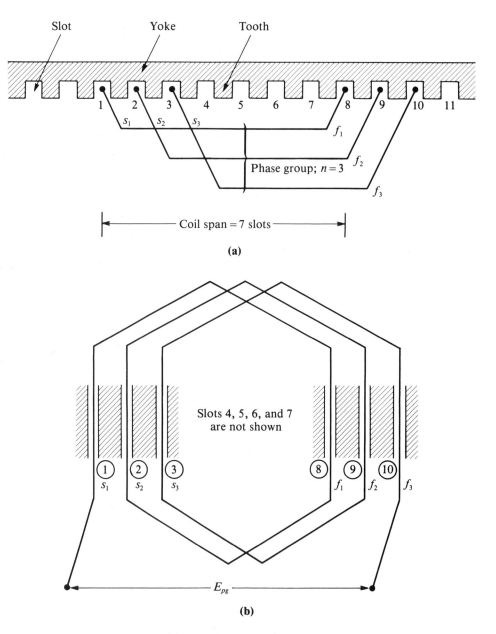

FIGURE 6.5
Placement of
Coils in Slots: (a)
Developed
Diagram; (b)
Series
Connection of n
Coils in a Phase
Group

group is the same. If terminal 4 is connected to terminal 7 and we are left with
terminals 1 and 10, then we have established a *series connection* as shown in Fig.
6.8a. The total voltage now available between terminals 1 and 10 equals P times
the voltage induced in each phase group, where P is the number of poles. Similar
series connections will also be established for the remaining two phases. Terminal

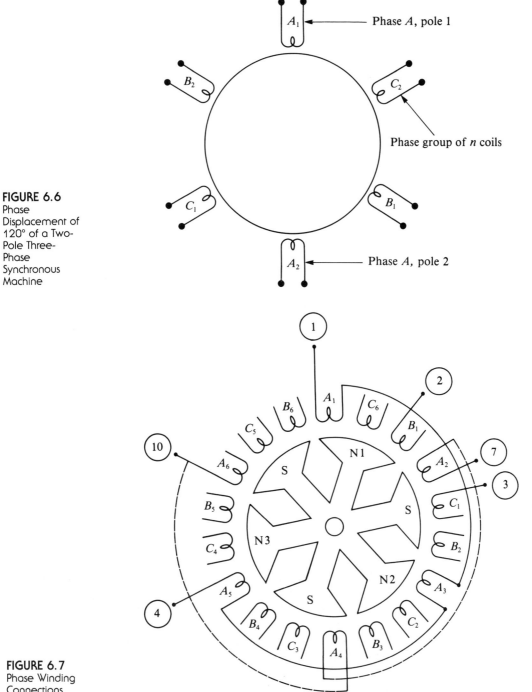

Phase A, pole 1

C_2

Phase group of n coils

B_1

Phase A, pole 2

FIGURE 6.6
Phase
Displacement of
120° of a Two-
Pole Three-
Phase
Synchronous
Machine

FIGURE 6.7
Phase Winding
Connections

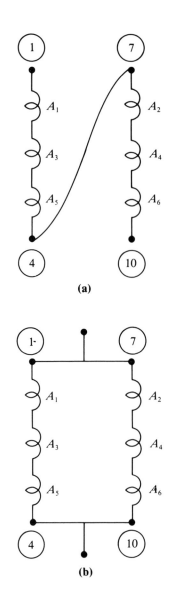

FIGURE 6.8
(a) Series
Connection and
(b) Parallel
Connection for
Phase *A*
Windings

1 for phase *A* will be brought outside, while terminal 10 may be internally connected to its counterparts for the other two phases.

The two winding groups with terminals 1 and 4 under north poles and terminals 7 and 10 under south poles can also be connected in parallel. Such a parallel connection can be established by connecting terminals 1 and 7 and terminals 4 and 10, as shown in Fig. 6.8b. As is now obvious, the available output voltage in a parallel connection is one-half the available output voltage in a series connection. Thus, a generator can satisfy dual-voltage requirements just by changing its connections. When all the phase groups are connected in series, there is only one path for the current to take. However, the parallel connection, as discussed previously, provides two parallel paths. The current capabilities in a parallel connection have been doubled as compared to a series connection.

6.2 INDUCED VOLTAGE EQUATION

Let us consider a round rotor synchronous generator rotating at an angular velocity of ω_s. The flux density emanating from the surface of a pole on the rotor due to the spatial distribution of the field winding, as shown in Figure 6.2, can be approximated to vary as

$$B = B_m \cos \theta \tag{6.3}$$

where B_m is the flux density per pole as shown in Fig. 6.9.

The magnetic flux linking the coil AB (Fig. 6.10) on the stator is

$$\phi = \int \vec{B} \cdot \vec{ds} \tag{6.4}$$

$$ds = Lrd\theta_m = \frac{2rLd\theta}{P} \tag{6.5}$$

$$L = \text{axial length of the rotor}$$

$$r = \text{radius of the rotor}$$

If the coil span is ρ, the maximum flux linking the coil will be

$$\phi_{ct} = \int_{-\rho/2}^{\rho/2} B_m \cos \theta \, 2Lr \frac{d\theta}{P}$$

$$= \frac{4LrB_m}{P} \sin \frac{\rho}{2} \tag{6.6}$$

$$= \phi_p \sin \frac{\rho}{2} = \phi_p k_p$$

where $k_p = \sin(\rho/2)$ is known as the pitch factor and ϕ_p is the total flux linking a full-pitch coil, $\phi_p = (4LrB_m)/P$.

As this flux revolves with an angular frequency ω such that

$$\omega = \frac{P}{2} \omega_s \tag{6.7}$$

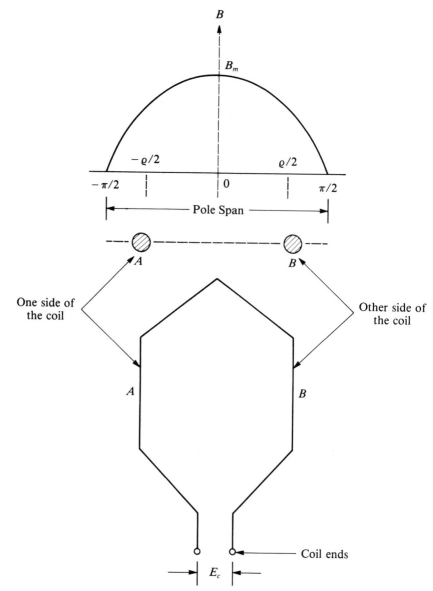

FIGURE 6.9
Flux Density
Distribution and
the Placement
of a Coil

the flux linking the coil at any time t can then be expressed as

$$\phi_c(t) = \phi_{ct} \cos \omega t \qquad (6.8)$$

or

$$= \phi_p k_p \cos \omega t$$

If the coil has N_c turns, the voltage induced in the coil will be

$$e = -N_c \frac{d\phi_c(t)}{dt}$$

$$= N_c k_p \omega \phi_p \sin \omega t \qquad (6.9)$$

The maximum value of the induced voltage is

$$E_m = N_c k_p \omega \phi_p \qquad (6.10)$$

and its rms value is

$$E_c = \frac{1}{\sqrt{2}} E_m$$

$$(6.11)$$

or

$$E_c = 4.44 f N_c k_p \phi_p$$

Since a phase group may have more than one coil connected in series and each coil is displaced by a slot pitch, the voltage induced in the phase group is the phasor sum of the voltages induced in individual coils. The space displacement of the coils is viewed as if it further reduces the induced voltage in each coil by a factor k_d, which is known as *distribution factor*.

Let n be the number of coils in a phase group and γ be the slot pitch; then the distribution factor is given as (from Appendix D)

$$k_d = \frac{\sin(n\gamma/2)}{n \sin(\gamma/2)} \qquad (6.12)$$

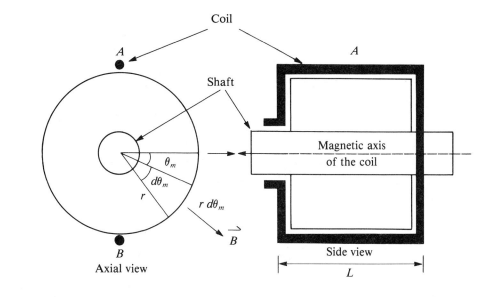

FIGURE 6.10
Magnetic Flux
Linking a Coil
AB

In terms of the distribution factor, the induced voltage for n coils in a phase group is given as

$$E_{pg} = nk_dE_c$$

or
$$E_{pg} = 4.44fnN_ck_pk_d\phi_p \qquad (6.13)$$

If all the phase groups are connected in series, the total voltage induced per phase is

$$E_{\phi s} = PE_{pg}$$

or
$$E_{\phi s} = 4.44fPnN_ck_pk_d\phi_p \qquad (6.14)$$

However, if the windings are connected in parallel and there are a parallel paths, the induced voltage per phase is

$$E_\phi = \frac{4.44fPnN_ck_pk_d\phi_p}{a} \qquad (6.15)$$

The product of pitch factor and distribution factor is called the *winding factor*. That is,

$$k_w = k_pk_d \qquad (6.16)$$

The product PnN_c gives the total number of turns per phase irrespective of the type of connection. However, taking into account the winding factor, k_w, and the number of parallel paths, a, we can define the effective turns per phase as

$$N_e = \frac{PnN_ck_w}{a} \qquad (6.17)$$

and the induced voltage per phase now becomes

$$E_\phi = 4.44fN_e\phi_p \qquad (6.18)$$

Note that this expression is very similar to the one obtained for the transformer. In the case of a transformer, the winding factor is unity because each winding is made of one coil that embraces the total flux in the core.

■ EXAMPLE 6.2

A three-phase synchronous generator has 16 poles and 144 slots, which are wound with 10-turn coils using double-layer winding. The rotor is driven by a dc motor at a speed of 600 rpm. The rotor flux per pole is 0.025 Wb. Determine (a) the frequency of the induced voltage, (b) the induced voltage per phase, and (c) the open-circuit line voltage for a Y-connected generator.

Solution

Since double layer winding is used, $C = 144$.

(a) The frequency is $f = 600 \times 16/120 = 80$ Hz.

(b) The total number of turns in the generator is $144 \times 10 = 1440$. Thus, turns per phase is $1440/3 = 480$. There are $144/16 = 9$ slots per pole. Thus $\gamma = 180/9 = 20°$ (electrical). The number of coils per pole is $144/16 = 9$. The number of coils per pole per phase is $n = 9/3 = 3$.

Since there are 9 slots per pole and there are 3 coils in a phase group, it can be found with the help of a developed diagram that each coil must span 7 slots. Thus the coil span is $\rho = 20 \times 7 = 140°$. The pitch factor is $k_p = \sin(140/2) = 0.94$. The distribution factor is

$$k_d = \frac{\sin(3 \times 20/2)}{3 \sin(20/2)} = 0.96$$

The winding factor is

$$k_w = 0.94 \times 0.96 = 0.9024$$

Since there is no mention of the winding connections, we will assume that all phase groups are connected in series for each of the three phases. In other words, there is only one path for the phase current to take. In that case, the effective number of turns per phase is equal to the total turns per phase times the winding factor. That is,

$$N_e = 480 \times 0.9024 = 433.152$$

The voltage generated per phase is

$$E_\phi = 4.44 \times 80 \times 433.152 \times 0.025 = 3846.4 \text{ V}$$

(c) For a Y-connected generator, the line voltage under no load is

$$V_t = \sqrt{3} \times 3846.4 = 6662 \text{ V}$$

■ EXAMPLE 6.3

A 440-V, Y-connected, three-phase synchronous generator supplies the rated voltage at 1800 rpm and a field current of 0.9 A. If the speed of the prime mover falls to 1200 rpm, what must the field current be to produce the rated voltage?

Solution

Let $E_{\phi 1}$ and $E_{\phi 2}$ be the induced voltages at 1800 rpm and 1200 rpm, respectively. Then, from Eq. (6.18),

$$\frac{E_{\phi 1}}{E_{\phi 2}} = \frac{f_1 \phi_{p1}}{f_2 \phi_{p2}}$$

But the flux produced is proportional to the field current under linear conditions; therefore,

$$\frac{\phi_{p1}}{\phi_{p2}} = \frac{I_{f1}}{I_{f2}}$$

Since the frequency is proportional to the rated speed, the ratio of the frequencies is equal to the ratio of the speeds. That is,

$$\frac{f_1}{f_2} = \frac{N_1}{N_2}$$

Thus,

$$\frac{E_{\phi 1}}{E_{\phi 2}} = \frac{I_{f1} N_1}{I_{f2} N_2}$$

But $E_{\phi 1} = E_{\phi 2}$; therefore,

$$I_{f2} = I_{f1} \frac{N_1}{N_2} = 0.9 \times \frac{1800}{1200} = 1.35 \text{ A}$$

6.3 EQUIVALENT CIRCUIT DEVELOPMENT

A synchronous generator can be represented by an equivalent circuit under steady-state conditions, as shown in Fig. 6.11. The stator is shown having three identical phases, which are connected internally to form a Y connection. A delta connection could also have been used. Each phase has its own winding resistance, R_1. Whenever there is a current flow in a phase winding, it produces its own flux. Part of this flux links the phase winding only and is lost as the leakage flux. This leakage flux can be modeled by introducing a leakage reactance X_1 in series with the winding resistance. The remaining flux crosses the air gap and tends to oppose the flux set up by the rotor.

Since all three phases are identical, it is a common practice to draw an equivalent circuit on a per phase basis. Such a simple circuit, which takes into account the winding resistance and the leakage reactance, is shown in Fig. 6.12. This is an exact equivalent circuit as long as we neglect the effect of armature

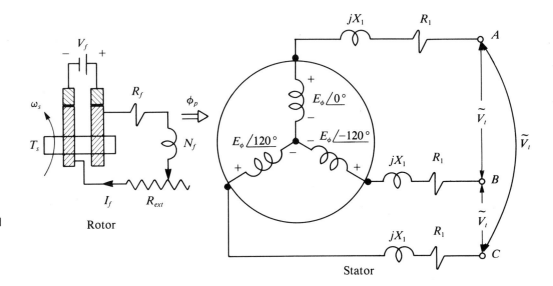

FIGURE 6.11
Equivalent
Circuit of a
Synchronous
Generator

reaction. In the case of dc machines, the overall effect of armature reaction is to reduce the total flux in the machine. A very similar phenomenon takes place in synchronous machines as outlined next.

The armature reaction in synchronous machines is caused by the current in the phase windings on the stator. Let us study it by considering a single full-pitch coil representing a phase. The flux produced by the rotor induces a maximum voltage when the magnetic axis of the rotor field is 90° to the magnetic axis of the coil, because this is the condition when the coil is experiencing maximum change in the flux linkages. The induced voltage, $\tilde{E}_\phi$, however, would be lagging the flux, $\tilde{\phi}_p$, by 90°, as shown in Fig. 6.13. Under no-load condition, the induced voltage would appear as the terminal voltage.

Let us now assume that a resistive load is connected to the output terminals and a current is flowing in the phase winding. This current sets up its own magnetic field, which induces a voltage of its own in the stator phase winding. Let us call the stator flux $\tilde{\phi}_{ar}$ and the voltage induced in the stator phase winding $\tilde{E}_{ar}$. The induced voltage $\tilde{E}_{ar}$ must lag $\tilde{\phi}_{ar}$ by 90°. Since $\tilde{\phi}_{ar}$ is produced by the current $\tilde{I}_1$

FIGURE 6.12
Per Phase
Equivalent
Circuit of a
Synchronous
Generator
without
Armature
Reaction

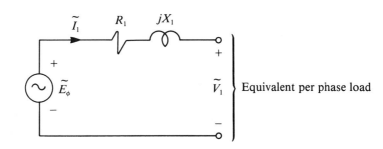

Equivalent per phase load

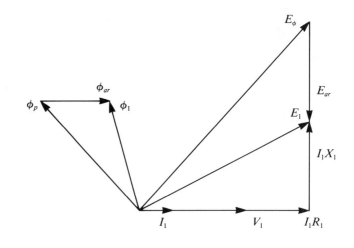

FIGURE 6.13
Phasor Diagram
of a
Synchronous
Generator
Including
Armature
Reaction

in the stator phase winding, $\tilde{\phi}_{ar}$ and $\tilde{I}_1$ must be in phase. The sum of the flux set up by the rotor, $\tilde{\phi}_p$, and the flux, $\tilde{\phi}_{ar}$, produced by the armature current, $\tilde{I}_1$, results in the net flux, $\tilde{\phi}_1$, in the air gap of the machine. The net flux is then responsible for the net voltage induced in each of the phase windings by taking into account the armature reaction. The net voltage must also be the phasor sum of the voltage $\tilde{E}_\phi$ induced by the flux $\tilde{\phi}_p$ and the voltage $\tilde{E}_{ar}$ induced by the flux $\tilde{\phi}_{ar}$. Furthermore, the net induced voltage, $\tilde{E}_1$, must lag the net flux by 90°. The phasor diagram depicting these relationships is shown in Fig. 6.13. Mathematically, we can express these relationships as

$$\tilde{\phi}_1 = \tilde{\phi}_p + \tilde{\phi}_{ar} \tag{6.19}$$

and

$$\tilde{E}_1 = \tilde{E}_\phi + \tilde{E}_{ar} \tag{6.20}$$

An equivalent circuit for the synchronous generator representing armature reaction voltage $\tilde{E}_{ar}$ is shown in Fig. 6.14. However, from Fig. 6.13 it is evident that $\tilde{E}_{ar}$ lags $\tilde{I}_1$ by 90°. From the preceding discussion it is also clear that $\tilde{E}_{ar}$ is directly proportional to $\tilde{I}_1$. Therefore, we can say that

$$\tilde{E}_{ar} = -j\tilde{I}_1 X_m \tag{6.21}$$

where X_m is the constant of proportionality. In synchronous machines, X_m is known as the *reactance of the armature reaction* or simply the *magnetizing reactance*.

It is rather difficult to separate one reactance from the other. For that reason, the two reactances are often represented by one reactance, which is known as

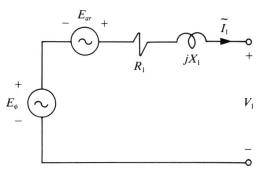

FIGURE 6.14
Equivalent
Circuit
Representing
Armature
Reaction in
Terms of
Voltage E_{ar}

the *synchronous reactance*, X_s. That is,

$$X_s = X_1 + X_m \tag{6.22}$$

The exact equivalent circuit can now be represented in terms of its synchronous reactance as shown in Fig. 6.15.

6.3.1 Phasor Diagrams

Depending on the load impedance, the current in each phase winding of a synchronous generator may be lagging, in phase, or leading the terminal voltage, $\tilde{V}_1$. For any load current,

$$\tilde{E}_\phi = \tilde{V}_1 + \tilde{I}_1(R_1 + jX_s) \tag{6.23}$$

and the corresponding phasor diagrams are shown in Fig. 6.16.

■ EXAMPLE 6.4

A 10-kW, 440-V, Δ-connected, three-phase synchronous generator has a stator winding resistance of 1.0 Ω per phase and synchronous reactance of 10.0 Ω per

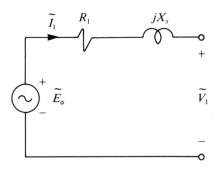

FIGURE 6.15
Equivalent
Circuit of a
Synchronous
Generator with
Armature
Reaction

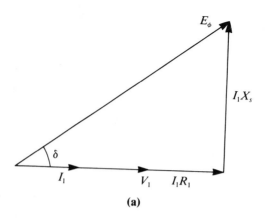

(a)

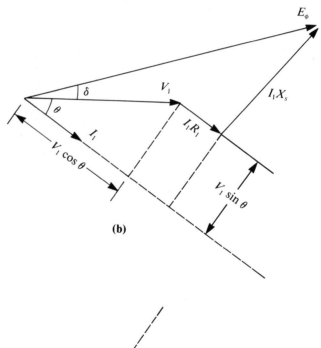

(b)

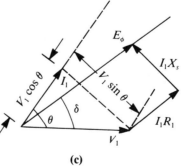

FIGURE 6.16
Phasor
Diagrams: (a)
Unity Power
Factor; (b)
Lagging Power
Factor; (c)
Leading Power
Factor

(c)

phase. What is the induced voltage when the generator is delivering full load at 0.8 power factor lagging?

Solution

The phasor diagram for a synchronous generator operating under lagging power factor is given in Fig. 6.16b.

Load voltage per phase: $V_1 = 440$ V

Power output per phase: $P_o = 10 \times 1000/3 = 3333.33$ W

Current per phase: $I_1 = 3333.33/(440 \times 0.8) = 9.47$ A

Thus, $\tilde{I}_1 = 9.47\underline{/-36.87°}$ A

The induced voltage per phase is

$$\tilde{E}_\phi = \tilde{V}_1 + \tilde{I}_1(R_1 + jX_s)$$
$$= 440 + (9.47\underline{/-36.87°}) \times (1.0 + j10)$$
$$= 440 + 95.17\underline{/47.42°}$$
$$= 509.24\underline{/7.91°} \text{ V}$$

Thus, the line voltage is 509.24 V at no load.

6.4 VOLTAGE REGULATION

The percent of voltage regulation of a generator is defined as the ratio of the change in voltage from no load to full load to the full-load voltage. The speed of rotation, the current in the field winding, and the temperature of the generator must be held constant. That is,

$$\text{VR\%} = 100 \frac{V_{nl} - V_{fL}}{V_{fL}}$$

where
V_{nL} = no-load voltage
V_{fL} = full-load voltage

The calculations for the voltage regulation can be carried out on a per phase or line to line basis. On a per phase basis, the no-load voltage is the same as the generated voltage, E_ϕ.

■ **EXAMPLE 6.5**

A 10-kVA, 230-V, three-phase Y-connected generator has a winding resistance of 0.3 Ω and synchronous reactance of 5.6 Ω on a per phase basis. Determine its voltage regulation when the power factor is (a) 65% lagging, (b) unity, and (c) 65% leading.

Solution

Since the voltage regulation is only defined at full load, the full-load voltage on a per phase basis is

$$V_1 = \frac{230}{\sqrt{3}} = 132.79 \text{ V}$$

The full-load current is

$$I_1 = \frac{10 \times 1000}{\sqrt{3} \times 230} = 25.1 \text{ A}$$

(a) The lagging power factor is $\cos \theta = 0.65 \Rightarrow \theta = -49.46°$.

$$\tilde{E}_\phi = \tilde{V}_1 + \tilde{I}_1(R_1 + jX_s) = 132.79 + (25.1\underline{/-49.46°}) \times (0.3 + j5.6)$$

$$= 244.51 + j85.63 = 259.07\underline{/19.3°} \text{ V}$$

$$\text{VR\%} = 100 \times \frac{259.07 - 132.79}{132.79} = 95.1\%$$

(b) The unity power factor is $\cos \theta = 1 \Rightarrow \theta = 0°$.

$$\tilde{E}_\phi = 132.79 + 25.1(5.608\underline{/86.93°}) = 198.62\underline{/45.05°} \text{ V}$$

$$\text{VR\%} = 100 \times \frac{198.62 - 132.79}{132.79} = 49.57\%$$

(c) The leading power factor is $\cos \theta = 0.65 \Rightarrow \theta = 49.46°$.

$$\tilde{E}_\phi = 132.79 + (25.1\underline{/49.46°}) \times (5.608\underline{/86.93°}) = 101.88\underline{/72.36°} \text{ V}$$

$$\text{VR\%} = 100 \times \frac{101.88 - 132.79}{132.79} = -23.28\%$$

6.5   POWER RELATIONSHIPS

The rotor of the synchronous machine is connected to a prime mover, which may be a dc motor, steam turbine, diesel engine, or the like, in order to rotate the field winding at its synchronous speed. If T_s is the available torque at the shaft and ω_s is the speed at which it is being rotated, the mechanical power input to the synchronous machine will be $T_s\omega_s$. In addition, the dc power supplied to the field winding is V_fI_f. Thus, total power input is

$$P_{\text{in}} = T_s\omega_s + V_fI_f \tag{6.24}$$

Since the power input is mostly mechanical in nature, we must take care of all the rotational losses in the machine in a similar way as we did for dc generators. The remaining mechanical power is converted into electrical power. From this developed electrical power we must now subtract all the electrical losses to obtain power output, as indicated in the power-flow diagram of Fig. 6.17.

The power output can, however, be expressed as

$$P_o = 3V_1I_1 \cos\theta \tag{6.25}$$

Most often the prime mover is directly coupled to the shaft of the rotor and there is no way we can measure its torque. The input power, therefore, is usually expressed in terms of the output power and the losses, such that

$$P_{\text{in}} = P_o + P_{\text{elec}} + P_{\text{rot}}$$

or

$$P_{\text{in}} = 3V_1I_1 \cos\theta + 3I_1^2R_1 + V_fI_f + P_{\text{rot}} \tag{6.26}$$

If the generator is rotating at a fixed speed, the rotational losses will become constant. Also, for the constant flux in the machine, the power lost in the field winding is constant. In this case we can represent the losses as

Fixed losses:
$$P_c = P_{\text{rot}} + V_fI_f \tag{6.27}$$

and variable losses:
$$3I_1^2R_1 \tag{6.28}$$

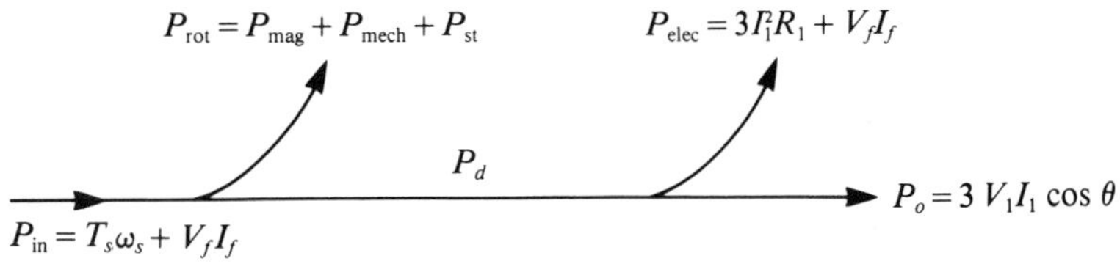

FIGURE 6.17
Power-Flow Diagram of a Synchronous Generator

$P_{\text{rot}} = P_{\text{mag}} + P_{\text{mech}} + P_{\text{st}}$

$P_{\text{elec}} = 3I_1^2R_1 + V_fI_f$

P_d

$P_o = 3V_1I_1 \cos\theta$

$P_{\text{in}} = T_s\omega_s + V_fI_f$

The efficiency of the generator is

$$\eta = \frac{3V_1I_1 \cos \theta}{3V_1I_1 \cos \theta + 3I_1^2R_1 + P_c} \tag{6.29}$$

The efficiency of the synchronous generator is maximum when the fixed losses are equal to the variable losses. That is,

$$P_c = 3I_1^2R_1 \tag{6.30}$$

6.5.1 Approximate Power Relationships

The armature winding resistance of a synchronous generator is usually quite small and can be neglected in comparison with its synchronous reactance, as shown in Fig. 6.18. The phasor diagram is also given for a lagging power factor. The terminal voltage is

$$\tilde{V}_1 = \tilde{E}_\phi - j\tilde{I}_1X_s$$

or

$$\tilde{I}_1 = \frac{\tilde{E}_\phi - \tilde{V}_1}{jX_s}$$

$$= \frac{E_\phi \cos \delta + jE_\phi \sin \delta - V_1}{jX_s}$$

$$= \frac{E_\phi \sin \delta}{X_s} - j\frac{E_\phi \cos \delta - V_1}{X_s} \tag{6.31}$$

or

$$I_1 \cos \theta = \frac{E_\phi \sin \delta}{X_s} \tag{6.32}$$

and

$$I_1 \sin \theta = -\frac{E_\phi \cos \delta - V_1}{X_s} \tag{6.33}$$

Since the power output is $P_o = 3V_1I_1 \cos \theta$, the approximate power output becomes

$$P_o \cong \frac{3V_1E_\phi \sin \delta}{X_s} \tag{6.34}$$

For a synchronous generator running at constant speed with a constant field current, X_s and E_ϕ are constant. V_1 is the terminal voltage, which is usually held constant. Therefore, the power output depends on the angle δ, which is known as the *power angle*. A synchronous generator is usually operated at a power angle

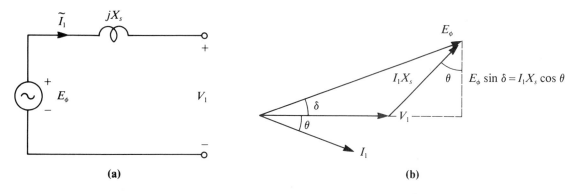

FIGURE 6.18
(a) Approximate Equivalent Circuit of a Synchronous Generator; (b) Phasor Diagram

varying from 15° to 25°. Since we have neglected the resistance of the phase winding, the power output is also equal to the power developed by the machine. The power developed is maximum when $\sin \delta = 1$ or $\delta = 90°$. This corresponds to the static stability limit of the generator. In this case, the maximum power developed or output is

$$P_{dm} = \frac{3V_1 E_\phi}{X_s} \tag{6.35}$$

The torque developed is

$$T_d = \frac{3V_1 E_\phi \sin \delta}{X_s \omega_s} \tag{6.36}$$

and the maximum torque developed is

$$T_{dm} = \frac{3V_1 E_\phi}{X_s \omega_s} \tag{6.37}$$

■ EXAMPLE 6.6

A 10-kVA, 230-V, three-phase, Y-connected synchronous generator has a field-winding resistance of 4.5 Ω, the armature-winding resistance per phase is 0.3 Ω, and the excitation currents are 10, 7, and 11 A for 0.85 pf lagging, unity pf, and 0.85 pf leading, respectively. Determine the efficiency of the generator in all cases when the core losses are 300 W and the mechanical losses are 200 W.

Solution

Per phase load voltage: $\quad V_1 = \dfrac{230}{\sqrt{3}} = 132.79$ V

Per phase load current: $\quad I_1 = \dfrac{10 \times 1000}{\sqrt{3} \times 230} = 25.12$ A

Rotational loss: $\quad P_{rot} = 300 + 200 = 500$ W

Stator copper loss: $\quad$ SCL $= 3 \times 25.1^2 \times 0.3 = 567$ W

Lagging power factor:

Power output: $\quad P_o = 10 \times 1000 \times 0.85 = 8500$ W

Field copper loss: $\quad I_f^2 R_f = 10^2 \times 4.5 = 450$ W

Power input: $\quad P_{in} = 8500 + 450 + 500 + 567 = 10{,}017$ W

Efficiency: $\quad \eta = \dfrac{8500}{10{,}017} = 0.849$ or 84.9%

Unity power factor:

Power output: $\quad P_o = 10 \times 1000 \times 1 = 10{,}000$ W

Field copper loss: $\quad I_f^2 R_f = 7^2 \times 4.5 = 221$ W

Power input: $\quad P_{in} = 10{,}000 + 221 + 567 + 500 = 11{,}288$ W

Efficiency: $\quad \eta = \dfrac{10{,}000}{11{,}288} = 0.886$ or 88.6%

Leading power factor:

Power output: $\quad P_o = 10 \times 1000 \times 0.85 = 8500$ W

Field copper loss: $\quad I_f^2 R_f = 11^2 \times 4.5 = 545$ W

Power input: $\quad P_{in} = 8500 + 545 + 567 + 500 = 10{,}112$ W

Efficiency: $\quad \eta = \dfrac{8500}{10{,}112} = 0.841$ or 84.1%

6.6 OPERATING CHARACTERISTICS OF SYNCHRONOUS GENERATORS

In this section we will discuss the operating characteristics of synchronous generators, that is, the no-load, short-circuit, and load characteristics, because it is important to know the capabilities of the machine when it runs under various modes of operation.

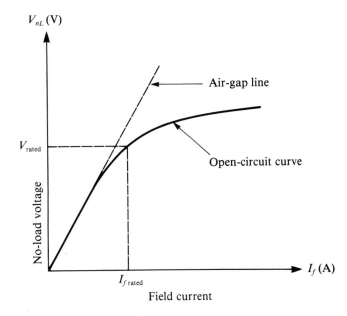

FIGURE 6.19
No-Load
Characteristic of
a Synchronous
Generator

No-load Characteristic   The no-load characteristic is the variation of the output voltage as a function of the excitation current when the generator operates at its rated synchronous speed with the armature terminals open-circuited. Figure 6.19 shows the no-load characteristic of a synchronous generator on a per phase basis.

Short-circuit Characteristic   The short-circuit characteristic of a synchronous generator shows the variation of the armature current as a function of the excitation current when the generator is operating at its synchronous speed by establishing a short circuit between the armature terminals. The short-circuit characteristic provides information about the current capabilities of the generator. Figure 6.20 shows the short-circuit characteristic of the synchronous generator on a per phase basis.

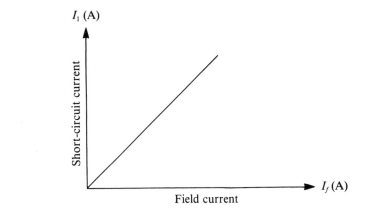

FIGURE 6.20
Short-Circuit
Characteristic of
a Synchronous
Generator

It is a common practice to plot the no-load and short-circuit characteristics on the same graph for the purpose of calculating the synchronous impedance of the generator, as shown in Fig. 6.21. For a given field current, the corresponding no-load voltage and the short-circuit current can be determined from the no-load and open-circuit curves. The synchronous impedance, on a per phase basis, is

$$Z_s = \frac{V_{nL}}{I_{sc}}$$
(6.38)

The synchronous reactance of the generator is

$$X_s = \sqrt{Z_s^2 - R_1^2}$$
(6.39)

In the linear region of the no-load characteristic the synchronous impedance is fairly constant and is referred to as the unsaturated synchronous impedance of the synchronous generator. On the other hand, the saturated synchronous impedance of the synchronous generator is the ratio of the no-load voltage in the saturated region to the armature current for the same excitation current. Since the no-load voltage does not increase proportionally with the increase in the excitation current in the saturated region, the saturated synchronous impedance is smaller than the unsaturated. At the rated no-load voltage, the unsaturated and saturated values of the synchronous impedance can be determined from Fig. 6.21.

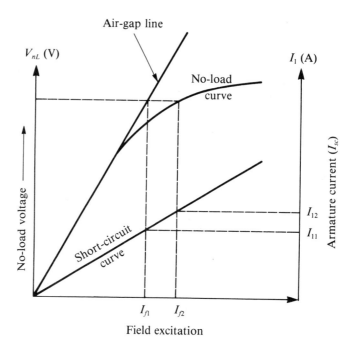

FIGURE 6.21
Open-Circuit and Short-Circuit Characteristics of a Y-Connected Synchronous Generator

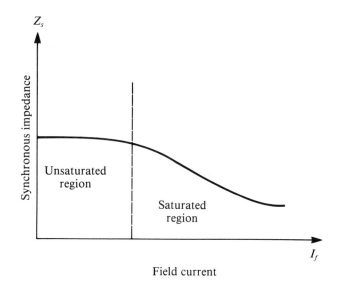

FIGURE 6.22
Variations in
Synchronous
Impedance as a
Function of Field
Current

The unsaturated synchronous impedance is

$$Z_{su} = \frac{V_{nL}}{I_{11}} \qquad (6.40)$$

and the saturated synchronous impedance is

$$Z_{ss} = \frac{V_{nL}}{I_{12}} \qquad (6.41)$$

The variation in the synchronous impedance as a function of the excitation current is given in Fig. 6.22.

Another important aspect of the no-load and short-circuit characteristics is the short-circuit ratio. It is defined as the ratio of the excitation current to obtain the rated voltage under no-load conditions to the excitation current required to establish the rated armature current under short-circuit conditions.

6.6.1 External Characteristics

The external characteristics of a synchronous generator highlight the variation in line voltage as the load current increases from no load to full load. In a given synchronous generator operating at its rated speed, the excitation current can be held constant. The synchronous impedance remains the same under different loads in the linear operating region. The variation in line voltage now depends on the power factor of the load and the armature current. From the phasor diagram for the approximate equivalent circuit (Fig. 6.18), it is evident that

$$\tilde{E}_\phi = \tilde{V}_1 + j\tilde{I}_1 X_s \tag{6.42}$$

In a synchronous generator with constant excitation current, the magnitude of the induced voltage, $\tilde{E}_\phi$, is fixed. However, the power angle of $\tilde{E}_\phi$ is free to change with changes in the loading conditions. It is, therefore, obvious that the locus of $\tilde{E}_\phi$ is a circle. Let us now consider the loads with unity, lagging, and leading power factors.

Unity Power Factor   Figure 6.23 shows the two possibilities with small and large armature currents. As the armature current increases, the drop across the synchronous reactance increases, thereby reducing the terminal voltage, as shown

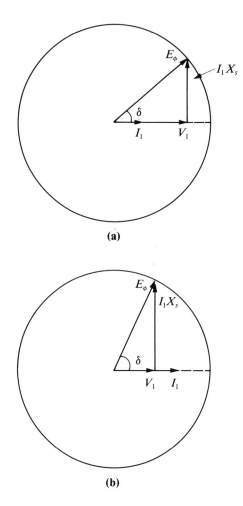

FIGURE 6.23
Resistive
Loading Effect
on Terminal
Voltage when
(a) the
Armature
Current is
Relatively Small
(Nearly No
Load); and (b)
the Armature
Current is
Relatively Large
(Full Load)

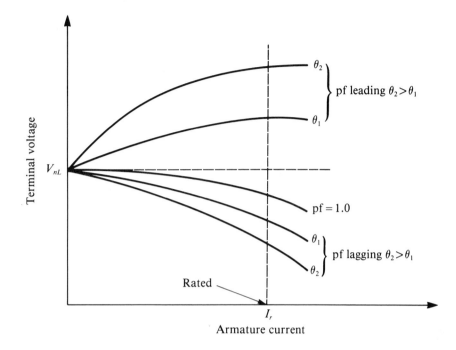

FIGURE 6.24
External
Characteristics of
a Synchronous
Generator

in the figure. Notice the increase in the power angle with the increase in the armature current. Also, the terminal voltage is smaller in magnitude than the no-load voltage. The variation in terminal voltage with the load current is shown in Fig. 6.24.

Lagging Power Factor   Let us consider two cases with lagging power factor angles θ_1 and θ_2 such that θ_2 is greater than θ_1. For the lagging power factor angle θ_1, the effects of small and large armature currents on the terminal voltage are shown in Figs. 6.25a and 6.25b. It is obvious that the terminal voltage decreases with the increase in the armature current. Figures 6.25c and 6.25d are for the large power factor angle θ_2 for small and large armature currents. A direct comparison of Fig. 6.25a with Fig. 6.25c or Fig. 6.25b with Fig. 6.25d shows that for the same armature current the terminal voltage decreases even further with the increase in power factor angles, as shown in Fig. 6.24.

Leading Power Factor   Once again, consider the loading conditions with two leading power factor angles θ_1 and θ_2 such that θ_1 is smaller than θ_2. The phasor diagrams for the two power factor angles for small and large armature currents are given in Fig. 6.26. From these diagrams, it is obvious that the terminal voltage increases with the increase in load current. Also, the terminal voltages are higher for large (leading) power factor angles for a given load current. The variations in the terminal voltage for the two power factor angles as a function of armature current are shown in Fig. 6.24.

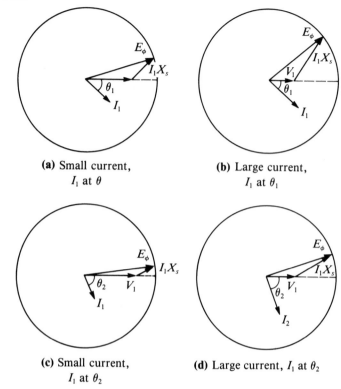

(a) Small current,
I_1 at θ

(b) Large current,
I_1 at θ_1

(c) Small current,
I_1 at θ_2

(d) Large current, I_1 at θ_2

FIGURE 6.25
Effect of
Inductive
Loading on the
Terminal
Voltage: θ_1 is
Smaller than θ_2.

In the determination of the preceding characteristic, we held the field current constant to obtain the same no-load voltage. The terminal voltage either increased or decreased from its no-load value depending on whether the load was capacitive or inductive. There is another external characteristic that can be determined by adjusting the field current to obtain the rated voltage at the rated line current for a given load. This characteristic is more desirable because a given synchronous generator is usually operated at its rated values. However, if we now reduce the line current without making any adjustments in the excitation current, we will notice the following:

1. For a resistive load, the terminal voltage increases with the decrease in the line current.

2. For an inductive load, the terminal voltage also increases with the decrease in the line current. The increase, however, is larger than that for the resistive load.

3. For a capacitive load, the terminal voltage decreases with the decrease in the line current.

These observations can be explained with the help of the phasor diagrams given in Fig. 6.16 and are plotted in Fig. 6.27.

■ **EXAMPLE 6.7**

The test data on a three-phase, Y-connected synchronous generator are as follows:

Short-circuit test: Field current = 1.2 A; short-circuit current = 25 A

Open-circuit test: Field current = 1.2 A; open-circuit voltage = 440 V

Both tests are done at the rated speed of the generator. The winding resistance is 1.2 Ω per phase. Determine the synchronous reactance of the generator.

Solution

Open-circuit voltage per phase: $E_\phi = 440/\sqrt{3} = 254.03$ V

Short-circuit current per phase: $I_{sc} = 25$ A

Synchronous impedance: $Z_s = 254.03/25 = 10.16\,\Omega$

Synchronous reactance: $X_s = \sqrt{Z_s^2 - R_1^2} = \sqrt{10.16^2 - 1.2^2} = 10.09\,\Omega$

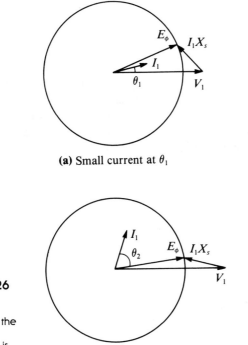

(a) Small current at θ_1

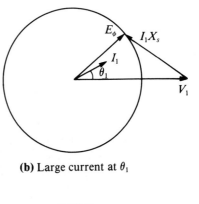

(b) Large current at θ_1

FIGURE 6.26
Effect of
Capacitive
Loading on the
Terminal
Voltage: θ_2 is
Larger than θ_1.

(c) Small current at θ_2

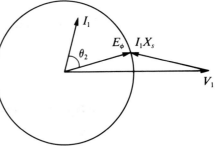

(d) Large current at θ_2

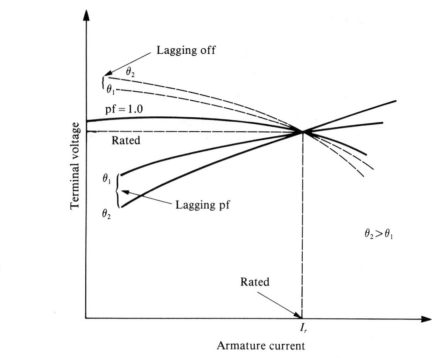

FIGURE 6.27
External
Characteristics of
a Synchronous
Generator
Adjusted to
Operate at
Rated Voltage
and Current

■ **EXAMPLE 6.8**

A 2-MVA, 4400-V, Y-connected, three-phase synchronous generator rated at 1200 rpm is tested as follows: The open-circuit voltage and the short-circuit current are 5200 V and 260 A, respectively, when the field winding current is held at 20 A. The generator is designed to result in a resistive voltage drop of 2.5% of its line voltage. Determine (a) the winding resistance, (b) the synchronous reactance, and (c) the induced voltage per phase at unity power factor.

Solution

(a) The rated voltage per phase is

$$V_1 = \frac{4400}{\sqrt{3}} = 2540.34 \text{ V}$$

The rated current per phase is

$$I_1 = \frac{2 \times 10^6}{3 \times 2540.34} = 262.43 \text{ A}$$

The voltage drop across the winding resistance is 2.5% of the load voltage. That is,

$$I_1 R_1 = \frac{2540.34 \times 2.5}{100} = 63.51 \text{ V}$$

Therefore, the resistance per phase is $R_1 = 63.51/262.43 = 0.242 \ \Omega$.

(b) The synchronous impedance is

$$Z_s = \frac{5200/\sqrt{3}}{260} = 11.547 \ \Omega$$

The synchronous reactance is

$$X_s = \sqrt{Z_s^2 - R_1^2} = \sqrt{11.547^2 - 0.241^2} = 11.544 \ \Omega$$

(c) Operation at the rated load is

$$\tilde{E}_\phi = \tilde{V}_1 + \tilde{I}_1 \hat{Z}_s = 2540.34 + (262.43\underline{/0°}) \times (0.241 + j11.544)$$

$$= 2540.34 + 3030.28\underline{/88.8°}$$

$$= 3994.8\underline{/49.32°} \text{ V}$$

6.7 SALIENT-POLE SYNCHRONOUS GENERATOR

The round rotor construction is commonly used for synchronous machines operating at 1800 or 3600 rpm. These high-speed generators are driven by steam turbines and are known as turbogenerators. The cylindrical rotor structure helps to reduce the windage loss, provides quiet operation, and enables the rotor to withstand high centrifugal forces. For low-speed operation, synchronous machines having four or more poles usually employ salient-pole rotors. Low-speed generators are normally driven by water turbines and are known as hydroelectric generators.

In the foregoing analysis of a synchronous generator, the effect of saliency was not taken into consideration. Due to uniform air gap, the reluctance and thereby the reactances were considered as constant parameters. In the salient-pole rotor, the region midway between the poles (quadrature or q axis) has a larger air gap as compared to the region between the pole centers (direct or d axis), as shown in Fig. 6.28. Therefore, a given armature mmf produces more flux along the d axis as compared to that along the q axis. Thus, we can define two armature reactances corresponding to the d and q axes of the machine. The reactance corresponding to the d axis is known as the direct-axis synchronous

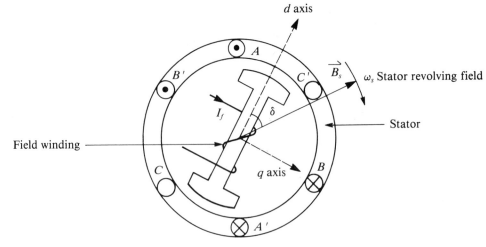

FIGURE 6.28
Salient-Pole
Synchronous
Generator
Showing *d* and
q Axes

reactance, x_d, while that corresponding to the q axis is termed the quadrature-axis synchronous reactance, x_q. The armature current, $\tilde{I}_a$, can be resolved into two components, $\tilde{I}_d$ and $\tilde{I}_q$. $\tilde{I}_d$ is responsible for the production of the armature field along the d axis, while $\tilde{I}_q$ establishes the armature field along the q axis.

In the absence of the armature current, the voltage induced in the armature winding due to the flux produced by the field current I_f is $\tilde{E}_\phi$ on a per phase basis. The armature reaction due to the armature current $\tilde{I}_a$ can now be taken into consideration by decomposing it along the d and q axes. Let $\tilde{E}_d$ and $\tilde{E}_q$ be the voltages induced in the stator winding by the armature current components $\tilde{I}_d$ and $\tilde{I}_q$, respectively. Thus, the per phase voltage of the generator is

$$\tilde{V}_1 = \tilde{E}_\phi + \tilde{E}_d + \tilde{E}_q - \tilde{I}_1 R_1 \tag{6.43}$$

Just as in the cylindrical rotor theory, the armature reaction voltages can be expressed as

$$\tilde{E}_d = -j\tilde{I}_d x_d \tag{6.44}$$

and

$$\tilde{E}_q = -j\tilde{I}_q x_q \tag{6.45}$$

From Eq. (6.43), the generated voltage of the salient-pole generator is

$$\tilde{E}_\phi = \tilde{V}_1 + \tilde{I}_1 R_1 + j\tilde{I}_d x_d + j\tilde{I}_q x_q \tag{6.46}$$

where R_1 is the armature-winding resistance. The corresponding phasor diagram is shown in Fig. 6.29.

To make the calculations using the preceding equations, we must know $\tilde{I}_d$, $\tilde{I}_q$, x_d, and x_q. The value of x_d can be determined from Eq. (6.39). The quadrature-

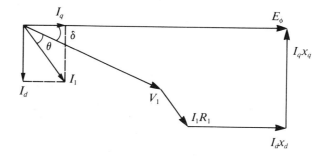

FIGURE 6.29
Phasor Diagram
of a Salient-Pole
Synchronous
Generator

axis synchronous reactance can be determined by means of standard tests as described in *IEEE Test Code for Synchronous Machines*. To determine $\tilde{I}_d$ and $\tilde{I}_q$, we must know θ and δ. Usually, the angle θ is given in the specifications and the power angle δ is unknown. However, to complete the phasor diagram, the power angle δ must be known. From the phasor diagram (Fig. 6.29), it can be shown that

$$V_1 \sin \delta + I_1 R_1 \sin(\theta + \delta) = I_q x_q$$

But

$$I_q x_q = I_1 \cos(\theta + \delta) x_q$$

Thus,

$$\tan \delta = \frac{I_1 x_q \cos \theta}{V_1 + I_1(R_1 \cos \theta + x_q \sin \theta)} \tag{6.47}$$

The power output of a three-phase synchronous generator is

$$P_o = 3V_1 I_1 \cos \theta$$

$$= 3V_1 I_q \cos \delta + 3V_1 I_d \cos(90 - \delta)$$

$$= 3V_1 I_q \cos \delta + 3V_1 I_d \sin \delta \tag{6.48}$$

Neglecting the armature resistance, the power developed must be equal to the power output. Thus,

$$P_d = 3V_1 I_q \cos \delta + 3V_1 I_d \sin \delta \tag{6.49}$$

Moreover,

$$V_1 \sin \delta = I_q x_q$$

or

$$I_q = \frac{V_1 \sin \delta}{x_q} \tag{6.50}$$

and

$$V_1 \cos \delta = E_\phi - I_d x_d$$

or

$$I_d = \frac{E_\phi - V_1 \cos \delta}{x_d} \tag{6.51}$$

Substituting for I_d and I_q in Eq. (6.49), the power developed can be expressed as

$$P_d = \frac{3V_1E_\phi}{x_d} \sin \delta + \frac{3}{2} V_1^2 \sin 2\delta \left(\frac{x_d - x_q}{x_d x_q}\right) \tag{6.52}$$

The first term is the same as the power developed by a cylindrical rotor synchronous machine. The second term highlights the effect of saliency. It is the power developed by reluctance action in a salient-pole machine. The torque developed can be expressed as

$$T_d = \frac{3V_1E_\phi}{\omega_s x_d} \sin \delta + \frac{3V_1^2}{2\omega_s} \sin 2\delta \left(\frac{x_d - x_q}{x_d x_q}\right) \tag{6.53}$$

The torque developed by a salient-pole synchronous generator as a function of torque angle δ is plotted in Fig. 6.30. The reluctance torque is independent of the excitation. Furthermore, for a cylindrical or nonsalient machine, the synchronous reactances along the d and q axes are equal to each other, and the contribution due to reluctance torque disappears.

■ **EXAMPLE 6.9**

A 70-MVA, 13.8-kV, 60-Hz, salient-pole, Y-connected, three-phase synchronous generator has $R_1 = 0.0125 \ \Omega$, $x_d = 1.83 \ \Omega$, and $x_q = 1.21 \ \Omega$. It is operating at full load at a 0.8 power factor lagging. Determine the voltage regulation of the generator.

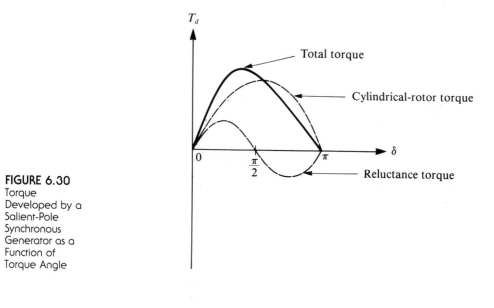

FIGURE 6.30
Torque
Developed by a
Salient-Pole
Synchronous
Generator as a
Function of
Torque Angle

Solution

The phase voltage is

$$V_1 = \frac{13,800}{\sqrt{3}} = 7967.434 \text{ V}, \qquad \theta = \cos^{-1}(0.8) = 36.87°$$

The per phase current is

$$I_1 = \frac{70 \times 10^6}{\sqrt{3} \times 13,800} = 2928.588 \text{ A}$$

From Eq. (6.47),

$$\tan \delta = \frac{2928.588 \times 1.21 \cos(36.87°)}{7967.434 + 2928.588[0.0125 \cos(36.87°) + 1.21 \sin(36.87°)]}$$

$$= 0.28$$

or $\delta = 15.64°$

With E_ϕ as a reference phasor, the phase current I_1 lags by an angle of 15.64 + 36.87 = 57.51°. Thus,

$$\tilde{I}_1 = 2928.588\underline{/-52.51°}$$

From the phasor diagram (Fig. 6.29),

$$E_\phi = 7967.434 \cos(15.64°) + 0.0125 \times 2928.588 \cos(52.51°)$$
$$+ 1.83 \times 2928.588 \sin(52.51°)$$
$$= 11947.117 \text{ V}$$

The percent of voltage regulation is

$$VR\% = 100 \frac{11,947.117 - 7967.434}{7967.434} = 49.95\%$$

6.8 PARALLEL OPERATION OF SYNCHRONOUS GENERATORS

The generation of electric power, its transmission, and distribution have to be conducted in an efficient and reliable way at a reasonable cost with the least amount of interruptions. Consequently, in a large power system many synchronous generators are connected in parallel to a common line known as an *infinite bus.* An infinite bus has a fixed voltage, constant frequency, and predetermined phase sequence. Thus, a number of requirements have to be satisfied prior to connecting a synchronous generator to the infinite bus.

1. The operating frequency must be the same as that of the infinite bus.
2. The line voltage must be equal to the constant voltage of the bus.
3. The phase sequence must be identical with the phase sequence of the infinite bus.

To meet the increased load demand, let us bring a second generator into service as shown in Fig. 6.31. The speed of the incoming generator is adjusted to obtain the frequency of the induced voltage equal to that of the infinite bus. Now the field current can be raised to a level at which the induced voltage of the

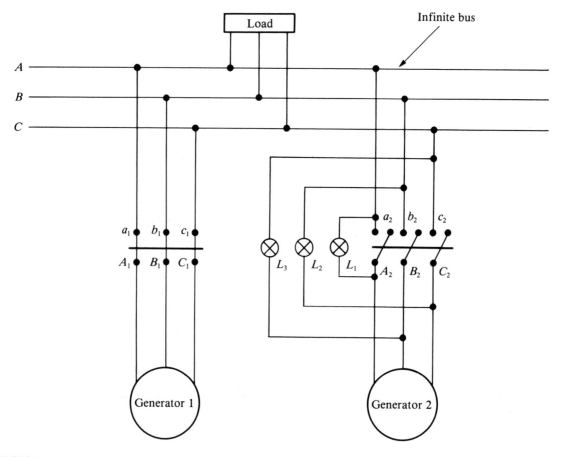

FIGURE 6.31
Parallel Operation of Synchronous Generators

second generator is identical to the voltage of the infinite bus. To verify the phase sequence, three lamps are connected asymmetrically as shown. When the phase sequence is also proper, lamp L_1 will be dark, while the other two lamps will glow brightly. If the phase sequence is not proper, all the lamps will glow or become dark simultaneously. When all the above conditions are met, the second generator can then be connected to the infinite bus by closing the breaker switch S.

In addition to the lamps to check the conditions for synchronism, a device called a synchroscope is also used to perform a precise switching when the above conditions are met. The synchroscope provides the information on the incoming generator as to whether it is running at a lower or higher speed. At the proper speed and thereby the proper phase sequence, the synchroscope indicates zero. That is the time at which the switch must be closed.

Once the breaker switch is closed, the second generator is on line. At this moment, it is neither receiving nor delivering any power. This is referred to as the *floating stage* of the generator. If we now increase the field current, the second generator will deliver reactive power and help to improve voltage regulation. On the other hand, if an attempt is made to increase the speed of the rotor, the torque developed will increase and the second generator will begin to supply active power to the load.

PROBLEMS

6.1. The stator of a three-phase synchronous generator has 48 teeth. If it is to be wound for 8-pole applications, what must be the speed of the rotor for a supply frequency of 60 Hz? Determine the number of coils per pole per phase for a double-layer winding. What are the pole pitch, coil pitch, and slot pitch?

6.2. For Problem 6.1, determine the pitch factor, distribution factor, and winding factor. How many electrical degrees are there in one revolution?

6.3 The rotor of the synchronous generator discussed in Problems 6.1 and 6.2 has a radius of 40 cm. The pole length is 80 cm and the maximum flux density is 1.2 T. Each coil has 6 turns. What is the flux per pole? Determine the induced voltage if (a) all phase groups are connected in series and (b) there are two parallel paths. What is the no-load line-to-line voltage for (c) a Y-connection, and (d) a Δ-connection.

6.4. A three-phase, 12-pole, synchronous generator is required to produce a no-load voltage of 5600 V per phase at 50 Hz. The rotor flux per pole is 0.0185 Wb. The stator has 108 slots, which are wound using double-layer winding. What is the number of turns per coil for a series connection?

6.5. A three-phase synchronous generator is designed to produce 25 Hz when operated at 1500 rpm. A double-layer winding is to be placed in such a way that there are 4 coils per pole per phase. Determine the number of slots in the stator. How many poles must be there?

6.6. A 6-pole, three-phase synchronous generator has 72 slots. It is to be wound using double-layer winding. Determine the coil pitch, pitch factor, distribution factor, and winding factor.

6.7. A 12-pole, Y-connected, three-phase synchronous generator is designed to produce 400 Hz with 3 coils per phase group. Each coil has 2 turns and the rotor flux is 0.023 Wb/pole. The stator uses double-layer winding and the coils are connected in two parallel groups. Determine the (a) speed of the rotor, (b) total number of slots, (c) coil pitch, (d) pitch factor, (e) distribution factor, (f) winding factor, (g) effective coils per phase, (h) generated voltage per phase, and (i) line-to-line voltage.

6.8. A three-phase, Δ-connected synchronous generator supplies a rated current of 13 A at 2300 V. The power delivered to the balanced load is 40 kW. Determine the phase voltage, phase current, power per phase, and power factor.

6.9. A 100-kVA, three-phase, Y-connected synchronous generator is designed to supply a full load at 480 V. The power factor of the load is 0.9 lagging. Determine the current and total power supplied by the generator.

6.10. In the synchronous generator of Problem 6.9 the per phase resistance is 0.15 Ω and the synchronous reactance is 1.3 Ω. Determine the voltage regulation of the generator. What is the power developed?

6.11. A three-phase, Y-connected synchronous generator delivers a load current of 50 A at 230 V with 0.8 power factor leading. Under no-load conditions, the generated voltage is 280 V leading the full-load voltage by 30°. Determine the effective resistance and synchronous reactance. What is the voltage regulation of the generator?

6.12. A three-phase, Y-connected, 8-pole, 48-slot, double-layer-wound, 6000-rpm synchronous generator capable of generating 440 V from line to line when connected in series is to be connected in such a way that its output line voltage is 110 V. Each of its coil has 12 turns. If one side of the coil is in slot 1, the other side is in slot 6. The full-load voltage regulation is 5%. What is the flux per pole to satisfy the requirements?

6.13. A three-phase, 2-pole, Y-connected synchronous generator rated at 10 kVA, 380 V (line to line), and 60 Hz operates at its rated values with a lagging power factor of 0.8. The per phase synchronous reactance and resistance are 4 and 2 Ω, respectively. Calculate the synchronous speed of the generator. Determine the per phase induced voltage, power developed, and the efficiency of the generator when the fixed loss is 1 kW.

6.14. A 100-kW, 1-kV, Δ-connected, three-phase synchronous generator has a winding resistance of 1.5 Ω per phase and synchronous reactance of 15 Ω per phase. Determine the generated line voltage under full load at 0.8 power factor lagging. What is the voltage regulation of the generator?

6.15. A Y-connected, three-phase synchronous generator is tested under short-circuit and open-circuit conditions while varying the field current. At the rated armature current of 80 A, the per phase voltage is 120 V. The dc resistance measurements between the three line terminals are 0.25, 0.29, and 0.22 Ω. Calculate the synchronous impedance of the generator.

6.16. A 600-V, 60-kVA, Y-connected, three-phase synchronous generator has a synchronous reactance of 1.2 Ω per phase. It is supplying full load at 0.9 leading power factor. Determine (a) the generated voltage, (b) the power angle, and (c) the voltage regulation of the generator. Draw the phasor diagram. Neglect the winding resistance.

6.17. A 600-kVA, 5000-V, Δ-connected, three-phase synchronous generator is delivering full load at a unity power factor. The synchronous impedance is $1.5 + j12$ Ω per phase. Determine the generated voltage. If the field current is unchanged and the generator is now supplying the same current at a 0.8 power factor lagging, what is the terminal voltage?

6.18. A 300-kVA, 500-V, Δ-connected, three-phase synchronous generator draws a field current of 2 A to maintain the rated current under short-circuit conditions. For the same field current, the open-circuit voltage is 572 V. Determine the synchronous reactance of the generator if it has negligible winding resistance. What are the voltage regulation and the efficiency of the generator at 0.707 pf lagging?

6.19. A 5-MVA, three-phase, Y-connected generator rated at 6600 V at 60 Hz has the open-circuit characteristic given by the following data:

I_f: 10   15   20   25   30 A
E_ϕ: 4500 6600 7500 8250 8950 V

A field current of 20 A is found necessary to circulate full-load current on short circuit of the armature. Determine the full-load power angle and the efficiency at 0.8 pf lagging when $R_1 = 1.2$ Ω, $Rf = 500$ Ω, and $P_{\text{rot}} = 250$ kW.

6.20. A 4160-V, 3.5-MVA, 60-Hz, three-phase, Y-connected, salient-pole generator has a direct-axis reactance of 2.75 Ω and a quadrature-axis reactance of 1.8 Ω per phase. If the generator is delivering full load at a 0.8 pf lagging, determine the induced voltage and torque angle. Neglect the armature-winding resistance.

6.21. A salient-pole synchronous generator has a direct-axis synchronous reactance of 0.75 per unit and a quadrature-axis synchronous reactance of 0.5 per unit. Determine the voltage regulation if the generator delivers full load at 0.8 pf lagging at the rated voltage. What is the per unit power delivered by the generator? Assume that the armature-winding resistance can be neglected.

CHAPTER 7

Synchronous Motors

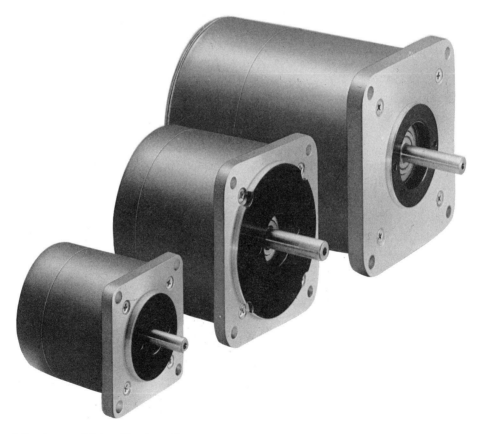

Slow-Speed Synchronous Motors (Courtesy of
Bodine Electric Company)

A synchronous motor, as the name implies, runs at its synchronous speed dictated by the frequency of the applied voltage and the number of poles on the stator. Most large-size synchronous motors have three-phase windings on the stator and a field winding on the rotor. The stator winding is connected to a balanced three-phase system, while the dc voltage is impressed on the rotor winding. Therefore, these motors are also a doubly fed type. Often single-phase synchronous motors do not have a field winding on the rotor. However, the polyphase synchronous motors with no dc excitation are not that common.

7.1 CONSTRUCTION AND OPERATION

In construction a synchronous motor looks the same as a synchronous generator. In fact, a synchronous motor can be made to operate as a generator, and vice versa. However, a synchronous motor must have an additional winding, which is referred to as the *damper or amortisseur winding*. This winding serves two basic purposes. First, it helps to start the motor because a synchronous motor is not a self-starting motor. Second, the damper winding tends to minimize motor hunting. Hunting is the term that is used to define successive overshoots and undershoots in speed due to sudden variations in the load. Whenever the load is changed suddenly, a synchronous motor cannot adjust its power angle instantaneously to account for the variation in load. As it tries to adjust its power angle, due to its inertia, it may go over the needed adjustment. Then it will try to correct itself and may overdo again. Therefore, it may swing back and forth many times before reaching the proper value. A damper winding minimizes these swings and helps the motor to adjust its power angle promptly due to changes in the load. In a synchronous motor without the damper winding, hunting may become intolerably severe.

7.1.1 Starting a Three-phase Synchronous Motor

A three-phase synchronous motor is not self-starting if a damper winding is not provided. A damper winding is an auxiliary winding located in slots near the surface of the rotor. The damper winding is usually of the squirrel-cage type. For synchronous motors for which very large starting torques are required, the damper winding can be of the wound-rotor type. As soon as the three-phase voltage is impressed on the stator windings, revolving flux is produced, which moves around the periphery of the rotor at synchronous speed. Since the rotor is at standstill, in one-half the cycle of the stator revolving field the rotor polarity may be such that it experiences a force of attraction and tends to go along with the field. Due to the heavy mass of the rotor, it takes time before it can start moving, but by then the stator flux has already reversed. Now a force of repulsion acts on the rotor pole and tends to rotate the rotor in the opposite direction. As the rotor now tries to rotate in the opposite direction, the stator field changes again. Thus the rotor pole is acted on by a rapidly reversing force of equal magnitude in both directions. The net torque is therefore zero, as shown in Fig. 7.1 for a 2-pole motor. The revolving field is assumed to be rotating in the clockwise direction. In Fig. 7.1a, the alignment of like poles results in a force of repulsion that tends to rotate the rotor in the counterclockwise direction. However, half a period later (Fig. 7.1(b)), the force of attraction tends to rotate the rotor in the clockwise direction due to locking between the north and south poles of the rotor and stator. Thus the rotor stays in its original position and no motion is imparted to it. If the stator field flux is strong enough, the motor will draw very high current and produce a growling noise. Thus, in the absence of the damper windings, the rotor must be driven at nearly its synchronous speed by another prime mover and then synchronized by applying the dc voltage to the rotor winding.

If the motor has damper windings, the revolving field set up by the stator winding will induce a voltage in the damper winding. The rotor will now tend to rotate along with the revolving field of the stator, as explained in Chapter 8. As the motor attains nearly 95% of its synchronous speed, the current is then applied to the insulated dc field winding on the rotor. The dc field winding now produces strong magnetic poles, and therefore the force of attraction between the poles of the rotor and the stator flux helps move the rotor into step with the stator flux. At that time, the motor is said to be *synchronized*. During the rotor acceleration period, the rotor field windings must be shorted through an appropriate bank of resistors. The rotor should not be left open because it can develop high voltage due to its rotation in the magnetic field of the stator.

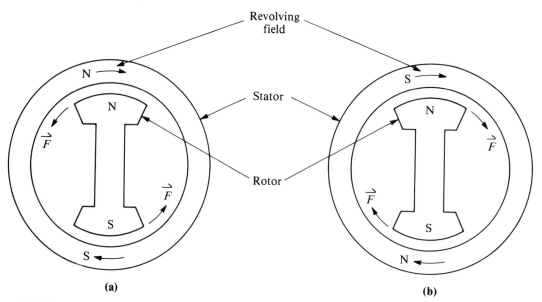

FIGURE 7.1
(a) Force of Repulsion Tending to Cause Rotation in Counterclockwise Direction; (b) Force of Attraction Tending to Cause Rotation in Clockwise Direction

As the rotor is now rotating at its synchronous speed, there is an induced voltage in the stator windings owing to generator action. The expression for the induced voltage is the same as that for the synchronous generator. It must, however, be noted that, when the rotor of a synchronous motor is rotating at its sychronous speed, the damper winding does not play any role. But the presence of the damper winding still helps control the hunting of the motor due to changes in the load. If the load suddenly increases, the rotor of the synchronous motor tends to slow down due to increased torque. As soon as it tries to slow down, voltage will be induced in the damper winding conductors, which will try to accelerate the rotor. On the other hand, if the changes on the load suddenly reduce the torque on the rotor, the rotor will at first try to accelerate. In this case, a voltage is again induced in the damper winding, but this voltage is of the opposite polarity as compared to the situation when the rotor tried to slow down. Therefore, the damper winding creates a torque in the direction opposite to the rotation, which tends to slow the rotor.

7.2   EQUIVALENT CIRCUIT OF A SYNCHRONOUS MOTOR

The equivalent circuit of a synchronous motor is the same as that of a synchronous generator except for the direction of the stator winding current. Such an equivalent circuit is given in Fig. 7.2 on a per phase basis. It is obvious that

$$\tilde{V}_1 = \tilde{E}_\phi + \tilde{I}_1 R_1 + j\tilde{I}_1 X_s \tag{7.1}$$

or
$$\tilde{I}_1 = \frac{\tilde{V}_1 - \tilde{E}_\phi}{R_1 + jX_s} \tag{7.2}$$

Let θ be the phase angle between $\tilde{V}_1$ and $\tilde{I}_1$. The power factor of a three-phase synchronous motor can be unity, leading, or lagging. Taking the applied voltage as a reference, we can draw the phasor diagram for any power factor, as shown in Fig. 7.3.

FIGURE 7.2
Per Phase
Equivalent
Circuit of a
Synchronous
Motor

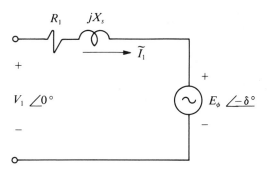

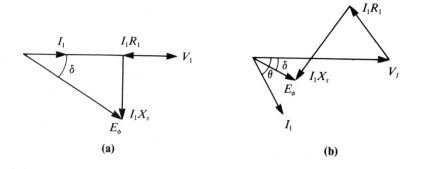

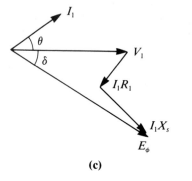

FIGURE 7.3
Phasor Diagrams
of a
Synchronous
Motor with (a)
Unity; (b)
Lagging; and
(c) Leading
Power Factors

The power input is

$$P_{in} = 3V_1I_1 \cos \theta + V_fI_f \qquad (7.3)$$

The electrical loss in the machine takes place in the stator winding resistance, as well as in the field winding of the rotor. Thus, the total electrical loss is

$$P_{elec} = 3I_1^2R_1 + I_fV_f \qquad (7.4)$$

where I_f is the current in the field winding of the rotor and V_f is the dc voltage impressed on it.

The power developed by the motor is

$$P_d = 3V_1I_1 \cos \theta - 3I_1^2R_1 \qquad (7.5)$$

The torque developed by the motor is

$$T_d = \frac{P_d}{\omega_s} \qquad (7.6)$$

The power output is obtained by subtracting the rotational losses from the power developed. The power-flow diagram for a synchronous motor is given in Fig. 7.4.

■ EXAMPLE 7.1

A 10-hp, 230-V, 60-Hz, three-phase, Y-connected synchronous motor delivers full load at a power factor of 0.707 leading. The synchronous reactance is 5 Ω, the rotational loss is 230 W, and the field loss is 70 W. Determine (a) the armature current, (b) the motor efficiency, and (c) the power angle. Neglect the stator winding resistance.

Solution

Power output:
$$P_o = 10 \times 746 = 7460 \text{ W}$$

Total losses:
$$P_{\text{loss}} = 230 + 70 = 300 \text{ W}$$

Power input:
$$P_{\text{in}} = 7460 + 300 = 7760 \text{ W}$$

(a) Since
$$P_{\text{in}} = \sqrt{3} \, V_l I_l \cos \theta,$$

$$I_l = \frac{7760}{\sqrt{3} \times 230 \times 0.707} = 27.55 \text{ A}$$

(b) The efficiency is

$$\eta = \frac{7460}{7760} = 0.9613 \quad \text{or} \quad 96.13\%$$

(c) Since the motor is Y-connected, the phase current is the same as the line

FIGURE 7.4
Power-Flow Diagram of a Synchronous Motor Including Field-Winding Loss. For Approximate Calculations, Field-Winding Loss Can Be Ignored

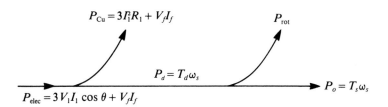

$$P_{\text{Cu}} = 3I_1^2 R_1 + V_f I_f \qquad P_{\text{rot}}$$

$$P_d = T_d \omega_s$$

$$P_{\text{elec}} = 3V_1 I_1 \cos \theta + V_f I_f \qquad P_o = T_s \omega_s$$

current. From the equivalent circuit of a three-phase synchronous motor on a per phase basis, we have

$$\tilde{V}_1 = \tilde{E}_\phi + j\tilde{I}_1 X_s$$

Thus
$$\tilde{E}_\phi = \frac{230}{\sqrt{3}} - j5 \times 27.55\underline{/45°} = 249.35\underline{/-22.93°} \text{ V}$$

The power angle is 22.93° and $\tilde{E}_\phi$ lags the applied voltage.

■ EXAMPLE 7.2

A 2-hp, 120-V, 60-Hz, 4-pole, three-phase, Y-connected synchronous motor has a stator impedance of $0.2 + j10 \ \Omega$. The friction and windage loss is 20 W, the core loss is 35 W, and the field winding loss is 30 W. Determine the stator current, the efficiency, and the power angle of the motor when it operates at full-load with unity power factor.

Solution

Power output: $P_o = 2 \times 746 = 1492 \text{ W}$

Power developed: $P_d = 1492 + 20 + 35 = 1547 \text{ W}$

However, $P_d = \sqrt{3} \ V_l I_l \cos \theta - 3I_1^2 R_1$

where $\cos \theta$ is the power factor and is given to be unity. Thus

$$1547 = \sqrt{3} \times 120 I_l - 3 \times 0.2 \ I_l^2, \qquad \because I_l = I_1$$

or $I_1^2 - 346.41 I_1 + 2578.33 = 0$

$$I_1 = \frac{346.41 \pm \sqrt{346.41^2 - 10{,}313.33}}{2} = \frac{346.41 - 331.19}{2} = 7.61 \text{ A}$$

$$\text{SCL} = 3I_1^2 R_1 = 3 \times 7.61^2 \times 0.2 = 34.75 \text{ W}$$

The power input is

$$P_{\text{in}} = 1547 + 34.75 + 30 = 1611.75 \text{ W}$$

Note that the power loss in the field winding is added to the total power input to the motor.
The efficiency is

$$\eta = \frac{1492}{1611.75} = 0.9257 \quad \text{or} \quad 92.57\%$$

Since $\tilde{V}_1 = \tilde{E}_\phi + \tilde{I}_1(R_1 + jX_s)$,

$$\tilde{E}_\phi = \frac{120}{\sqrt{3}} - 7.61 \times (0.2 + j10) = 67.76 - j76.1$$

$$= 101.9 \underline{/-48.32°} \text{ V}$$

Thus, the power angle is 48.32° lagging.

7.3 POWER EXPRESSION

Assume that the current in a stator winding lags the applied voltage by an angle θ. The case when the current leads the applied voltage is left as an exercise for the student to verify that the power developed is the same.

From the exact equivalent circuit, as given in Fig. 7.2, we have

$$\tilde{V}_1 = \tilde{E}_\phi + \tilde{I}_1(R_1 + jX_s) \tag{7.7}$$

or

$$\tilde{I}_1(R_1 + jX_s) = \tilde{V}_1 - \tilde{E}_\phi \tag{7.8}$$

which can also be written as

$$(I_1 \cos\theta - jI_1 \sin\theta)(R_1 + jX_s) = V_1 - E_\phi \cos\delta + jE_\phi \sin\delta \tag{7.9}$$

Equating the real and imaginary parts, we get

$$I_1 R_1 \cos\theta + I_1 X_s \sin\theta = V_1 - E_\phi \cos\delta \tag{7.10}$$

and

$$-I_1 R_1 \sin\theta + I_1 X_s \cos\theta = E_\phi \sin\delta \tag{7.11}$$

Squaring and adding these two equations and after some simplifications, we get

$$I_1^2 = \frac{(V_1 - E_\phi \cos\delta)^2 + E_\phi^2 \sin^2\delta}{R_1^2 + X_s^2} \tag{7.12}$$

or

$$I_1^2 = \frac{V_1^2 + E_\phi^2 - 2V_1 E_\phi \cos\delta}{Z_s^2} \tag{7.13}$$

where

$$Z_s^2 = R_1^2 + X_s^2 \tag{7.14}$$

The copper loss in the stator winding on a per phase basis is

$$\text{SCL}_\phi = I_1^2 R_1 = \frac{V_1^2 R_1 + E_\phi^2 R_1 - 2V_1 E_\phi R_1 \cos\delta}{Z_s^2} \tag{7.15}$$

Multiplying Eq. (7.10) by R_1 and Eq. (7.11) by X_s and adding them together, we obtain

$$I_1 \cos \theta = \frac{V_1 R_1 - E_\phi R_1 \cos \delta + E_\phi X_s \sin \delta}{Z_s^2} \tag{7.16}$$

The per phase power input is

$$
\begin{aligned}
P_{in} &= V_1 I_1 \cos \theta \\
&= \frac{V_1^2 R_1 - V_1 E_\phi R_1 \cos \delta + E_\phi V_1 X_s \sin \delta}{Z_s^2}
\end{aligned}
\tag{7.17}
$$

The per phase power developed is

$$
\begin{aligned}
P_{d\phi} &= P_{in} - SCL_\phi \\
&= \frac{V_1 E_\phi R_1 \cos \delta + V_1 E_\phi X_s \sin \delta - E_\phi^2 R_1}{Z_s^2}
\end{aligned}
$$

or
$$P_{d\phi} = \frac{V_1 E_\phi}{Z_s^2} (R_1 \cos \delta + X_s \sin \delta) - \frac{E_\phi^2 R_1}{Z_s^2} \tag{7.18}$$

Equation (7.18) can be simplified by neglecting the stator winding resistance and can be expressed, on a three-phase basis, as

$$P_d = \frac{3 V_1 E_\phi \sin \delta}{X_s} \tag{7.19}$$

and the torque developed is

$$T_d = \frac{3 V_1 E_\phi \sin \delta}{X_s \omega_s} \tag{7.20}$$

■ EXAMPLE 7.3

A 440-V, three-phase, Δ-connected synchronous motor draws 8660 W when operating under a certain load at 188.5 rad/s with an induced voltage of 560 V. The synchronous reactance is 25 Ω and the winding resistance is negligible. Determine the line current and the torque developed by the motor.

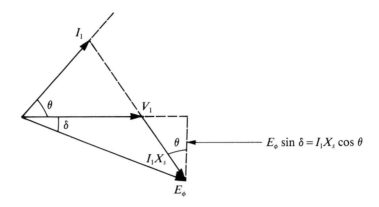

$E_\phi \sin \delta = I_1 X_s \cos \theta$

FIGURE 7.5
Phasor Diagram
for Example 7.3

Solution

Since the induced voltage is greater in magnitude then the applied voltage, the power factor of the motor must either be unity or leading. For general development, let us assume that the power factor is leading, as shown in Fig. 7.5.

Power input per phase: $\qquad P_{in} = 8660/3 = 2886.67$ W

Therefore, $\qquad I_1 \cos \theta = 2886.67/440 = 6.56$ A

But, from Fig. 7.5, $E_\phi \sin \delta = I_1 X_s \cos \theta = 6.56 \times 25 = 164$ V. Thus

$$\sin \delta = \frac{164}{560} = 0.293 \quad \text{or} \quad \delta = 17.03°$$

However, $\qquad E_\phi \cos \delta = V_1 + I_1 X_s \sin \theta$

or $\qquad I_1 X_s \sin \theta = E_\phi \cos \delta - V_1$

or $\qquad \tan \theta = \dfrac{E_\phi \cos \delta - V_1}{E_\phi \sin \delta}$

$$= \frac{560 \cos(17.03) - 440}{164} = 0.58$$

$$\theta = 30.1°$$

The phase current is

$$I_1 = \frac{6.56}{\cos(30.1°)} = 7.58 \text{ A}$$

The line current is

$$I = \sqrt{3} \times 7.58 = 13.13 \text{ A}$$

Since the stator winding resistance is negligible, the power developed is the same as the power input. That is,

$$P_d = 8660 \text{ W}$$

The torque developed is

$$T_d = \frac{P_d}{\omega_s} = \frac{8660}{188.5} = 45.94 \text{ N-m}$$

■ **EXAMPLE 7.4**

The full-load stator winding current in a 460-V, three-phase, Y-connected synchronous motor is maintained at 12.5 A. The synchronous impedance of the motor is $1.0 + j12 \ \Omega$. What is the nominal kVA rating of the motor? Determine the induced voltage and the power angle of the rotor when the motor is fully loaded at (a) unity power factor, (b) 0.707 power factor leading, and (c) 0.707 power factor lagging.

Solution

Since the motor is Y-connected, the per phase current in the stator winding is $I_1 = 12.5$ A. The per phase voltage is

$$V_1 = \frac{460}{\sqrt{3}} = 265.58 \text{ V}$$

and the rating of the motor is

$$S = 3V_1 I_1 = 3 \times 265.58 \times 12.5 = 9959 \text{ W}$$

Thus the nominal rating of the motor would be 10 kVA.

(a) Unity power factor:

$$\tilde{E}_\phi = \tilde{V}_1 - \tilde{I}_1(R_1 + jX_s) = 265.58 - 12.5 \times (1 + j12) = 294.19\underline{/-30.65°} \text{ V}$$

Thus, the induced line voltage is $\sqrt{3} \times 294.19 = 509.56$ V and the power angle is 30.65° when the motor is operating at unity power factor.

(b) Leading power factor:

$$\tilde{E}_\phi = \tilde{V}_1 - \tilde{I}_1(R_1 + jX_s)$$

$$= 265.58 - (12.5\underline{/45°}) \times (1 + j12) = 380.54\underline{/-17.57°} \text{ V}$$

Thus, the induced line voltage is 659.12 V and the power angle is 17.57°.

(c) Lagging power factor:

$$\tilde{E}_\phi = \tilde{V}_1 - \tilde{I}_1(R_1 + jX_s)$$

$$= 265.58 - (12.5\underline{/-45°}) \times (1 + j12) = 179.31\underline{/-32.83°} \text{ V}$$

Thus, the induced line voltage is 310.58 V and the power angle is 32.83°.

7.3.1 Condition for Maximum Power

The power developed by the synchronous motor depends on the applied voltage, induced voltage, synchronous impedance, and power angle. If we assume that the synchronous impedance, applied voltage, and induced voltage are constant, the power developed is dependent on the power angle. Since the induced voltage is maintained constant, the locus of the $\tilde{E}_\phi$ phasor is a circle, as shown in Fig. 7.6. The power developed is maximum when

$$\frac{dP_d}{d\delta} = 0 \tag{7.21}$$

Differentiating the expression for the power developed with respect to δ and setting it equal to zero, we get

$$-R_1 \sin \delta + X_s \cos \delta = 0$$

or
$$\tan \delta = \frac{X_s}{R_1} \tag{7.22}$$

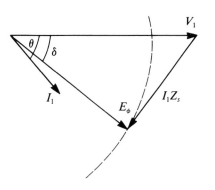

FIGURE 7.6
Induced Voltage Traces a Circle as θ Changes

which clearly indicates that the power angle will approach 90° when the winding resistance becomes negligible. The maximum power developed is

$$P_{dm} = \frac{V_1 E_\phi}{Z_s^2} \left(X_s \frac{X_s}{Z_s} + R_1 \frac{R_1}{Z_s} \right) - \frac{E_\phi^2 R_1}{Z_s^2}$$

$$= \frac{V_1 E_\phi}{Z_s^2} \frac{R_1^2 + X_s^2}{Z_s} - \frac{E_\phi^2 R_1}{Z_s^2}$$

or
$$P_{dm} = \frac{V_1 E_\phi}{Z_s} - \frac{E_\phi^2 R_1}{Z_s^2} \tag{7.23}$$

If the maximum power developed by a three-phase synchronous motor is known, we can determine the necessary induced voltage in the motor. The maximum power developed equation can be rewritten as

$$E_\phi^2 - V_1 \frac{Z_s}{R_1} E_\phi + \frac{Z_s^2}{R_1} P_{dm} = 0$$

Thus,
$$E_\phi = \frac{V_1 Z_s}{2R_1} \pm \frac{1}{2} \sqrt{\left(\frac{V_1 Z_s}{R_1}\right)^2 - \frac{4 P_{dm} Z_s^2}{R_1}}$$

or
$$E_\phi = \frac{Z_s}{2R_1} [V_1 \pm \sqrt{V_1^2 - 4R_1 P_{dm}}] \tag{7.24}$$

Equation (7.24) gives us two values for E_ϕ and these are the excitation limits for any load.

We can also determine the criterion for the maximum power developed as a function of the induced voltage E_ϕ by differentiating Eq. (7.18) with respect to E_ϕ and setting it equal to zero. By doing so, we obtain

$$V_1[R_1 \cos \delta + X_s \sin \delta] = 2E_\phi R_1$$

or
$$E_\phi = \frac{V_1}{2R_1} [R_1 \cos \delta + X_s \sin \delta] \tag{7.25}$$

Thus, the maximum power developed is

$$P_{dm} = \frac{V_1^2}{4R_1 Z_s^2} [R_1 \cos \delta + X_s \sin \delta]^2$$

$$= \frac{V_1^2}{4R_1} \left[\frac{R_1}{Z_s} \cos \delta + \frac{X_s}{Z_s} \sin \delta \right]^2 \tag{7.26}$$

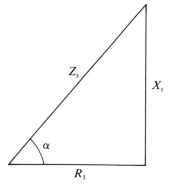

FIGURE 7.7
Impedance
Diagram of a
Synchronous
Motor

But from the impedance diagram (Fig. 7.7),

$$R_1 = Z_s \cos \alpha \quad \text{and} \quad X_s = Z_s \sin \alpha$$

Therefore, the maximum power developed becomes

$$P_{dm} = \frac{V_1^2}{4R_1} \cos^2(\alpha - \delta) \tag{7.27}$$

It is obvious that the power is maximum when $\alpha = \delta$. Therefore, the maximum power developed by a three-phase synchronous motor when E_ϕ is a variable becomes

$$P_{dm} = \frac{V_1^2}{4R_1} \tag{7.28}$$

and the corresponding value of the induced voltage is

$$E_\phi = \frac{V_1 Z_s}{2R_1}$$

This is the value of the induced voltage that gives the maximum power but is not the maximum value that can be induced in the stator winding.

■ EXAMPLE 7.5

A 120-V, 60-Hz, three-phase, Y-connected synchronous motor has a stator impedance of $0.2 + j3$ Ω per phase. When operating on full load, the stator line current is 10 A at a leading power factor of 0.8. Determine the maximum power developed by the motor if the field excitation is unchanged.

Solution

The per phase voltage is

$$V_1 = \frac{120}{\sqrt{3}} = 69.28 \text{ V}$$

The induced emf is

$$\begin{aligned}
\tilde{E}_\phi &= \tilde{V}_1 - \tilde{I}_1(R_1 + jX_s) \\
&= 69.28 - (10\underline{/36.87°}) \times (0.2 + j3) \\
&= 89.31\underline{/-16.39°} \text{ V}
\end{aligned}$$

The maximum power developed per phase is

$$\begin{aligned}
P_{dm} &= \frac{V_1 E_\phi}{Z_s} - \frac{E_\phi^2 R_1}{Z_s^2} \\
&= \frac{69.28 \times 89.31}{3.007} - \frac{89.31^2 \times 0.2}{3.007^2} \\
&= 1881.2 \text{ W per phase}
\end{aligned}$$

The total maximum power developed by the motor is $3 \times 1881.2 = 5643.7$ W.

■ EXAMPLE 7.6

A 440-V, 60-Hz, 4-pole, three-phase, Δ-connected synchronous motor has a synchronous impedance of $1.2 + j8$ Ω per phase. The induced line voltage is 560 V. Determine the line current, power factor, and torque angle when the motor is delivering maximum power. What is the maximum torque developed by the motor?

Solution

For the motor to operate at maximum power

Torque angle: $\delta = \tan^{-1}\left(\dfrac{X_s}{R_1}\right) = \tan^{-1}\left(\dfrac{8}{1.2}\right) = 81.47°$

Therefore, $\tilde{E}_\phi$ lags $\tilde{V}_1$ by 81.47°.

$$\tilde{V}_1 = \tilde{E}_\phi + \tilde{I}_1(R_1 + jX_s)$$

or
$$\tilde{I}_1 = \frac{\tilde{V}_1 - \tilde{E}_\phi}{R_1 + jX_s}$$

$$= \frac{440 - 560\underline{/-81.47°}}{1.2 + j8}$$

$$= 81.44\underline{/-24.27°} \text{ A}$$

Thus the line current is $I_l = \sqrt{3} \times 81.44 = 141.06$ A.

Power factor:

$$\cos 24.27° = 0.912$$

Total power input:

$$P_{\text{in}} = \sqrt{3}V_lI_l \cos\theta = \sqrt{3} \times 440 \times 141.06 \times \cos(24.27) = 98,000 \text{ W}$$

Stator copper loss:

$$\text{SCL} = 3I_1^2R_1 = 3 \times 81.44^2 \times 1.2 = 23,877 \text{ W}$$

Maximum power developed:

$$P_{dm} = 98,000 - 23,877 = 74,123 \text{ W}$$

The synchronous speed is

$$N_s = \frac{120f}{p} = \frac{120 \times 60}{4} = 1800 \text{ rpm}$$

or
$$\omega_s = \frac{2\pi N_s}{60} = \frac{2\pi \times 1800}{60} = 188.5 \text{ rad/s}$$

Thus, the maximum torque developed by the motor is

$$T_{dm} = \frac{P_{dm}}{\omega_s} = \frac{74,122}{188.5} = 393.2 \text{ N-m}$$

■ EXAMPLE 7.7

A 440-V, 60-Hz, 6-pole, three-phase, Δ-connected synchronous motor draws 86.6 A on full load. The friction, windage, and core losses are 3 kW, while the field excitation loss is 1 kW. The field current is adjusted to obtain an open-circuit voltage equal to the applied voltage. The synchronous reactance is 5 Ω per phase, and the winding resistance is negligible. Determine the (a) torque angle, (b) power factor, (c) power input, (d) power output, (e) efficiency, and (f) net torque developed by the motor.

Solution

Since the induced voltage is the same as the applied voltage, the power factor angle of the motor must be lagging. The phase current is

$$I_1 = \frac{86.6}{\sqrt{3}} = 50 \text{ A}$$

$$\tilde{V}_1 = \tilde{E}_\phi + j\tilde{I}_1 X_s$$

or

$$\tilde{I}_1 = \frac{\tilde{V}_1 - \tilde{E}_\phi}{jX_s}$$

Thus, $$I_1^2 = \frac{V_1^2 + E_\phi^2 - 2V_1 E_\phi \cos \delta}{X_s^2}$$

$$50^2 = \frac{440^2 + 440^2 - 2 \times 440 \times 440 \cos \delta}{5^2} \Rightarrow \delta = 33°$$

(a) Therefore, the power angle is $\delta = 33°$. Thus,

$$\tilde{I}_1 = \frac{\tilde{V}_1 - \tilde{E}_\phi}{jX_s}$$

$$= \frac{440 - 440\underline{/-33°}}{5\underline{/90°}} = 50\underline{/-16.46°} \text{ A}$$

(b) Thus, the power factor is pf $= \cos(16.46) = 0.96$ (lag).

(c) The power input is

$$P_{in} = \sqrt{3}V_l I_l \cos \theta$$

$$= \sqrt{3} \times 440 \times 86.6 \times 0.96 = 63,358 \text{ W}$$

(d) Since the stator winding resistance is negligible, power developed is equal to the power input. That is,

$$P_d = 63,358 \text{ W}$$

(e) The power output is

$$P_o = 63,358 - 3000 = 60,358 \text{ W}.$$

(f) The motor speed is

$$N_s = \frac{120f}{P} = \frac{120 \times 60}{6} = 1200 \text{ rpm}$$

or

$$\omega_s = \frac{2\pi N_s}{60} = \frac{2\pi \times 1200}{60} = 125.66 \text{ rad/s}$$

The net torque developed is

$$T_s = \frac{P_o}{\omega_s} = \frac{60,358}{125.66} = 480.33 \text{ N-m}$$

(g) For the purpose of computing the efficiency of the motor, let us add the field excitation loss to the power input of the motor. Thus, the modified power input is

$$P_{\text{in}} = 63,358 + 1000 = 64,358 \text{ W}$$

Thus, efficiency is

$$\eta = \frac{P_o}{P_{\text{in}}} = \frac{60,358}{64,358} = 0.938 \quad \text{or} \quad 93.8\%$$

■ EXAMPLE 7.8

A 50-hp, 1200-V, three-phase, Δ-connected synchronous motor is 95% efficient on full load with a power angle of 20°. The synchronous reactance of the motor is 15 Ω per phase, while the winding resistance is negligible. Determine the line current and the power factor of the motor.

Solution

The power input is

$$P_{\text{in}} = \frac{P_o}{\eta} = \frac{50 \times 746}{0.95} = 39,263 \text{ W}$$

Since the winding resistance is negligible, the power developed is equal to the power input. Thus,

$$P_d = 39,263 \text{ W}$$

However, the power developed by the motor is given as

$$P_d = \frac{V_1 E_\phi \sin \delta}{X_s}$$

Thus,

$$E_\phi = \frac{P_d X_s}{V_1 \sin \delta} = \frac{39{,}263 \times 15}{1200 \times \sin 20} = 1435 \text{ V}$$

However,

$$\tilde{I}_1 = \frac{\tilde{V}_1 - \tilde{E}_\phi}{jX_s} = \frac{1200 - 1435\underline{/-20°}}{15\underline{/90°}}$$

$$= 34.18\underline{/16.8°} \text{ A}$$

The line current is

$$I_l = \sqrt{3} \times I_1 = \sqrt{3} \times 34.18 = 59.2 \text{ A}$$

and the power factor is

$$\text{pf} = \cos(16.8) = 0.957 \text{ (lead)}$$

7.4 EFFECT OF EXCITATION

No-load Condition Let us consider a synchronous motor with negligible arma-ture-winding resistance operating under no-load conditions. Under ideal and nor-mal excitation, without any losses, the current in the stator winding must be zero. Under such normal conditions, the induced voltage must be equal to the applied voltage, and the power angle must be zero because the power output is zero. The normal excitation condition is shown in Fig. 7.8a. In this case, the motor is said to be idling because there is no power input and there is no power output. How-ever, if we now try to increase the field current in the rotor winding, the induced voltage will also increase. In this case, the motor is said to overexcited, which will result in a current flow as given by

$$\tilde{I}_1 = j\left(\frac{\tilde{E}_\phi - \tilde{V}_1}{X_s}\right) \tag{7.29}$$

The current I_1 is leading the phasor, $\tilde{E}_\phi - \tilde{V}_1$, by 90°. The magnitude of the current depends on the level of overexcitation. Since the motor is running at no load, the power angle is still zero. Figure 7.8b shows the corresponding phasor diagram. The active power input is still zero because $\theta = 90°$. However, the motor now has some reactive power, which is equal to $3V_1 I_1$. These are the character-

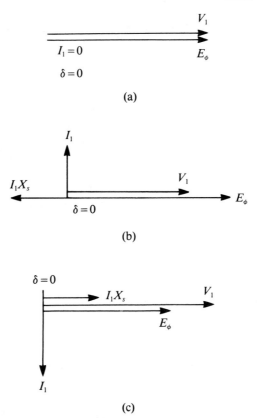

FIGURE 7.8
No-Load
Operation of a
Synchronous
Motor under
Ideal Conditions
for (a) Normal
Excitation; (b)
Overexcitation;
and (c)
Underexcitation

istics of a capacitor. Thus, a synchronous motor behaves like a capacitor when it has a leading power factor. A synchronous motor operating as a capacitor is known in the industry as a *synchronous condenser*.

On the other hand, if the field current is smaller than its normal value, the synchronous motor is said to be underexcited. In this case, the induced voltage is smaller than the applied voltage and there is some current in the stator winding. That is,

$$\tilde{I}_1 = -j \frac{\tilde{V}_1 - \tilde{E}_\phi}{X_s} \tag{7.30}$$

The presence of $-j$ indicates that the current is lagging by 90°. The real power input is still zero. The reactive power is again $3V_1I_1$. The synchronous motor now behaves as an inductor. The phasor diagram for an underexcited synchronous motor operating at no load is given in Fig. 7.8c.

It is therefore obvious that a change in the level of excitation affects the reactive power only. There is no change in the real power flow. The changes in the real power flow can only be accomplished by changing the load on the motor. These simple conclusions are drawn on the basis of no-load conditions when the power (torque) angle is essentially zero.

Load Condition   Let us now look at the condition when the motor is supplying some constant power. Since the power output is constant, the power input to the motor must also be constant. However, the power input is $3V_1I_1 \cos \theta$, and V_1 is constant; thus, the product $I_1 \cos \theta$ must be constant. But $I_1 \cos \theta$ is a projection of the current phasor I_1 on the voltage phasor; a line perpendicular to the voltage phasor, as shown in the phasor diagram (Fig. 7.9) will result in a constant magnitude of $I_1 \cos \theta$. In other words, when I_1 changes, the power factor must adjust itself in such a way that $I_1 \cos \theta$ is constant for constant power input. However, the current I_1 is

$$\tilde{I}_1 = \frac{\tilde{V}_1 - \tilde{E}_\phi}{jX_s} \tag{7.31}$$

Thus, as I_1 changes its magnitude when it moves from underexcitation to overexcitation, the phasor $\tilde{V}_1 - \tilde{E}_\phi$ should also change its magnitude and direction because I_1 lags it by 90°. Since $I_1X_s \cos \theta = E_\phi \sin \delta$ and $I_1 \cos \theta$ is constant, $E_\phi \sin \delta$ should also be constant. Therefore, the projection of the phasor $\tilde{V}_1 - \tilde{E}_\phi$ onto the voltage phasor must also be constant. In other words, the tip of the phasor $\tilde{E}_\phi$ must trace a line parallel to the voltage phasor as it changes its magnitude from one excitation level to another. Thus, the torque angle δ must also change. Such changes for three different excitation levels are depicted in Fig. 7.9.

V-Curves   As the current phasor moves from a lagging power factor caused by underexcitation to leading power factor caused by overexcitation, its magnitude first decreases until the power factor becomes unity and then increases again. If

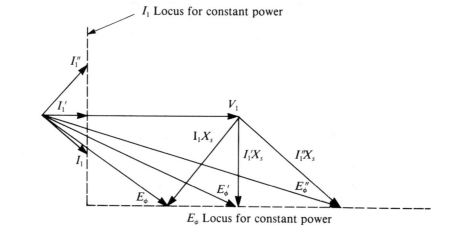

FIGURE 7.9
Loci of I_1 and E_ϕ
for Constant
Power
Operation

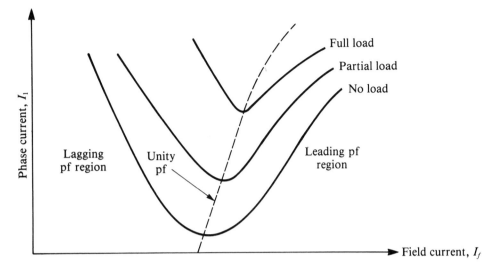

FIGURE 7.10
V-Curves for a
Synchronous
Motor

this magnitude of the stator current is plotted as a function of the field winding current for various loading conditions, a set of curves as shown in Fig. 7.10 will be obtained. Such curves are known as V-curves for the synchronous motor because they are shaped like the letter V. From the curves it is apparent that, as the load on the motor increases, the unity power factor line slants to the right. For the motor, on the right of the unity power factor line are the values of the currents when the motor is operating with leading power factor. The power factor is lagging on the left of the unity power factor line. We can also plot the variation in the power factor as a function of field current under different loads. Figure 7.11 shows the nature of these curves. These curves also indicate that the syn-

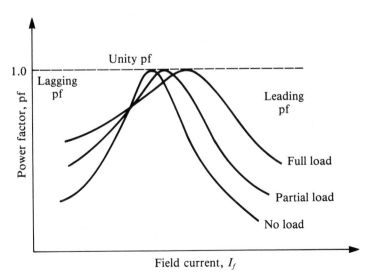

FIGURE 7.11
Power Factor
Characteristic of
a Synchronous
Motor

chronous motor can be operated at leading power factors under overexcited conditions. However, an increase in the field current also results in an increase in the stator current. Therefore, the operation under the overexcited condition must be below the maximum current ratings of the motor.

Such curves can also be obtained for synchronous generators. In this case, normal excitation results in a unity power factor. Overexcitation, the right side of the curve, results in a lagging power factor, and underexcitation, the left side of the curve, corresponds to a leading power factor.

7.5 POWER FACTOR CORRECTION

It was pointed out in the preceding sections that a synchronous motor with a leading power factor acts like a capacitor. This basic characteristic of the motor is exploited to improve the overall power factor of a plant or an industry. Most often, due to the use of induction motors, the overall power factor of a plant is lagging. If the power factor falls below a certain limit established by the power company, the plant will be assessed some surcharge for the use of power at those low power factors. To improve the low power factor, a synchronous motor with a leading power factor can be employed. A high-efficiency synchronous motor operating under no load as a synchronous condenser requires very little real power and supplies high reactive power with a power factor that is almost leading by 90°. However, in many cases an induction motor with a lagging power factor can be replaced by a synchronous motor with a leading power factor.

Let us assume that the total plant load is $\hat{S}$ with a lagging power factor angle θ as shown in Fig. 7.12. The real power requirement of the load is $P = S \cos \theta$, and the reactive power supplied to the load is $Q = S \sin \theta$.

Let us now add a synchronous motor with apparent power requirements of $\hat{S}_m$ at a leading power factor angle, θ_m. The real power intake by the synchronous motor is $P_m = S_m \cos \theta_m$, and the quadrature power delivered to the power

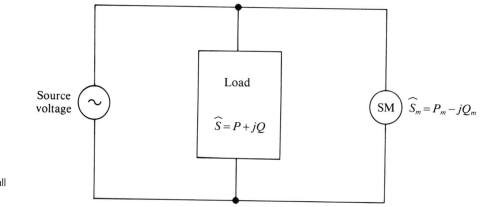

FIGURE 7.12
Connecting Synchronous Motor to Improve Overall Power Factor

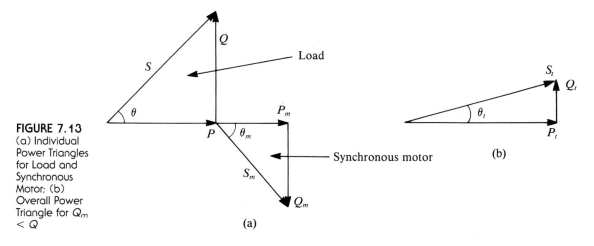

FIGURE 7.13
(a) Individual
Power Triangles
for Load and
Synchronous
Motor; (b)
Overall Power
Triangle for Q_m
$< Q$

system by the motor is $Q_m = S_m \sin \theta_m$. The overall power requirements are now

Real power: $$P_t = P + P_m \qquad (7.32)$$

Reactive power: $$Q_t = Q - Q_m \qquad (7.33)$$

Thus the apparent power requirements are now

$$\hat{S}_t = P_t + jQ_t \qquad (7.34)$$

and the new power factor is

$$\text{pf} = \cos \theta_t = \frac{P_t}{S_t} \qquad (7.35)$$

as indicated in Fig. 7.13 for $Q_m < Q$.

Note that if Q is less than Q_m the overall power factor will be leading. This requirement does not need to be satisfied. However, if Q_m is equal to Q, the power factor for the entire load will be unity. This is the ideal requirement for minimum energy consumption. An inductive load with lower power factor imposes high kVA requirements on the synchronous motor. Therefore, in most cases Q_m is usually smaller than Q. The overall power factor, although still lagging, is improved by adding a synchronous motor.

■ **EXAMPLE 7.9**

A manufacturing plant uses 10 kVA at 0.6 power factor lagging under normal operation. To improve the overall power factor, a synchronous motor is added to the system. The motor draws 1 kW. Determine the new power factor of the plant if the motor operates at (a) unity power factor, and (b) 0.8 leading power factor. What must the power factor of the motor be to improve the overall power factor of the plant to 0.85 lagging?

Solution

The plant requirements under operation are

$$P = S \cos \theta = 10 \times 0.6 = 6 \text{ kW}$$

$$Q = S \sin \theta = 10 \times \sqrt{1 - 0.6^2} = 8 \text{ kVAR}$$

(a) Synchronous motor operation at unity power factor is

$$P_m = 1 \text{ kW} \quad \text{and} \quad Q_m = 0$$

Thus, the overall plant requirements are

$$P_t = 6 + 1 = 7 \text{ kW}$$

and

$$Q_t = 8 - 0 = 8 \text{ kVAR}$$

$$S_t = 7 + j8 = 10.63\underline{/48.8°} \text{ kVA}$$

and the power factor is $\cos(48.8) = 0.66$ (lagging).

(b) Motor operation at 0.8 power factor leading is

$$P_m = 1 \text{ kW}$$

$$S_m = \frac{P_m}{\cos \theta_m} = \frac{1 \text{ kW}}{0.8} = 1.25 \text{ kVA}$$

$$Q_m = 1.25\sqrt{1 - 0.8^2} = 0.75 \text{ kVAR}$$

Thus, overall plant requirements are

$$P_t = 6 + 1 = 7 \text{ kW} \quad \text{and} \quad Q_t = 8 - 0.75 = 7.25 \text{ kVAR}$$

$$S_t = 7 + j7.25 = 10.08\underline{/46.0°} \text{ kVA} \quad \text{and} \quad \text{pf} = \cos(46) = 0.695$$

Overall plant pf of 0.85 lagging (Fig. 7.13):

$$\cos \theta_t = 0.85 \text{ (lag)}, \qquad \theta_t = 31.79°, \qquad \sin \theta_t = \sin(31.79°) = 0.527$$

$$P_t = 7 \text{ kW}$$

$$Q_t = P_t \tan(31.79) = 7 \times 0.62 = 4.34 \text{ kVAR}$$

Thus,
$$Q_m = Q - Q_t = 8 - 4.34 = 3.67 \text{ kVAR}$$
$$S_m = 1 - j3.67 = 3.804\underline{/-74.75°} \text{ kVA}$$

The power factor of the motor is $\cos(74.75°) = 0.263$ leading.

PROBLEMS

7.1. A 100-hp, 2300-V, 50-Hz, three-phase, Δ-connected synchronous motor delivers full load at a power factor of 0.8 leading. The synchronous reactance is 10 Ω per phase, the rotational loss is 10 kW, and the field excitation loss is 5 kW. Determine (a) the armature current, (b) the motor efficiency, and (c) the power angle. Assume that the stator-winding resistance is negligible.

7.2. A 10-hp, 230-V, 60-Hz, 6-pole, three-phase, Y-connected synchronous motor has a stator impedance of $0.5 + j10$ Ω per phase. The friction and windage loss is 450 W, the core loss is 250 W, and the field excitation loss is 100 W. Determine the stator current, efficiency, power angle, and net torque developed by the motor when it runs on full load with unity power factor.

7.3. A 1200-V, three-phase, Y-connected synchronous motor takes 110 kW (exclusive of field-winding loss) when operated under a certain load at 1200 rpm. The back emf of the motor is 2000 V. The synchronous reactance is 10 Ω per phase with negligible winding resistance. Calculate the line current and the torque developed by the motor.

7.4. A 230-V, three-phase, Δ-connected synchronous motor takes 86.6 A on full load. The synchronous impedance of the motor is $0.5 + j6$ Ω per phase. What is the nominal rating of the motor? Determine the back emf and the torque angle of the motor when it operates at (a) unity power factor, (b) 0.8 leading power factor, and (c) 0.8 lagging power factor.

7.5. A 440-V, 50-Hz, three-phase, Y-connected synchronous motor has a synchronous impedance of $0.4 + j4$ Ω per phase. Under full-load conditions the motor draws 25 A at a leading power factor of 0.707. Determine the maximum power developed by the motor if the field excitation is held constant.

7.6. A 1000-V, 50-Hz, 4-pole, three-phase, Y-connected synchronous motor has a synchronous impedance of $0.2 + j3$ Ω per phase. The induced line voltage is 1200 V. Determine the line current, power factor, and power angle when the motor delivers maximum power. What torque is developed by the motor?

7.7. A 460-V, 50-Hz, 4-pole, three-phase, Y-connected synchronous motor takes 40 A on full load. The rotational losses are 2500 W, and the field excitation loss is 1500 W. The field current is adjusted to obtain an open-circuit line voltage of 460 V. The synchronous reactance is 4 Ω per phase, and the winding resistance is negligible. Determine the efficiency and the net torque developed by the motor.

7.8. A 10-hp, 460-V, three-phase, Y-connected synchronous motor is 96% efficient on full load with a power angle of 12° (electrical). The synchronous reactance is 10 Ω per phase, while the resistance is negligible. Determine the line current and the power factor of the motor. Neglect the loss in the field winding.

7.9. A 1200-V, 50-Hz, 8-pole, three-phase, Y-connected synchronous motor has a synchronous impedance of 2 + j10 Ω per phase. The field excitation is adjusted to obtain an open-circuit line voltage of 1000 V. What is the maximum torque developed by the motor?

7.10. A 230-V, 2-hp, three-phase, Δ-connected synchronous motor has a synchronous impedance of 0.5 + j5 Ω per phase. The motor efficiency is 90% on full load. If the field excitation is adjusted in such a way that the motor draws the minimum current on full load, determine the induced voltage.

7.11. The per phase impedance of a 600-V, three-phase, Y-connected synchronous motor is 5 + j50 Ω. The motor takes 24 kW at a leading power factor of 0.707. Determine the induced voltage and the power angle of the motor.

7.12. The energy consumption in an industry is 100 kVA at a 0.45 lagging power factor. A synchronous motor with a 0.2 leading power factor is added to the load. The power intake of the synchronous motor is 10 kW. Determine the total power consumption and the overall power factor.

7.13. A manufacturing plant uses 20 kVA at 0.5 power factor lagging. It is desired to add a synchronous motor to improve the overall factor to 0.707 lagging. The input power to the synchronous motor is 2 kW. Determine the power factor of the motor.

CHAPTER 8

Polyphase Induction Motor

Three-Phase Induction Motor (Courtesy of
Franklin Electric)

An induction motor, as the name suggests, transfers the electrical power from the stationary member, the stator, to the rotating member, the rotor, completely by electromagnetic induction. An induction motor, therefore, is a singly fed machine as compared to doubly-fed dc and synchronous motors. Both dc and synchronous motors require the power input to both stationary and rotating members, while in the case of an induction motor the rotor is completely isolated from the stator as far as direct electrical connections are concerned. The currents are induced in the rotor windings by electromagnetic induction only. This is exactly the same principle we used in the study of the transformer. Therefore, an induction motor may be labeled as a transformer with a rotating secondary winding. In fact, we will make use of this similarity in developing the equivalent circuit of an induction motor.

Two basic types of induction motors are in regular use: (1) single-phase induction motors, and (2) polyphase induction motors. Single-phase induction motors are usually favored for domestic applications. These induction motors are normally available in fractional-horsepower range. On the other hand, the polyphase induction motors cover the entire spectrum of horsepower ratings and are preferably installed at locations where a polyphase power source is easily accessible.

Why is a polyphase induction motor more desirable than a single-phase motor? We will try to answer this question quantitatively later. At this time, it will suffice to say that a polyphase induction motor is superior than its single-phase counterpart because:

1. It is self-starting.
2. It usually has higher starting torque.
3. It is more reliable.
4. It is more efficient.

A polyphase induction motor may be equipped with a wound rotor or a squirrel-cage-type rotor. Later, we will examine the advantages and disadvantages of both types of rotors.

8.1 ESSENTIAL CONSTRUCTION DETAILS

A three-phase induction motor mainly consists of the following:

1. A stationary part, called the stator
2. A rotating part, called the rotor
3. The end plates or brackets, which are fastened to the stator by means of screws or bolts

Stator   The stator of a three-phase induction motor is similar to that of a three-phase synchronous machine. For further information, refer to Chapter 6.

Rotor The rotor is composed of thin-slotted steel laminations that are pressed together onto a shaft. The rotor laminations, especially for the die-cast rotors, are usually skewed in order to reduce cogging and electrical noise in the motor, as explained later.

For wound rotors, the winding must have as many poles and phases as the stator. For example, a three-phase, 4-pole induction motor must have a rotor that is also wound for 4 poles, and each pole must have three phases. The three-phase windings are usually symmetrical, and one end of each winding is internally connected to form a common connection. This is known as the *internal neutral* connection. The remaining three terminals can also be connected together. However, to control the performance of the motor, the remaining three terminals may be connected to the slip rings on the shaft of the rotor. With the help of brushes that ride on the slip rings, external resistances can be added to the rotor circuit. In this way the total resistance in the rotor circuit can be controlled. This technique provides a means to obtain the desired torque from a motor because, as we will see later, the torque developed by a three-phase induction motor is proportional to its rotor resistance.

8.2 PRINCIPLE OF OPERATION

When the windings of a three-phase induction motor are properly connected to a three-phase power source, a revolving field is produced by these windings. For details on how the revolving field is produced, see Appendix B. Let N_s be the speed at which the stator field revolves. This is also known as the synchronous speed of the motor. Due to flux-cutting action by the conductors on the rotor, the voltages are induced in them. Since the conductors on the rotor form a closed loop, the induced voltage results in the flow of current in those conductors. Whenever a current-carrying conductor is immersed in a magnetic field, it experiences a force that tends to rotate it. Based on this principle, the rotor of an induction motor starts rotating and achieves a speed that is very nearly equal to the synchronous speed of the stator under no-load conditions. However, the rotor of an induction motor can never rotate at the same speed as the field set up by the stator. If the rotor were to rotate at the synchronous speed, the rotor conductors would appear stationary with respect to the stator magnetic field and there would be no induced voltage in them. In the absence of the induced voltage, there would be no rotor current and thereby no force acting on the rotor conductors. Thus, the rotor will tend to slow down as a result of rotational losses due to friction and windage. Due to this reason, induction motors are also referred to as *asynchronous motors*.

Let N_m (ω_m) be the speed of rotation of the rotor. With respect to the rotating conductors on the rotor, the field is moving ahead at a relative speed of

$$N_r = N_s - N_m$$

or

$$\omega_r = \omega_s - \omega_m \qquad (8.1)$$

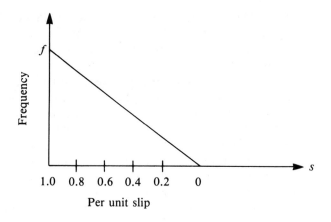

FIGURE 8.1
Slip-Frequency
Relationship for
a Rotor in a
Three-Phase
Induction Motor

Most often the relative speed is defined on a per unit basis. In that case we usually refer to it as slip or per unit slip, which is defined as the ratio of relative speed to the synchronous speed. That is,

$$s = \frac{N_r}{N_s} = \frac{\omega_n}{\omega_s}$$

or

$$s = \frac{N_s - N_m}{N_s} = \frac{\omega_s - \omega_n}{\omega_s} \tag{8.2}$$

In terms of the synchronous speed, the rotor speed is

$$N_m = (1 - s)N_s$$

or

$$\omega_m = (1 - s)\omega_s \tag{8.3}$$

Thus, s is the per unit slip with which the rotor is slipping behind the stator magnetic field. With respect to a stationary frame of axes, the stator field revolves at a constant speed, N_s or ω_s. The rotor is also rotating at a speed of N_m (ω_m). Since the stator magnetic field links the rotor, the magnetic field linking the rotor also rotates at the synchronous speed. However, with respect to the rotor, the rotating field produced by the rotor rotates at a relative speed of N_r or ω_r. This difference in speed is responsible for the induced voltage in the rotor conductors and thereby the production of electromagnetic torque.

Let P be the number of poles for the three-phase induction machine and f be the frequency of the three-phase power source; then the synchronous speed with which the stator flux revolves around the rotor conductor may be computed from

$$f = \frac{PN_s}{120}$$

or
$$f = \frac{P\omega_s}{4\pi} \tag{8.4}$$

These are the same equations we developed for the synchronous machines. If N_r (or ω_r) is the relative speed of the induced volltage in the rotor conductors, the frequency of the induced voltage will be

$$\begin{aligned} f_r &= \frac{PN_r}{120} \\ &= \frac{P}{120}(N_s - N_m) \\ &= \frac{P}{120}\frac{(N_s - N_m)}{N_s}N_s \\ &= sf \end{aligned} \tag{8.5}$$

The per unit slip in induction motors is always greater than zero but less than or equal to 1, because the rotor speed can change from zero to almost synchronous speed. Figure 8.1 shows the slip–frequency characteristic of the rotor. The rotor frequency is usually low as compared to the frequency of the applied voltage at rated speed. Hence, the core losses in the rotor magnetic circuit are most often ignored.

■ EXAMPLE 8.1

A 4-pole, three-phase, 230-V, 60-Hz induction motor runs at a speed of 1725 rpm under full-load conditions. Calculate the per unit slip and the rotor frequency at its rated speed.

Solution

The synchronous speed of the motor is

$$N_s = \frac{120f}{P} = \frac{120 \times 60}{4} = 1800 \text{ rpm}$$

The per unit slip is

$$s = \frac{N_s - N_m}{N_s} = \frac{1800 - 1725}{1800} = 0.0417$$

Thus, the rotor frequency is

$$f_r = sf = 0.0417 \times 60 = 2.5 \text{ Hz.}$$

8.3 DEVELOPMENT OF AN EQUIVALENT CIRCUIT

In a balanced three-phase motor, all the phase windings under each pole are exactly the same. Therefore, such a motor can be completely represented by its equivalent circuit on a per phase basis. When the rotor is held stationary (blocked rotor or standstill condition) by applying an external torque to the shaft that is greater than the starting torque developed by the motor, the per unit slip of the motor is unity. In that case, the relative frequency of the rotor with which the voltages and thereby the currents are induced in the rotor conductors is the same as that of the applied voltage to the stator windings. With this understanding, the basic equivalent circuit of an induction motor should be no different than that of the transformer. Such an equivalent circuit is shown in Fig. 8.2, where the secondary winding has been shorted to represent the closed circuit for the rotor windings. In this equivalent circuit,

V_1 = supply voltage on a per phase basis in volts

r_1 = per phase resistance of the stator winding

x_1 = per phase leakage reactance of the stator winding

f = frequency of the applied voltage in hertz

r_b = per phase rotor resistance, blocked rotor condition

x_b = per phase rotor reactance, blocked rotor condition

E_1 = per phase induced voltage in stator windings

E_b = per phase voltage induced in rotor circuit

f_b = rotor frequency, blocked rotor condition = f

X_m = per phase magnetizing reactance

R_c = per phase equivalent core-loss resistance

The induced voltages in the rotor and the stator windings are determined in the same fashion as for synchronous machines. If N_1 and N_2 are the total turns and $k_{\omega 1}$ and $k_{\omega 2}$ are the effective winding factors for the stator and the rotor windings on a per phase basis, the induced voltages will be

In stator: $$E_1 = 4.44 f N_1 k_{\omega 1} \phi_m \tag{8.6}$$

In rotor: $$E_b = 4.44 f_b N_2 k_{\omega 2} \phi_m \tag{8.7}$$

Since the rotor frequency is directly proportional to the the slip, Eq. (8.5), at any slip s the induced voltage in the rotor and its leakage reactance will be sE_b

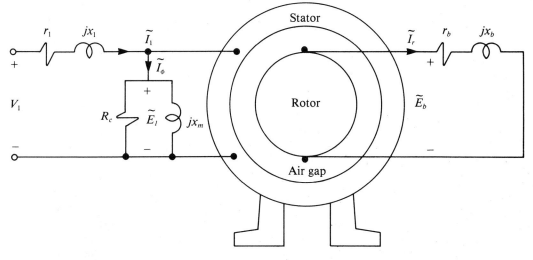

FIGURE 8.2
Schematic Representation of an Induction Motor on a Per Phase Basis under a Blocked-Rotor Condition

and sx_b, respectively. With these changes, we can now represent the equivalent circuit of a three-phase induction machine on a per phase basis when the rotor is rotating with per unit slip, s, as given in Fig. 8.3.

The current in the rotor circuit is

$$\tilde{I}_r = \frac{s\tilde{E}_b}{r_b + jsx_b} \tag{8.8}$$

However, we can express Eq. (8.8) as follows:

$$\tilde{I}_r = \frac{\tilde{E}_b}{(r_b/s) + jx_b} \tag{8.9}$$

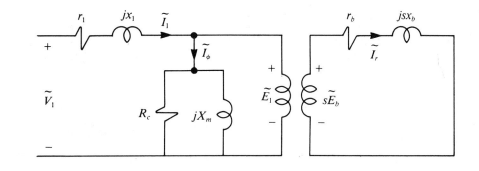

FIGURE 8.3
Per Phase
Equivalent
Circuit of an
Induction Motor
at any Slip s

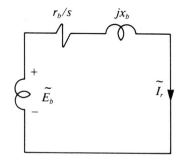

FIGURE 8.4
Modified
Equivalent
Circuit of the
Rotor at any
Slip s

Based on Eq. (8.9), we can develop another equivalent circuit of the rotor, as shown in Fig. 8.4. In this equivalent circuit, the induced voltage is now independent of the slip and so is the leakage reactance. However, we did introduce a hypothetical resistance, which is now inversely proportional to the slip. The hypothetical resistance will also be referred to as the effective resistance in the rotor circuit. As the slip approaches zero under no-load conditions, the effective resistance goes to infinity, which is electrically equivalent to an open circuit. On the other hand, under blocked rotor conditions the effective resistance is the same as the actual rotor resistance because the slip is unity. With these changes, the equivalent circuit of the motor is shown in Fig. 8.5.

Let us define a quantity similar to the transformation ratio, the a ratio, as the ratio of the effective turns in the stator to the effective turns in the rotor on a per phase basis. In other words, the a ratio is

$$a = \frac{N_1 k_{\omega 1}}{N_2 k_{\omega 2}} \tag{8.10}$$

We can now transform the equivalent circuit of the rotor as referred to the stator. Figure 8.6 shows such a transformed circuit, where

$$r_2 = a^2 r_b \tag{8.11}$$

$$x_2 = a^2 x_b \tag{8.12}$$

and $$\tilde{I}_2 = \tilde{I}_r/a \tag{8.13}$$

Phasor Diagram Figure 8.7 shows a complete phasor diagram for a three-phase induction motor on a per phase basis. In the phasor form, the basic equations used for the drawing of the phasor diagram are (1) for the rotor circuit,

$$\tilde{E}_2 = \tilde{I}_2 r_2/s + j\tilde{I}_2 x_2 \tag{8.14}$$

and (2) in the magnetization branch,

$$\tilde{I}_\phi = \tilde{I}_c + \tilde{I}_m \tag{8.15}$$

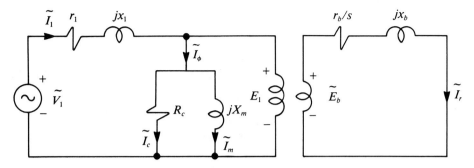

FIGURE 8.5
Modified
Equivalent
Circuit of an
Induction Motor

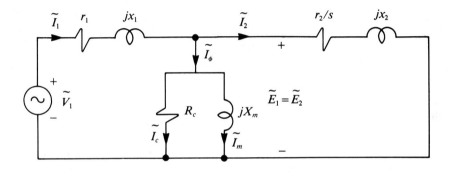

FIGURE 8.6
Exact Equivalent
Circuit of an
Induction Motor
as Referred to
the Stator

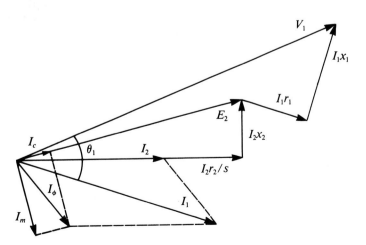

FIGURE 8.7
Phasor Diagram
for an Induction
Motor

where
$$\tilde{I}_c = \frac{\tilde{E}_2}{R_c} \quad \text{and} \quad \tilde{I}_m = \frac{\tilde{E}_2}{jX_m}$$

$\tilde{I}_c$ is in phase with $\tilde{E}_2$ and $\tilde{I}_m$ lags $\tilde{E}_2$ by 90°. The stator winding current is

$$\tilde{I}_1 = \tilde{I}_\phi + \tilde{I}_2 \tag{8.16}$$

Thus the applied voltage must be equal to

$$\tilde{V}_1 = \tilde{E}_2 + \tilde{I}_1 r_1 + j\tilde{I}_1 x_1 \tag{8.17}$$

8.3.1   Further Development of the Equivalent Circuit

In the preceding equivalent circuit for the induction motor, we introduced a hypothetical resistance in the equivalent rotor circuit. The actual resistance of the rotor as referred to the stator is r_2. The hypothetical resistance, since it is inversely proportional to the slip, must somehow or other represent the load on the motor in addition to the actual resistance of the rotor, because all the active power in the rotor circuit is dissipated by that resistance. The effective resistance can be rewritten as

$$\frac{r_2}{s} = r_2 + r_2 \frac{1-s}{s} \tag{8.18}$$

The additional resistance, $r_2[(1 - s)/s]$, is known as the *dynamic resistance or load resistance*. It depends on the speed of the motor and represents the load on the motor. In other words, it is the electrical equivalent of the mechanical load on the motor. The induction motor can now be modeled to incorporate the dynamic resistance as well. Such an equivalent-circuit representation of an induction machine is given in Fig. 8.8. In all these equivalent circuits, we have represented core loss by an equivalent resistance in parallel with the magnetizing reactance.

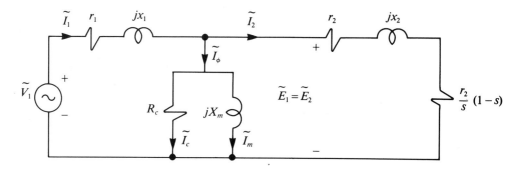

FIGURE 8.8
Equivalent Circuit of an Induction Motor in Terms of Load Resistance

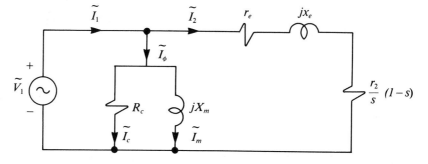

FIGURE 8.9
Approximate
Equivalent
Circuit of a
Balanced Three-
Phase Motor

The excitation branch, which is also known as the parallel branch of the equivalent circuit, is sometimes shifted so that it is directly across the supply. The stator-winding impedance and the equivalent rotor impedance can now be grouped together as illustrated in Fig. 8.9, where $r_e = r_1 + r_2$ and $x_e = x_1 + x_2$. This circuit is known as an approximate equivalent circuit of the three-phase induction motor on a per phase basis. The inaccuracy introduced is almost negligible under full-load conditions, but the calculations to determine the motor performance are immensely simplified. Sometimes the equivalent core-loss resistance is omitted and the core-loss is treated as a part of the rotational loss.

8.4 POWER RELATIONSHIPS

Under load, the motor adjusts its speed in such a way that the power loss in the dynamic resistance is equivalent to the torque developed by the motor. As the load on the motor changes, so does the speed and thereby the slip of the motor. The performance of the motor under load can be easily calculated from its equivalent circuit once the speed of rotation is known. Since we are considering only balanced three-phase motors, the calculations can be made on a per phase basis.

From the equivalent circuit (Fig. 8.6), the power input is

$$P_{\text{in}} = 3V_1I_1 \cos \theta_1 \tag{8.19}$$

Since the power input is electrical in nature, we must subtract the electrical losses first. The immediate electrical loss is the stator copper loss, or SCL for short. The total stator copper loss is

$$\text{SCL} = 3I_1^2 r_1 \tag{8.20}$$

If we subtract the stator copper loss from the input power, we are left with the power that is crossing the air gap of the motor and is transferred to the rotor by electromagnetic induction. The air-gap power is

$$P_g = 3V_1I_1 \cos \theta_1 - 3I_1^2 r_1 \tag{8.21a}$$

If the core-loss is modelled by an equivalent resistance, R_c, the power loss in that resistance must also be subtracted to obtain the air gap power. In this case,

$$P_g = 3V_1I_1 \cos \theta_1 - 3I_1^2 r_1 - 3I_c^2 R_c \tag{8.21b}$$

The air-gap power must be lost in the effective resistance of the rotor circuit. Thus,

$$P_g = 3I_2^2 \frac{r_2}{s} \tag{8.22}$$

The power loss in the actual rotor resistance is

$$RCL = 3I_2^2 r_2 = sP_g \tag{8.23}$$

The power developed by the motor is

$$P_d = P_g - RCL \qquad 3I_2^2 r_2 \qquad \frac{3I_2^2 r_2}{s} - 3I_2^2 R_2$$

$$r_2 \frac{1-s}{s}$$

$$= 3I_2^2 r_2 \left(\frac{1-s}{s}\right) = (1-s)P_g = SP_g \tag{8.24}$$

where $S = 1 - s = N_m/N_s$ is known as the per unit speed of the motor.

The torque developed by the motor is

$$T_d = \frac{P_d}{\omega_m}$$

$$= \frac{P_g}{\omega_s} = 3I_2^2 \frac{r_2}{s\omega_s} \tag{8.25}$$

It is evident that the torque developed is directly proportional to the square of the current and the equivalent effective resistance of the rotor. This is a very important relationship for the developed torque, which ties together the two quantities in the rotor circuit. These quantities are, in a sense, inversely related to each other. For example, if the rotor resistance is increased, the torque developed must increase linearly. But any increase in the rotor circuit resistance decreases the current in the rotor circuit and thereby causes a reduction in the torque developed by the motor. Whether the overall torque increases or decreases depends on which parameter plays a major role.

Let us try to examine the situation from standstill to a no-load condition. At standstill, the rotor slip is unity and the effective rotor resistance is r_2. As the rotor starts rotating, the slip starts decreasing, resulting in an increase in the hypothetical rotor resistance. Since other circuit parameters are unchanged, the

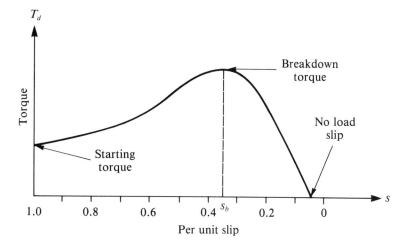

FIGURE 8.10
Typical Speed–
Torque Curve of
a Three-Phase
Induction Motor

overall increase in the total circuit impedance is small and results in very small reduction in the rotor current. Thus, in this speed range, the overall torque depends mainly on the rotor resistance and, therefore, will increase accordingly. As the slip keeps on decreasing further, the torque developed by the motor also keeps on increasing until the rotor resistance has greater influence on the current in the circuit.

Any further increase in the slip creates a rotor resistance many times higher than any other circuit parameters and it becomes the dominating factor. Under this condition we can simply neglect other circuit parameters in comparison to the hypothetical rotor resistance. The rotor current now is inversely proportional to the rotor resistance. The product of the square of the rotor current and the hypothetical rotor resistance becomes inversely proportional to the rotor resistance. Thus, the torque developed is now inversely proportional to the hypothetical rotor resistance. As this resistance increases, the developed torque falls. Under no-load conditions, the hypothetical rotor resistance is almost infinity and the torque developed by the motor becomes zero. It is therefore evident that at certain slip the torque developed by the motor becomes maximum; this is known as the *breakdown or maximum torque* of the motor. The slip at which the breakdown torque occurs is known as the *breakdown slip*. The speed–torque curve of a typical motor is shown in Fig. 8.10.

From the power developed, we can subtract the rotational losses to obtain the net power available at the shaft of the motor. The rotational losses include friction and windage, stray load loss if there is any, and may also include the core losses in the motor if these losses are not modelled by an equivalent resistance. The power output is

$$P_o = P_d - P_{rot} \tag{8.26}$$

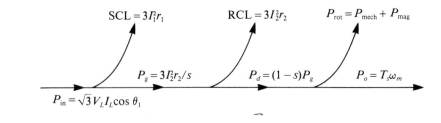

FIGURE 8.11
Power-Flow
Diagram of a
Three-Phase
Induction Motor

The power-flow diagram of the motor is shown in Fig. 8.11. The efficiency of the motor is

$$\eta = \frac{P_o}{P_{in}} \tag{8.27}$$

In a more elaborated fashion, the efficiency expression can be written as

$$\eta = \frac{3V_1I_1 \cos \theta_1 - 3I_1^2r_1 - 3I_2^2r_2 - P_{rot}}{3V_1I_1 \cos \theta_1} \tag{8.28}$$

■ **EXAMPLE 8.2**

A 10-hp, 4-pole, 440-V, 60-Hz, three-phase induction motor runs at 1725 rpm on full load. The stator copper loss is 212 W and the rotational loss is 340 W. Determine the (a) power developed, (b) air-gap power, (c) rotor copper loss, (d) total power input, and (e) efficiency of the motor.

Solution
Sychronous speed:

$$N_s = \frac{120f}{P} = \frac{120 \times 60}{4} = 1800 \text{ rpm}$$

Per unit slip:

$$s = \frac{N_s - N_m}{N_s} = \frac{1800 - 1725}{1800} = 0.0417$$

Power output:

$$P_o = 10 \times 746 = 7460 \text{ W}$$

Rotational power loss:

$$P_{rot} = 340 \text{ W (given)}$$

(a) The power developed is $P_d = P_o + P_{rot} = 7460 + 340 = 7800$ W.

(b) Since $P_d = (1 - s)P_g$, the air-gap power is

$$P_g = \frac{P_d}{1 - s} = \frac{7800}{1 - 0.0417} = 8139.41 \text{ W}$$

(c) The rotor copper loss is RCL = sP_g = 0.0417 × 8139.41 = 339.413 W, and the stator copper loss is SCL = 212 W (given).

(d) The power input is

$$P_{in} = P_g + SCL = 8139.41 + 212 = 8351.41 \text{ W}$$

(e) The efficiency is

$$\eta = \frac{P_o}{P_{in}} = \frac{7460}{8351.41} = 0.893 \quad \text{or} \quad 89.3\%$$

■ EXAMPLE 8.3

A 2-hp, 120-V, 60-Hz, 4-pole, three-phase induction motor operates at 1620 rpm on full load. The rotor impedance on a per phase basis referred to the stator is 0.02 + j0.06 Ω. What is the rotor current if the rotational losses are 160 W?

Solution

Power output: $P_o = 2 \times 746 = 1492 \text{ W}$

Synchronous speed: $N_s = \dfrac{120f}{P} = \dfrac{120 \times 60}{4} = 1800 \text{ rpm}$

Per unit slip: $s = \dfrac{N_s - N_m}{N_s} = \dfrac{1800 - 1620}{1800} = 0.1$

Power developed: $P_d = P_o + P_{rot} = 1492 + 160 = 1652 \text{ W}$

Air-gap power: $P_g = \dfrac{P_d}{1 - s} = \dfrac{1652}{1 - 0.1} = 1835.56 \text{ W}$

Rotor copper loss: $RCL = sP_g = 0.1 \times 1835.56 = 183.556 \text{ W}$

Thus, $3I_2^2 r_2 = 183.556$

$$I_2 = \sqrt{\frac{183.556}{3 \times 0.02}} = 55.31 \text{ A}$$

8.5 MAXIMUM EFFICIENCY CRITERION

The slip of the motor adjusts itself with the change in the mechanical load coupled to its shaft. This therefore causes a change in the current in the rotor circuit. If we differentiate the efficiency equation of the motor with respect to the rotor

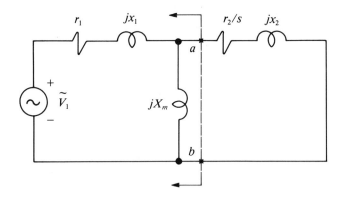

FIGURE 8.12
Equivalent
Circuit of an
Induction Motor
without
Equivalent Core-
Loss Resistance

current and set that equal to zero, we will arrive at a condition that should correspond to the maximum efficiency of the motor. However, the preceding equation for the efficiency of the motor also contains another variable current, $\tilde{I}_1$, which is equal to $\tilde{I}_2 + \tilde{I}_\phi$. As $\tilde{I}_2$ changes, so does the voltage across the magnetizing branch, thereby causing the change in the current in the magnetizing branch. We can, however, make the approximation that the change in I_ϕ is so small that it can be neglected or the magnetizing impedance is so high in comparison with all other circuit elements that it can be easily ignored. Instead of making such assumptions, let us take another look at the equivalent circuit of the motor.

Figure 8.12 shows an equivalent circuit of the motor where the magnetizing branch consists of nothing but the magnetizing reactance. It has been assumed in this circuit that the core losses are part of the rotational losses and therefore will be subtracted from the power developed by the motor. This equivalent circuit can be modified using Thevenin's theorem. Such a modified circuit is shown in Fig. 8.13. In this circuit,

$$V_{\text{Th}} = \frac{jV_1 X_m}{r_1 + j(x_1 + X_m)} \tag{8.29}$$

is the Thevenin's equivalent of the applied voltage, and $R_{\text{Th}} + jX_{\text{Th}}$ is the total impedance as viewed from terminals a and b of the equivalent circuit, where

$$Z_{\text{Th}} = R_{\text{Th}} + jX_{\text{Th}} = \frac{(r_1 + jx_1)jX_m}{r_1 + j(x_1 + X_m)} \tag{8.30}$$

Let $\tilde{I}_2$ be the current in the circuit. Then

$$\tilde{I}_2 = \frac{\tilde{V}_{\text{Th}}}{[R_{\text{Th}} + (r_2/s)] + j(X_{\text{Th}} + x_2)} \tag{8.31}$$

The total power input is

$$P_{in} = 3V_{Th}I_2 \cos \theta \qquad (8.32)$$

where θ is the angle between $\tilde{V}_{Th}$ and $\tilde{I}_2$. The power output becomes

$$P_o = 3V_{Th}I_2 \cos \theta - 3I_2^2(R_{Th} + r_2) - P_{rot} \qquad (8.33)$$

and the efficiency of the motor is

$$\eta = \frac{3V_{Th}I_2 \cos \theta - 3I_2^2(R_{Th} + r_2) - P_{rot}}{3V_{Th}I_2 \cos \theta} \qquad (8.34)$$

Now the only current variable is I_2. Thus, differentiating η with respect to I_2 and setting the derivative equal to zero, we obtain

$$3V_{Th}I_2 \cos \theta[3V_{Th} \cos \theta - 6I_2(R_{Th} + r_2)]$$
$$- 3V_{Th} \cos \theta[3V_{Th}I_2 \cos \theta - 3I_2^2(R_{Th} + r_2) - P_{rot}] = 0$$

or
$$3[I_2^2(R_{Th} + r_2)] = P_{rot} \qquad (8.35)$$

This expression states that the efficiency is maximum when the constant losses are equal to variable losses.

From Eq. (8.25), the torque developed can be expressed as

$$T_d = \frac{3V_{Th}^2(r_2/s)}{\omega_s\{[R_{Th} + (r_2/s)]^2 + (X_{Th} + x_2)^2\}} \qquad (8.36)$$

Thus, the torque developed by the motor is proportional to the square of the Thevenin's equivalent voltage. However, V_{Th} is proportional to V_1. Therefore, the electromagnetic torque developed by the motor is proportional to the square of the applied voltage.

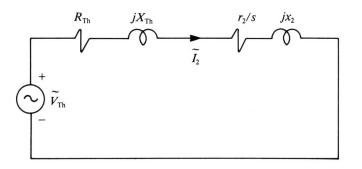

FIGURE 8.13
Thevenin's
Equivalent for
Figure 8.12

■ EXAMPLE 8.4

A 6-pole, 230-V, 60-Hz, Y-connected three-phase induction motor has the following parameters on a per phase basis: $r_1 = 0.5\ \Omega$, $r_2 = 0.25\ \Omega$, $x_1 = 0.75\ \Omega$, $x_2 = 0.5\ \Omega$, $X_m = 100\ \Omega$, and $R_c = 500\ \Omega$. The friction and windage loss is 150 W. Determine the stator current, magnetizing current, rotor current, power input, stator copper loss, rotor copper loss, power output, torque available at the shaft, and efficiency of the motor at its rated slip of 2.5%.

Solution

The synchronous speed is

$$N_s = \frac{120f}{P} = \frac{120 \times 60}{6} = 1200 \text{ rpm}$$

or

$$\omega_s = \frac{2\pi N_s}{60} = \frac{2\pi \times 1200}{60} = 125.66 \text{ rad/s}$$

Refer to the circuit shown in Fig. 8.14. The per phase voltage is

$$V_1 = \frac{230}{\sqrt{3}} = 132.79 \text{ V}$$

The effective rotor impedance as referred to the stator is

$$\hat{Z}_2 = \frac{r_2}{s} + jx_2 = \frac{0.25}{0.025} + j0.5 = 10 + j0.5 = 10.012\underline{/2.86°}\ \Omega$$

$$\frac{1}{\hat{Z}_g} = \frac{1}{R_c} + \frac{1}{jX_m} + \frac{1}{\hat{Z}_2}$$

$$= \frac{1}{500} + \frac{1}{j100} + \frac{1}{10.012\underline{/2.86°}}$$

$$= 0.103\underline{/-8.37°} \text{ siemens (S)}$$

Thus

$$\hat{Z}_g = 9.709\underline{/8.37°}\ \Omega$$

Total impedance:

$$\hat{Z} = r_1 + jx_1 + \hat{Z}_g = 0.5 + j0.75 + 9.709\underline{/8.37°} = 10.335\underline{/12.08°}\ \Omega$$

Stator current:

$$\tilde{I}_1 = \frac{\tilde{V}_1}{\hat{Z}} = \frac{132.79\underline{/0°}}{10.335\underline{/12.08°}} = 12.849\underline{/-12.08°} \text{ A}$$

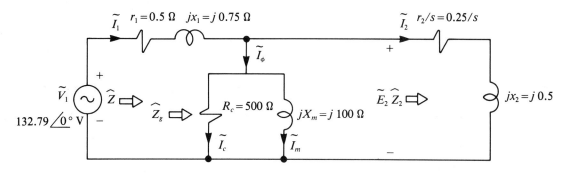

FIGURE 8.14
Equivalent Circuit for Example 8.4

Power factor:

$$\text{pf} = \cos(12.08) = 0.978 \text{ (lagging)}$$

Power input:

$$P_{\text{in}} = \sqrt{3} \, V_l I_l \cos \theta = \sqrt{3} \times 230 \times 12.849 \times 0.978 = 5006.06 \text{ W}$$

Stator copper loss:

$$\text{SCL} = 3I_1^2 r_1 = 3 \times 12.849^2 \times 0.5 = 247.7 \text{ W}$$

$$\tilde{E}_2 = \tilde{V}_1 - \tilde{I}_1(r_1 + jx_1) = 132.79 - (12.849\underline{/-12.08°}) \times (0.5 + j0.75)$$
$$= 124.76\underline{/-3.71°} \text{ V}$$

Core-loss current:

$$\tilde{I}_c = \frac{\tilde{E}_2}{R_c} = \frac{124.76\underline{/-3.71°}}{500} = 0.25\underline{/-3.71°} \text{ A}$$

Magnetization current:

$$\tilde{I}_m = \frac{\tilde{E}_2}{jX_m} = \frac{124.76\underline{/-3.71°}}{j100} = 1.248\underline{/-93.71°} \text{ A}$$

Total excitation current:

$$\tilde{I}_\phi = \tilde{I}_c + \tilde{I}_m$$
$$= 0.25\underline{/-3.71°} + 1.248\underline{/-93.71°}$$
$$= 1.272\underline{/-82.41°} \text{ A}$$

Rotor current:

$$\tilde{I}_2 = \tilde{I}_1 - \tilde{I}_\phi = 12.849\underline{/-12.08°} - 1.272\underline{/-82.41°} = 12.478\underline{/-6.57°} \text{ A}$$

Core loss:

$$P_c = 3I_c^2 R_c = 3 \times 0.25^2 \times 500 = 93.75 \text{ W}$$

Air-gap power:

$$P_g = P_{in} - SCL - P_c = 5006.06 - 247.65 - 93.75 = 4664.66 \text{ W}$$

Rotor copper Loss:

$$RCL = 3I_2^2 r_2 = 3 \times 12.478^2 \times 0.25 = 116.78 \text{ W}$$

Power developed:

$$P_d = P_g - RCL = 4664.66 - 116.78 = 4547.88 \text{ W}$$

Power output:

$$P_o = P_d - P_{mech} = 4547.88 - 150 = 4397.88 \text{ W}$$

Efficiency:

$$\eta = \frac{P_o}{P_{in}} = \frac{4397.88}{5006.06} = 0.879 \quad \text{or} \quad 87.9\%$$

Shaft torque:

$$T_s = \frac{P_o}{\omega_m} = \frac{P_o}{(1 - s)\omega_s} = \frac{4397.88}{(1 - 0.025) \times 125.66} = 35.9 \text{ N-m}$$

■ EXAMPLE 8.5

Analyze the motor discussed in Example 8.4 on the basis of an approximate equivalent circuit.

Solution

The approximate equivalent circuit of the motor is given in Fig. 8.15. The total series impedance is

$$\hat{Z} = r_1 + jx_1 + r_2/s + jx_2 = 0.5 + j0.75 + \frac{0.25}{0.025} + j0.5$$

$$= 10.5 + j1.25 = 10.574\underline{/6.79^\circ} \ \Omega$$

Core-loss current:

$$\tilde{I}_c = \frac{\tilde{V}_1}{R_c} = \frac{132.79\underline{/0^\circ}}{500} = 0.266\underline{/0^\circ} \text{ A}$$

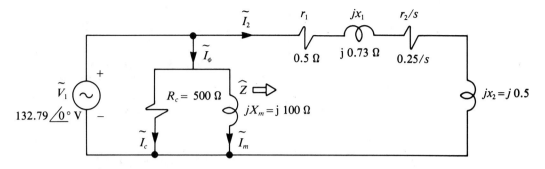

FIGURE 8.15
Approximate Equivalent Circuit for Example 8.5

Magnetization current:

$$\tilde{I}_m = \frac{\tilde{V}_1}{jX_m} = \frac{132.79\underline{/0^\circ}}{j100} = 1.328\underline{/-90^\circ}\ \text{A}$$

Rotor current:

$$\tilde{I}_2 = \frac{\tilde{V}_1}{\hat{Z}} = \frac{132.79\underline{/0^\circ}}{10.574\underline{/6.79^\circ}} = 12.558\underline{/-6.79^\circ}\ \text{A}$$

Total current:

$$\tilde{I}_1 = \tilde{I}_c + \tilde{I}_m + \tilde{I}_2 = 0.266 + 1.328\underline{/-90^\circ} + 12.558\underline{/-6.79^\circ}$$

$$= 12.736 - j1.328 = 13.043\underline{/-12.45^\circ}\ \text{A}$$

Power factor:

$$\text{pf} = \cos(12.45) = 0.976\ \text{lagging}$$

Stator copper loss:

$$\text{SCL} = 3I_2^2 r_1 = 3 \times 12.558^2 \times 0.5 = 236.56\ \text{W}$$

Rotor copper loss:

$$\text{RCL} = 3I_2^2 r_2 = 3 \times 12.558^2 \times 0.25 = 118.27\ \text{W}$$

Core loss:

$$P_c = 3I_c^2 R_c = 3 \times 0.266^2 \times 500 = 106.13\ \text{W}$$

Power input:

$$P_{\text{in}} = \sqrt{3} V_l I_l \cos\theta = \sqrt{3} \times 230 \times 13.043 \times 0.976 = 5071.26\ \text{W}$$

Air-gap power:

$$P_g = P_{in} - SCL - P_c$$

$$= 5071.26 - 236.56 - 106.13 = 4728.57 \text{ W}$$

Power developed:

$$P_d = P_g - RCL = 4728.57 - 118.27 = 4610.3 \text{ W}$$

Power output:

$$P_o = P_d - P_{mech} = 4610.3 - 150 = 4460.3$$

Efficiency:

$$\eta = \frac{P_o}{P_{in}} = \frac{4460.3}{5071.26} = 0.879 \quad \text{or} \quad 87.9\%$$

8.6 MAXIMUM TORQUE CRITERION

It is evident from Eq. (8.36) that for low values of slip the torque is directly proportional to the slip as there is no appreciable change in the value of the denominator. Thus, the speed–torque characteristic is basically a straight line for low values of the slip. For high-efficiency motors, this relationship is true for slips as high as 10%. Any further increase in slip will have a more and more pronounced effect on the value of the denominator, and the speed–torque curve will deviate from its straight-line relationship. At one particular value of the slip, which depends on the other parameters in the torque expression, the value of the denominator will vary directly with it. As the slip is further increased, the increase in the denominator will be faster than the numerator, and the torque developed by the motor will decrease. From now on the decrease in the torque will continue until the slip reaches unity. At unity slip, that is, under standstill or blocked-rotor conditions, a definite torque is developed by the motor, known as the *starting torque*. Therefore, for some value of slip the torque developed by the motor is maximum. The condition for maximum or breakdown torque can be obtained by differentiating the torque expression with respect to slip and setting it equal to zero. That is,

$$[(sR_{Th} + r_2)^2 + s^2(X_{Th} + x_2)^2]$$

$$- s[2R_{Th}(sR_{Th} + r_2) + 2s(X_{Th} + x_2)^2] = 0 \quad (8.37)$$

which, after simplifications, yields

$$r_2/s_b = \sqrt{R_{Th}^2 + (X_{Th} + x_2)^2} \quad (8.38)$$

where s has been replaced by s_b to indicate that this is the required slip at which the breakdown torque occurs.

The slip at the breakdown torque is

$$s_b = \frac{r_2}{\sqrt{R_{Th}^2 + (X_{Th} + x_2)^2}} \qquad (8.39)$$

This is the exact expression for the breakdown slip in terms of other motor circuit parameters. Substituting the preceding value of the slip in the torque equation, we can obtain the expression for the maximum torque developed as

$$T_{dm} = \frac{3V_{Th}^2}{2\omega_s} \left[\frac{1}{R_{Th} + \sqrt{R_{Th}^2 + (X_{Th} + x_2)^2}} \right] \qquad (8.40)$$

■ EXAMPLE 8.6

A three-phase, Y-connected, 120-V, 60-Hz, 4-pole induction motor has the following parameters: $r_1 = 10 \ \Omega$, $x_1 = 25 \ \Omega$, $r_2 = 3 \ \Omega$, $x_2 = 25 \ \Omega$, and $X_m = 75 \ \Omega$. Determine the breakdown slip and the maximum torque developed by the motor.

Solution

The synchronous speed of the motor is

$$N_s = \frac{120f}{P} = \frac{120 \times 60}{4} = 1800 \text{ rpm}$$

or

$$\omega_s = \frac{2\pi \times 1800}{60} = 188.5 \text{ rad/s}$$

The per-phase voltage is

$$V_1 = \frac{120}{\sqrt{3}} = 69.282 \text{ V}$$

Using Thevenin's equivalent circuit,

$$\tilde{V}_{Th} = \frac{j\tilde{V}_1 X_m}{r_1 + j(x_1 + X_m)} = \frac{69.282 \times 75\underline{/90°}}{10 + j(25 + 75)}$$

$$= 51.704\underline{/5.71°} \text{ V}$$

Thevenin's equivalent impedance is

$$\hat{Z}_{Th} = \frac{j(r_1 + jx_1)X_m}{r_1 + j(x_1 + X_m)} = \frac{j(10 + j25) \times 75}{10 + j(25 + 75)}$$

or $\qquad \hat{Z}_{Th} = 20.094\underline{/73.91°}\ \Omega$

Thus $\qquad R_{Th} = 5.569\ \Omega$ and $X_{Th} = 19.307\ \Omega$

The breakdown slip from Eq. (8.39) is

$$s_b = \frac{3}{\sqrt{5.569^2 + (19.307 + 25)^2}} = 0.067$$

The maximum torque developed, from Eq. (8.40), is

$$T_{dm} = \frac{3 \times 51.704^2}{2 \times 188.5[5.569 + \sqrt{5.569^2 + (19.307 + 25)^2}]} = 0.424\ \text{N-m}$$

8.6.1 An Approximate but Useful Relation

The preceding equations for breakdown slip and breakdown torque are exact. However, we can obtain a very useful approximate relationship by assuming that the voltage across the magnetizing reactance, $\hat{E}_\phi$, remains almost constant with slip. This can only be possible if the stator impedance on a per unit basis is very small as compared to other circuit parameters. In this case, the Thevenin's equivalent impedance can be assumed to be so small that it can be neglected. For negligible stator impedance, the expression for the breakdown slip simplifies to

$$s_b = \frac{r_2}{x_2} \tag{8.41}$$

which is simply the ratio of the actual rotor resistance to the rotor reactance. The approximate expression for the breakdown torque becomes

$$T_{dm} = \frac{3}{2} \frac{V_{Th}^2}{\omega_s} \left(\frac{1}{x_2}\right) = \frac{3}{2} \frac{V_{Th}^2}{\omega_s} \left(\frac{s_b}{r_2}\right) \tag{8.42}$$

For this approximate torque relationship to be true, the induction motor must operate at a very low slip. In fact, most three-phase induction motors operate below a slip of 5%, and the approximate relationship can be used to determine the maximum torque developed by the motor. As is obvious from the speed–torque curve, and now from the approximate equation for the breakdown torque, the speed–torque relationship is linear all the way from the maximum torque to zero torque under no-load condition.

We can expeditiously determine the torque developed by the motor at any other speed in terms of its maximum value using the approximate relationship. The developed torque at any slip, from Section 8.4, is

$$T_d = 3 \frac{I_2^2 \, r_2}{\omega_s \, s}$$

which becomes

$$T_d \cong \frac{3V_{\text{Th}}^2(r_2/s)}{\omega_s[(r_2/s)^2 + x_2^2]} \tag{8.43}$$

because

$$\tilde{I}_2 \cong \frac{\tilde{V}_{\text{Th}}}{r_2/s + jx_2} \tag{8.44}$$

ignoring the equivalent stator impedance. The ratio of the torque at any speed to the breakdown torque is

$$\frac{T_d}{T_{dm}} = \frac{\dfrac{3V_{\text{Th}}^2(r_2/s)}{\omega_s[(r_2/s)^2 + x_2^2)}}{\dfrac{3}{2}\dfrac{V_{\text{Th}}^2 s_b}{\omega_s r_2}} \tag{8.45}$$

Since $x_2 = r_2/s_b$, Eq. (8.45) reduces to

$$\frac{T_d}{T_{dm}} = \frac{2ss_b}{s^2 + s_b^2} \tag{8.46}$$

This is a very useful, although approximate, relation that can be used to determine the developed torque at any slip s once the breakdown torque and the corresponding slip are known. Since we have neglected the stator impedance with an understanding that it is very small, we can likewise assume that the stator copper losses are also negligible. Thus, the air-gap power must be equal to the power input to the motor. However, from the expression for the approximate maximum torque developed, the corresponding air-gap power is

$$P_{gm} = T_{dm}\omega_s$$

or

$$P_{gm} = \frac{3}{2} V_{\text{Th}}^2 \frac{s_b}{r_2} \tag{8.47}$$

Since SCL = 0, the power input is

$$P_{\text{in}} = P_g \tag{8.48}$$

For the efficiency to be theoretically maximum, total rotational loss should almost be negligible. In that case, $P_{rot} = 0$ and the power developed by the motor is also the power available at the shaft. The power developed by the motor, at any slip s is

$$P_o \cong P_d = (1 - s)P_g$$

Therefore, the efficiency of the motor is

$$\eta = \frac{P_o}{P_{in}} = 1 - s = S \tag{8.49}$$

That is, the maximum efficiency that can be obtained theoretically from a three-phase induction motor is simply equal to its per unit speed, S. This places the maximum limits on the efficiency of a three-phase induction motor. Actual efficiency, of course, will be lower than this maximum limit. Equation (8.49) only highlights the point that if a motor is operating at 60% of its synchronous speed, the maximum efficiency of that motor can never exceed 60%. Thus, the higher the speed of operation is, the higher is its efficiency. A motor operating at 5% slip can have a theoretically maximum efficiency of 95%. Therefore, it is to our advantage to design a three-phase induction motor to operate at a very low slip. It should, however, be noted that the expressions given are only approximate, but they do shed a great deal of light on the behavior of an induction machine. These expressions are usually used to make fast and relative comparisons among the same-sized motors. Once a motor is selected on this basis, it is usually a common practice to make an exact evaluation of its performance by making use of exact theoretical equations.

■ EXAMPLE 8.7

A 100-hp, 440-V, 60-Hz, 8-pole, three-phase induction motor has an equivalent rotor impedance of $0.02 + j0.08$ Ω on a per phase basis. What is its speed at the maximum torque? If the starting torque of the motor is 80% of its maximum value, determine the external resistance that must be added in series with the rotor.

Solution

Since the stator-winding impedance and the impedance of the exciting branch are unknown, we have to use the approximate expressions for the breakdown slip and the maximum torque. Thus,

$$s_b = \frac{r_2}{x_2} = \frac{0.02}{0.08} = 0.25$$

The synchronous speed of the motor is

$$N_s = \frac{120f}{P} = \frac{120 \times 60}{8} = 900 \text{ rpm}$$

The speed at which maximum torque occurs is

$$N_m = (1 - s_b)N_s = (1 - 0.25) \times 900 = 675 \text{ rpm}$$

The addition of resistance in series with the rotor circuit changes the slip at which the motor develops the maximum torque. Therefore, let us determine the breakdown slip such that the starting torque is 80% of the maximum torque developed at that slip. Since

$$\frac{T_s}{T_{bn}} = \frac{2ss_{bn}}{s^2 + s_{bn}^2}$$

where T_s and T_{bn} are the standstill and the new breakdown torques, respectively. At standstill, $s = 1$; hence

$$\frac{T_s}{T_{bn}} = \frac{2s_{bn}}{1 + s_{bn}^2}$$

However, $T_s/T_{bn} = 0.8$ implies that

$$s_{bn}^2 - 2.5s_{bn} + 1 = 0$$

Thus, $s_{bn} = 2$ or 0.5.

Since slip greater than 1 is not possible for motor operation, $s = 0.5$ is the new breakdown slip. The new rotor resistance must be

$$r_{2n} = s_{bn}x_2 = 0.5 \times 0.08 = 0.04 \text{ }\Omega$$

Thus, the additional resistance that should be added in the rotor circuit on a per phase basis is

$$r_2 = r_{2n} - r_2 = 0.04 - 0.02 = 0.02 \text{ }\Omega$$

8.7 MAXIMUM POWER OUTPUT CRITERION

In the preceding sections we examined the conditions for maximum efficiency and maximum torque for a three-phase induction motor. In this section we will

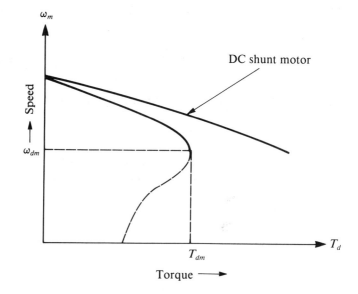

FIGURE 8.16
Speed–Torque
Characteristics of
an Induction
Motor and a dc
Shunt Motor for
Direct
Comparison

determine the condition that leads to the maximum power developed by the motor. When the motor is operating at no load, close to its synchronous speed, the torque developed by the motor is just sufficient to overcome the rotational losses, and the net power output is zero. On the other hand, the net power output is also zero for unity slip (blocked-rotor condition). Thus, the motor must provide maximum output at some speed between standstill and synchronous speed.

Let us examine the speed–torque characteristic of the motor. Under no load the motor operates very close to its synchronous speed. Any increase in the load decreases the speed of the motor in a manner that the load is balanced by the torque developed by the motor. The reduction in speed continues until the load equals the maximum torque developed by the motor. This is also the stable region of operation of the motor. The corresponding speed–torque curve is depicted in Fig. 8.16. Note that this characteristic is very similar to that of a dc shunt motor. Since the speed variations are small, an induction motor is also considered to be a constant-speed motor. However, if the load torque is more than the maximum torque developed by the motor, the motor will not be able to sustain the load and will stall. This is a remarkable difference between the operation of an induction motor and a dc shunt motor. At some speed in the stable operating region, the power output of the motor will be maximum.

The Thevenin's equivalent circuit of the motor (Fig. 8.13), can be redrawn as shown in Fig. 8.17, where $r_e = R_{Th} + r_2$ and $x_e = X_{Th} + x_2$. Thus, $\hat{z}_e = r_e + jx_e$.

The gross power developed by the motor, from Section 8.4, is

$$P_d = 3I_2^2 r_2 \frac{1-s}{s}$$

From Fig. 8.17,

$$\tilde{I}_2 = \frac{\tilde{V}_{Th}}{r_e + r_2[(1 - s)/s] + jx_e}$$

Thus,

$$P_d = \frac{3V_{Th}^2 r_2[(1 - s)/s]}{\{r_e + r_2[(1 - s)/s]\}^2 + x_e^2} \tag{8.50}$$

The condition for maximum power output can be found by differentiating Eq. (8.50) with respect to slip s and setting the derivative equal to zero. That is,

$$\left\{ \left[r_e + r_2 \left(\frac{1 - s}{s} \right) \right]^2 + x_e^2 \right\} \left(-\frac{1}{s^2} \right)$$

$$- 2 \left(\frac{1 - s}{s} \right) \left[r_e + r_2 \left(\frac{1 - s}{s} \right) \right] \left(-\frac{r_2}{s^2} \right) = 0$$

which simplifies to

$$r_e^2 + x_e^2 = \left[r_2 \left(\frac{1 - s_p}{s_p} \right) \right]^2 \tag{8.51}$$

where s has been replaced by s_p to indicate the slip at maximum power output. But

$$r_e^2 + x_e^2 = z_e^2$$

Thus,

$$z_e = r_2 \left(\frac{1 - s_p}{s_p} \right) \tag{8.52}$$

Hence, the power output of an induction motor is maximum when its dynamic resistance is equal to the leakage impedance of the motor at standstill.

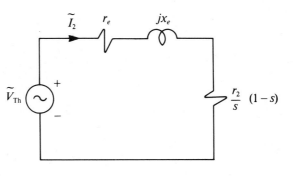

FIGURE 8.17
Simplified
Version of
Thevenin's
Equivalent
Circuit for an
Induction Motor

The slip at which the motor develops the maximum power is given as

$$s_p = \frac{r_2}{r_2 + z_e} \tag{8.53}$$

Let us compare this expression for slip with that of the breakdown slip. The denominator in Eq. (8.53) is larger than that of the breakdown slip. Thus, the motor develops maximum power at a slip lower than that at which it develops maximum torque. Substituting for s_p in the expression for the power output of the motor, we obtain an expression for the maximum power output as

$$P_{dm} = \frac{3}{2} \left(\frac{V_{Th}^2}{r_e + z_e} \right) \tag{8.54}$$

The net power output is, of course, less than the gross power output by the amount of rotational loss in the motor.

■ EXAMPLE 8.8

A 120-V, 60-Hz, 6-pole, Y-connected, three-phase induction motor has a stator impedance of $0.1 + j0.15$ Ω and an equivalent rotor impedance of $0.2 + j0.25$ Ω on a per phase basis. Calculate (a) the maximum gross power output and corresponding torque developed, and (b) the maximum torque developed and the corresponding power output of the motor.

Solution

Since the equivalent core-loss resistance, R_c, and the magnetization reactance, X_m, are not given, we will assume them to be quite large and replace R_{Th} with r_1 and X_{Th} with x_1.

(a) $z_e = r_e + jx_e = (r_1 + r_2) + j(x_1 + x_2)$

$$= (0.1 + 0.2) + j(0.15 + 0.25) = 0.3 + j0.4 = 0.5\underline{/53.13°}\,\Omega$$

The slip at which the motor provides maximum power output is

$$s_p = \frac{r_2}{r_2 + z_e} = \frac{0.2}{0.2 + 0.5} = 0.286 \quad \text{or} \quad 28.6\%$$

Thus, $P_{dm} = \frac{3}{2} \frac{V_{Th}^2}{r_e + z_e} = \frac{3}{2} \frac{(120/\sqrt{3})^2}{0.3 + 0.5} = 9000 \text{ W}$

The synchronous speed is

$$N_s = \frac{120f}{P} = \frac{120 \times 60}{6} = 1200 \text{ rpm}$$

or

$$\omega_s = \frac{2\pi N_s}{60} = \frac{2\pi \times 1200}{60} = 40\pi \text{ rad/s}$$

The speed of the motor is

$$\omega_m = (1 - s_p)\omega_s = (1 - 0.286) \times 40\pi = 89.724 \text{ rad/s}$$

The developed torque is

$$T_d = \frac{P_{dm}}{\omega_m} = \frac{9000}{89.724} = 100.3 \text{ N-m}$$

(b) The slip for the maximum torque is

$$s_b = \frac{r_2}{\sqrt{r_1^2 + (x_1 + x_2)^2}} = \frac{0.2}{\sqrt{0.1^2 + (0.15 + 0.25)^2}} = 0.485$$

The speed of the motor is

$$\omega_m = (1 - s_b)\omega_s = (1 - 0.485) \times 40\pi = 64.717 \text{ rad/s}$$

The maximum torque developed is

$$T_{dm} = \frac{3}{2} \frac{V_{Th}^2}{\omega_s} \left[\frac{1}{R_{Th} + \sqrt{R_{Th}^2 + (X_{Th} + x_2)^2}} \right]$$

$$= \frac{3}{2} \frac{(120/\sqrt{3})^2}{40\pi} \left[\frac{1}{0.1 + \sqrt{0.1^2 + (0.15 + 0.25)^2}} \right]$$

$$= 111.84 \text{ N-m}$$

The corresponding power output of the motor is

$$P_d = T_{dm}\omega_m = 111.84 \times 64.717 = 7238 \text{ W}$$

8.8   SPEED–TORQUE CHARACTERISTICS

In the foregoing sections we have carefully examined the speed–torque and speed–power characteristics of an induction motor by using the exact equations for the torque developed and the power output. As stated earlier, most induction motors operate on full load at a slip of usually less than 10%. In this range of slips, the effective rotor resistance is very high, and the speed–torque characteristic is practically a straight line. From the approximate equivalent circuit of an induction motor (Fig. 8.9), the torque developed by the motor is

$$T_d \cong \frac{3V_1^2}{\omega_s} \frac{s}{r_2}$$

We can now make the following observations:

1. If the applied voltage, V_1, and the actual rotor resistance, r_2, are constant, the torque developed by the motor is directly proportional to the slip. Thus, in this linear range, the ratio of the torques developed is equal to the ratio of the slips.

2. The torque developed is inversely proportional to the actual rotor resistance at a definite slip as long as the impressed voltage is held constant. Therefore, the torque produced by a motor can be adjusted by changing its rotor resistance. This can be easily accomplished in wound-rotor motors.

3. At a definite value of slip and the actual rotor resistance, the torque developed by the motor is directly proportional to the square of the impressed voltage.

4. For constant-torque operation under fixed impressed voltage, the motor slip is directly proportional to the actual rotor resistance.

■ **EXAMPLE 8.9**

A 2-hp, 230-V, 60-Hz, three-phase, Δ-connected induction motor operates at a full load speed of 1710 rpm. If the supply voltage fluctuates ±10%, determine the torque range of the motor at its rated speed.

Solution

Since the motor runs at 1710 rpm at full load, the synchronous speed of the motor must be 1800 rpm.

$$s = \frac{N_s - N_m}{N_s} = \frac{1800 - 1710}{1800} = 0.05$$

Operation at the rated voltage is

$$V_1 = 230 \text{ V}, \qquad P_o = 2 \times 746 = 1492 \text{ W}$$

$$\omega_m = \frac{2\pi N_m}{60} = \frac{2\pi \times 1710}{60} = 179.07 \text{ rad/s}$$

$$T_{s1} = \frac{P_o}{\omega_m} = \frac{1492}{179.07} = 8.33 \text{ N-m}$$

When the supply voltage is up by 10%

$$V_2 = 1.1 \times 230 = 253 \text{ V}$$

Since the slip is only 5%, the speed–torque characteristic can be assumed to be linear. Therefore,

$$\frac{T_{s2}}{T_{s1}} = \left(\frac{V_2}{V_1}\right)^2$$

$$T_{s2} = T_{s1} \left(\frac{V_2}{V_1}\right)^2 = 8.33 \left(\frac{253}{230}\right)^2 = 10.08 \text{ N-m}$$

When the supply voltage is down by 10%,

$$V_2 = 0.9 \times 230 = 207 \text{ V}$$

and

$$T_{s2} = 8.33 \left(\frac{207}{230}\right)^2 = 6.75 \text{ N-m}$$

Thus, the torque developed by the motor varies from 6.75 to 10.08 N-m.

■ EXAMPLE 8.10

A 10-hp, 440-V, 50-Hz, 4-pole induction motor runs at a slip of 5% on full load. The motor slip falls to 2% when lightly loaded. Determine the torque developed and the power output of the motor when the motor runs at (a) full load and (b) light load.

Solution

(a) *Full-load condition:* The power output is $P_o = 10 \times 746 = 7460$ W.

$$N_s = \frac{120f}{P} = \frac{120 \times 50}{4} = 1500 \text{ rpm} \quad \text{or} \quad \omega_s = 157.08 \text{ rad/s}$$

The full-load is speed

$$\omega_m = (1 - s)\omega_s = (1 - 0.05) \times 157.08 = 149.23 \text{ rad/s}$$

The full-load torque is

$$T_f = \frac{P_o}{\omega_m} = \frac{7460}{149.23} = 50 \text{ N-m}$$

(b) *Light-load condition:* Since the torque is directly proportional to the slip, the ratio of the torque under light load to full load becomes

$$\frac{T_L}{T_f} = \frac{s_L}{s_f}$$

$$T_L = T_f \left(\frac{s_L}{s_f}\right) = 50 \left(\frac{0.02}{0.05}\right) = 20 \text{ N-m}$$

The power output is $P_o = (1 - s)\omega_s T_L = (1 - 0.02) \times 157.08 \times 20 = 3079 \text{ W}$.

8.9 INDUCTION MOTOR TESTS

Four basic tests are used to determine the performance of an induction motor.

(1) Stator Resistance Test This test is performed to determine the stator-winding resistance on a per phase basis. Since the stator of a three-phase induction machine is wound exactly like the stator of a synchronous machine, the test data can be interpreted the same way as we did for the synchronous machine. Since the motor may be Y- or Δ-connected, the dc value of the measured resistance on a per phase basis is then

$$r_1 = 0.5R, \quad \text{for Y-connection}$$

$$r_1 = 1.5R, \quad \text{for Δ-connection}$$

where R is the measured resistance between any two lines of a three-phase induction motor. Often the calculated value of the stator-winding resistance is multiplied by a factor of 1.15 or so to convert it from its dc to an ac value. This is done to account for the *skin effect*. This factor may be debatable at power frequencies of 50 or 60 Hz, but it does become important for 400-Hz applications.

(2) No-load Test In this case the rated voltage is applied to the stator windings without any load connected to the rotor. This test is therefore similar to the open-circuit test on the transformer except for the friction and windage loss in an induction motor. If the core loss is considered as a part of the rotational loss, this test can be referred to as the *rotational loss test*. The schematic of the motor connections is shown in Fig. 8.18.

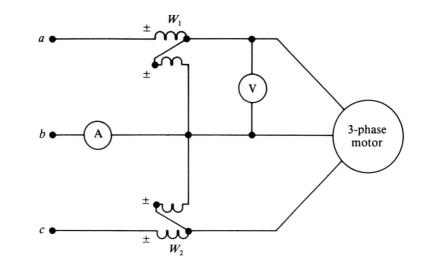

FIGURE 8.18
Typical
Connections to
Perform
Induction Motor
Tests

Under no-load conditions, the slip is nearly equal to zero. Thus, the rotor circuit may be considered an open circuit. However, if we make a similar approximation as we did for transformers that the series impedance is very small compared to the impedance of the magnetization branch, the induction motor can be represented as shown in Fig. 8.19.

Let W_{oc}, I_{oc}, and V_{oc} be the power input, input current, and rated voltage under no-load conditions on a per phase basis. Since we are trying to represent the core loss by an equivalent resistance R_c, we must subtract the loss due to friction and windage from W_{oc}. The loss due to friction and windage can be determined by coupling the motor under test to another motor with a calibrated output and running it at the no-load speed of the induction motor under test. Let this loss be P_{mech} on a per phase basis. Then the power loss in the core-loss resistance is

$$P_{oc} = W_{oc} - P_{\text{mech}} \tag{8.55}$$

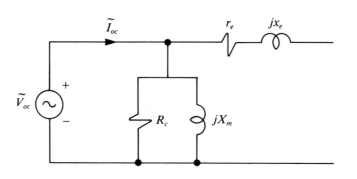

FIGURE 8.19
Approximate
Equivalent
Circuit under No
Load

The core-loss resistance is

$$R_c = \frac{V_{oc}^2}{P_{oc}} \tag{8.56}$$

The power factor is

$$\cos \theta = \frac{P_{oc}}{V_{oc}I_{oc}} \tag{8.57}$$

The magnetizing reactance is

$$X_m = \frac{V_{oc}}{I_{oc} \sin \theta} \tag{8.58}$$

As shown in Fig. 8.18, the total power input, W_{oc}, is being measured by the two-wattmeter method. If both wattmeters, as connected, show positive power indications, the total power is the sum of the two readings. On the other hand, if one wattmeter reads negative and its leads have to be reversed, the total power input is then the difference of the two readings.

(3) Blocked-rotor Test   This test is also known as the locked-rotor test and is very similar to the short-circuit test of the transformer. In this case, the rotor is held stationary by applying external torque to the shaft. The stator field windings are connected to a balanced three-phase power source. The voltage is carefully increased from zero so as to obtain the rated current (full load) in the windings. The circuit set up is the same as that for the no-load test. Since the rotor circuit impedance is small under standstill conditions and the applied voltage is considerably lower than the full-load voltage, the current in the magnetization branch would be very small and can be neglected. In this case, the approximate equivalent circuit of the induction motor is given in Fig. 8.20. The total series impedance is

$$\hat{z}_e = r_1 + r_2 + j(x_1 + x_2) = r_e + jx_e \tag{8.59}$$

Let V_b, I_b, and P_b be the applied voltage, the rated current, and the power input under the locked-rotor condition on a per phase basis. Since the rotor is stationary, there is no loss due to friction and windage. By neglecting the current flow in the magnetization branch, we have already assumed that the core loss is zero. Thus, the total power supplied on a per phase basis must be equal to the loss in the equivalent resistance, r_e, in the series circuit. Hence,

$$r_e = \frac{P_b}{I_b^2} \tag{8.60}$$

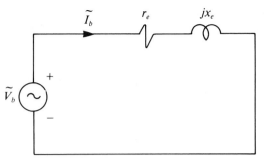

FIGURE 8.20
Approximate
Equivalent
Circuit for
Blocked-Rotor
Test

Since r_1 is already known from the resistance test, the rotor circuit resistance r_2 can now be determined as

$$r_2 = r_e - r_1 \tag{8.61}$$

However,

$$z_e = \frac{V_b}{I_b}$$

and $\tag{8.62}$

$$x_e = \sqrt{z_e^2 - r_e^2}$$

It is rather difficult to isolate the leakage reactances x_1 and x_2. For all practical purposes, it is usually assumed that these reactances are equal. That is,

$$x_1 = x_2 = 0.5x_e \tag{8.63}$$

Figure 8.21 shows the approximate equivalent circuit of an induction motor based on the test data. Once these parameters are known, the performance of the induction motor can be calculated at any speed. In making such calculations, the equivalent-circuit parameters are usually assumed to be constant. This is only true for those motors operating linearly under unsaturated conditions. Most three-phase induction motors for high-efficiency applications fall into this category.

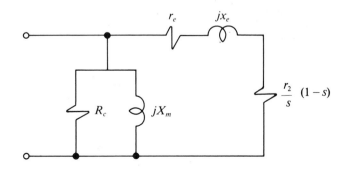

FIGURE 8.21
Approximate
Equivalent
Circuit of an
Induction Motor
as Determined
from Tests

(4) Load Test To experimentally determine the complete speed–torque curve of a three-phase induction motor, couple the shaft to the dynamometer and connect the stator windings to a balanced three-phase source. Starting from the no-load condition, the load is slowly increased and the corresponding readings for the rotor speed, winding current, power input, and so on, are recorded. Often the meters have the capability to be directly connected to X-Y recorders for plotting the graphs.

■ **EXAMPLE 8.11**

The test data on a 208-V, 60-Hz, 4-pole, Y-connected, three-phase induction motor rated at 1710 rpm are as follows:

Resistance between any two lines = 2.4 Ω

	No-load test	Blocked-rotor test
Total input power	450.0 W	59.4 W
Line current	1.562 A	2.77 A
Line-to-line voltage	208.0 V	27.0V
Friction and windage loss = 18 W		

Calculate the equivalent-circuit parameters of the motor.

Solution

Since the motor is Y-connected, $r_1 = 2.4/2 = 1.2$ Ω. *From the no-load test* on a per phase basis,

$$W_{oc} = \frac{450}{3} = 150 \text{ W} \quad \text{and} \quad P_{\text{mech}} = \frac{18}{3} = 6 \text{ W}$$

Thus,

$$P_{oc} = 150 - 6 = 144 \text{ W}, \qquad V_{oc} = \frac{208}{\sqrt{3}} = 120 \text{ V}, \qquad I_{oc} = 1.562 \text{ A}$$

The core-loss resistance is

$$R_c = \frac{V_{oc}^2}{P_{oc}} = \frac{120^2}{144} = 100 \text{ Ω}$$

The apparent power input is

$$S_{oc} = V_{oc}I_{oc} = 120 \times 1.562 = 187.44 \text{ VA}$$

The power factor is

$$\cos \theta = \frac{P_{oc}}{S_{oc}} = \frac{144}{187.44} = 0.768$$

and

$$\sin \theta = \sqrt{1. - 0.768^2} = 0.64$$

Thus, the magnetizing current is

$$I_m = I_{oc} \sin \theta = 1.562 \times 0.64 = 1.0 \text{ A}$$

The magnetizing reactance is

$$X_m = \frac{V_{oc}}{I_m} = \frac{120}{1} = 120 \ \Omega$$

From the blocked rotor test on a per phase basis,

$$V_b = \frac{27}{\sqrt{3}} = 15.588 \text{ V}, \qquad P_b = \frac{59.4}{3} = 19.8 \text{ W}, \qquad I_b = 2.77 \text{ A}$$

The equivalent series resistance is

$$r_e = \frac{P_b}{I_b^2} = \frac{19.8}{2.77^2} = 2.58 \ \Omega$$

Since $r_1 = 1.2 \ \Omega$,

$$r_2 = r_e - r_1 = 2.58 - 1.2 = 1.38 \ \Omega$$

$$z_e = \frac{V_b}{I_b} = \frac{15.588}{2.77} = 5.627 \ \Omega$$

Thus

$$x_e = \sqrt{z_e^2 - r_e^2} = \sqrt{5.627^2 - 2.58^2} = 5.0 \ \Omega$$

or

$$x_1 = x_2 = \frac{5}{2} = 2.5 \ \Omega$$

8.10 EFFECT OF ROTOR RESISTANCE ON MOTOR PERFORMANCE

Earlier we obtained the expression for the torque developed by an induction motor as

$$T_d = \frac{3V_{\text{Th}}^2}{\omega_s} \frac{r_2/s}{[R_{\text{Th}} + (r_2/s)]^2 + (X_{\text{Th}} + x_2)^2} \tag{8.64}$$

At standstill, the rotor slip is unity, the starting torque developed by the motor is

$$T_{ds} = \frac{3V_{Th}^2}{\omega_s} \left[\frac{r_2}{(R_{Th} + r_2)^2 + (X_{Th} + x_2)^2} \right] \tag{8.65}$$

and the starting current is

$$\tilde{I}_{2s} = \frac{\tilde{V}_{Th}}{(R_{Th} + r_2) + j(X_{Th} + x_2)} \tag{8.66}$$

At rated speed, the slip is low, and therefore the ratio of r_2/s is high. This high value of hypothetical rotor resistance controls the current in the rotor as well as in the stator windings. As $s = 1$, the rotor circuit resistance is very small. Thus, the starting current is many times higher than the rated full-load current. The starting current can, however, be controlled by adding additional resistance in the rotor circuit. In a wound-rotor machine, the additional resistance is added in series with the rotor windings through the slip rings. In die-cast rotors, the change in resistance from a high value at the time of starting to a low value at the rated speed is accomplished by the *double-cage* construction of the rotor, which will be studied later. The increase in rotor resistance also increases the starting torque of the motor. The starting torque is maximum when the rotor resistance is

$$r_2 = \sqrt{R_{Th}^2 + (X_{Th} + x_2)^2} \tag{8.67}$$

$$\cong \sqrt{r_1^2 + (x_1 + x_2)^2} \tag{8.68}$$

when X_m and R_c are high compared to r_1 and x_1.

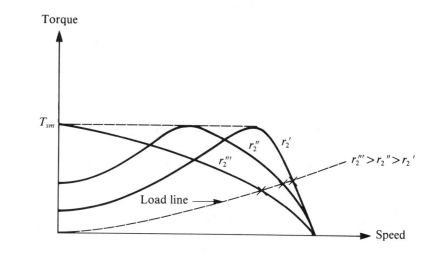

FIGURE 8.22
Effect of Rotor
Resistance on
Motor
Performance

How does the increase in the rotor resistance affect the maximum torque of the motor? An answer can be found in the expression for the maximum torque developed by the motor under approximate conditions when the hypothetical rotor resistance is large compared to the impedance of the stator winding. We obtained the expression for the maximum torque as

$$T_{dm} = \frac{3}{2\omega_s} \frac{V_{Th}^2}{x_2}$$

If the applied voltage is held constant and the equivalent-circuit parameter x_2 is assumed to be constant, the maximum torque developed by the motor will be constant. In other words, the maximum torque developed by the motor is independent of rotor resistance. However, we found that for the maximum torque to exist

$$x_2 = \frac{r_2}{s_b}$$

Therefore, the expression for the maximum torque can also be given as

$$T_{dm} = \frac{3}{2\omega_s} V_{Th}^2 \frac{s_b}{r_2}$$

which is in terms of the rotor resistance and the slip. For the maximum torque to be constant, it is evident from this expression that the ratio of r_2/s_b must be constant. Speed–torque curves for various values of actual rotor resistance are given in Fig. 8.22.

8.9.1 Multicage Rotors

In large squirrel-cage rotors, rotor resistance control can be obtained by incorporating more than one squirrel-cage in the rotor. Many different designs (Fig. 8.23) have been proposed to this end, and each allows the rotor circuit to see different values of rotor resistance at different speeds of operation.

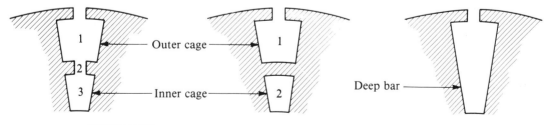

FIGURE 8.23
Triple-Cage, Double-Cage, and Deepbar Rotor Construction

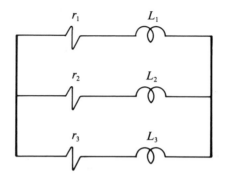

FIGURE 8.24
Equivalent
Circuit
Representation
of a Triple-Cage
Rotor

In each design, the underlying principle is to achieve a high resistance at starting and a low resistance at the rated speed. At starting, the frequency of the rotor current is the same as the frequency of the applied voltage to the stator winding. As the currents are induced in the rotor bars, they create a secondary magnetic field. Part of that field links only the rotor conductors and becomes the leakage flux. As is obvious, the leakage flux increases as we move radially toward the rotor shaft and away from the air gap. Thus, the inner cage presents high leakage reactance compared to the outer cage. The distribution of resistances and reactances for a triple-cage rotor is shown in Fig. 8.24. Note that $L_1 < L_2 < L_3$. Due to the high leakage reactance of the inner cages, the rotor current tends to go through the outer cage. If the cross-sectional area of the outer cage is smaller than the inner cage, it will present comparatively high resistance, which is needed for the starting torque. At the normal operating speed the rotor frequency is very low, causing the leakage reactances to decrease. In this case, the current can flow equally well in all the cages. In fact, the current density in each cage will be

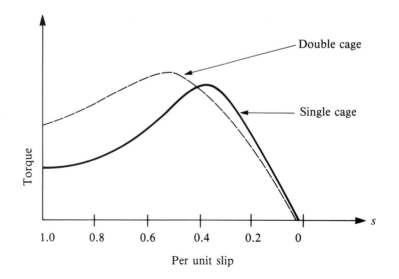

FIGURE 8.25
Speed–Torque
Characteristics
for Single- and
Double-Cage
Rotors

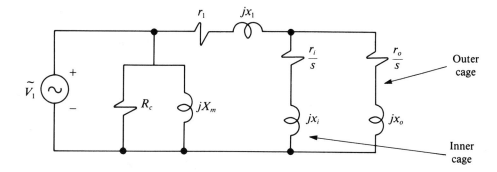

FIGURE 8.26
Equivalent
Circuit of a
Double-Cage
Induction Motor

practically the same. As a result, the total effective resistance of the rotor is low. The speed–torque curves for motors with single- and double-cage rotors are given in Fig. 8.25. An equivalent circuit for an induction motor with a double-cage rotor is shown in Fig. 8.26.

Another technique that is commonly used to increase the rotor resistance and minimize the effects of harmonics in the motor is known as *skewing*. With respect to the rotor shaft, the rotor bars are skewed at an angle as depicted in Fig. 8.27. Usually, skew is given in terms of rotor slots or rotor bars. The minimum skew must be one bar to avoid cogging. Skews of more than one bar are common. Skew also plays a role in the reduction of noise in the motor.

■ EXAMPLE 8.12

The impedances of the inner and outer cages of a double-cage, three-phase, 4-pole induction motor are $0.2 + j0.8\ \Omega$ and $0.6 + j0.2\ \Omega$, respectively. Determine the ratio of torques developed by the two cages (a) at standstill and (b) at 2% slip.

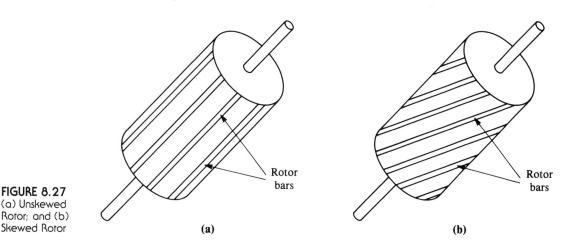

FIGURE 8.27
(a) Unskewed
Rotor; and (b)
Skewed Rotor

(a)   **(b)**

Solution

The approximate equivalent circuit, neglecting the stator and the magnetization branch impedances, is shown in Fig. 8.28. Let $\tilde{I}_i$ and $\tilde{I}_o$ be the currents in the inner and outer cages as shown. Then

$$\tilde{I}_i = \frac{\tilde{V}_1}{(0.2/s) + j0.8}$$

and

$$\tilde{I}_o = \frac{\tilde{V}_1}{(0.6/s) + j0.2}$$

For any rotor resistance r_2, the torque developed is

$$T_d = \frac{3I^2 r_2}{s\omega_s}$$

If T_i and T_o are the torques developed by the inner and outer cages, respectively, then

$$\frac{T_o}{T_i} = \frac{I_o^2 r_o}{I_i^2 r_i} = \frac{(0.2/s)^2 + (0.8)^2}{(0.6/s)^2 + (0.2)^2} \times \frac{0.6}{0.2}$$

(a) At standstill, $s = 1$, and

$$\frac{T_o}{T_i} = 5.1$$

Thus, the torque developed by the outer cage is five times as much as that developed by the inner cage at standstill.

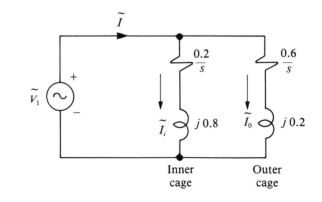

FIGURE 8.28
Equivalent
Circuit for a
Double-Cage
Induction Motor
for Example
8.12

(b) At 2% slip, $s = 0.02$, and

$$\frac{T_o}{T_i} = \frac{(0.2/0.02)^2 + 0.8^2}{(0.6/0.02)^2 + 0.2^2} \frac{0.6}{0.2} = 0.34$$

or
$$T_i = 2.98T_o$$

In other words, at the rated slip of 2%, the inner cage is most effective because it develops nearly three times as much torque as the outer cage.

■ EXAMPLE 8.13

The double-cage induction motor discussed in Example 8.12 has a stator impedance of $2.1 + j8.3\ \Omega$. Find the starting torque developed by the motor if the phase voltage is 110 V, 60 Hz.

Solution

The impedances of the inner and outer cages, respectively, at standstill are

$$\hat{z}_i = 0.2 + j0.8\ \Omega \quad \text{and} \quad \hat{z}_o = 0.6 + j0.2\ \Omega$$

The equivalent rotor impedance is

$$\hat{z}_r = \frac{\hat{z}_o \hat{z}_i}{\hat{z}_o + \hat{z}_i} = \frac{(0.2 + j0.8)(0.6 + j0.2)}{(0.2 + j0.8) + (0.6 + j0.2)} = 0.297 - j0.278\ \Omega$$

The total impedance is

$$\hat{z} = \hat{z}_s + \hat{z}_r = 2.1 + j8.3 + 0.297 + j0.278 = 8.906\underline{/74.39°}\ \Omega$$

The stator current is

$$\tilde{I}_2 = \frac{\tilde{V}_1}{\hat{z}} = \frac{110\underline{/0°}}{8.906\underline{/74.39°}} = 12.35\underline{/-74.39°}\ A$$

The torque developed is

$$T_d = 3I_2^2 \frac{r_2}{s} \frac{1}{\omega_s}$$

where r_2 is the effective rotor impedance and is equal to 0.336 Ω and ω_s is the synchronous speed of the motor.

$$\omega_s = \frac{4\pi f}{P} = \frac{4 \times \pi \times 60}{4} = 60\pi \text{ rad/s}$$

At standstill, $s = 1$. Thus,

$$T_d = \frac{3 \times 12.35^2 \times 0.297}{60\pi} = 0.72 \text{ N-m}$$

8.11   ROTOR IMPEDANCE TRANSFORMATION

Thus far we have tacitly assumed that whether the rotor is the wound or squirrel-cage (die-cast) type the rotor circuit elements can be transformed to the stator side in terms of an a ratio, which was defined on a per phase basis as the ratio of effective turns on the stator to the effective turns on the rotor. That is,

$$a = \frac{k_{\omega_1}N_1}{k_{\omega_2}N_2} = \frac{E_1}{E_b} \qquad (8.69)$$

The exact equivalent circuit based on the transformer analogy is given in Fig. 8.5. For a wound rotor that has the same number of poles and phases as that of the stator, the total turns per phase, N_2, and the winding factor, k_{ω_2}, can be calculated the same way as for the stator. However, the problem is somewhat more complex for the squirrel-cage rotor. Let us assume that there are P poles in the stator and Q bars on the rotor. Let us assume that one of these bars is under the middle of the north pole of the stator. There also exists another bar, which is in the middle of the adjacent south pole. The voltage induced in both bars is maximum but of opposite polarity. Thus, the maximum current will flow in these two bars. We can assume that these two bars make one turn. Thus, the total number of turns on the rotor is $Q/2$. The voltages are also induced in other bars. If the flux is distributed sinusoidally, the voltage induced will also follow the same pattern. However, the rms value of the voltage induced in each turn is the same. Since each turn is offset by one slot pitch on the rotor, the voltage induced in each bar is offset by that angle. Thus, we can say that each turn is equivalent to a phase group, and there are $Q/2$ phase groups in all. Since there are Q bars and P poles, the number of bars per pole is Q/P. Since each bar identifies a different phase group, the number of bars per pole is then equivalent to the number of phases on the rotor, m_2. That is,

$$m_2 = \frac{Q}{P} \qquad (8.70)$$

In other words, the number of bars per pole per phase is 1. Stated differently, the number of turns per pole per phase is ½. Now we can determine the total number of turns per phase by multiplying the number of turns per pole per phase by the number of poles. That is,

$$N_2 = \frac{P}{2} \tag{8.71}$$

Since the two bars that are 180° away form a turn, the winding factor is unity because (1) the pitch factor is unity as the turn is a full-pitch turn, and (2) the distribution factor is unity as there is only one turn in a phase group.

Since we are trying to refer the rotor circuit to the stator side, let m_1 be the number of phases, E_2 be the voltage, and I_2 be the current. For the equivalent representation to be valid, the apparent power of the original rotor must be the same as referred to the stator side. That is,

$$m_1 E_2 I_2 = m_2 E_b I_b \tag{8.72}$$

Since, on the stator side, the induced voltage is E_1, E_2 must then be equal to E_1. Thus,

$$I_2 = \frac{m_2 k_{\omega 2} N_2}{m_1 k_{\omega 1} N_1} I_b \tag{8.73}$$

The rotor resistance losses prior to and after the transformation must be equal. That is,

$$m_1 I_2^2 r_2 = m_2 I_b^2 r_b \tag{8.74}$$

Therefore,

$$r_2 = \frac{m_1}{m_2} \left(\frac{k_{\omega 1} N_1}{k_{\omega 2} N_2} \right)^2 r_b \tag{8.75}$$

Finally, the magnetic energy stored in the rotor leakage inductance before and after transformation must be the same. That is,

$$\frac{1}{2} m_1 I_2^2 \frac{x_2}{2\pi f} = \frac{1}{2} m_2 I_b^2 \frac{x_b}{2\pi f}$$

or

$$x_2 = \frac{m_1}{m_2} \left(\frac{k_{\omega 1} N_1}{k_{\omega 2} N_2} \right)^2 x_b \tag{8.76}$$

Equations (8.75) and (8.76) outline how the actual rotor circuit parameters for a squirrel-cage rotor can be transformed into parameters referred to the stator. r_2 and x_2 are the parameters that have been used in the equivalent circuit of an induction motor.

From the preceding equations, it is evident that the a ratio is

$$a = \sqrt{\frac{m_1}{m_2}} \left(\frac{k_{\omega_1} N_1}{k_{\omega_2} N_2} \right) \tag{8.77}$$

and for a wound rotor $m_1 = m_2$.

■ EXAMPLE 8.14

A 4-pole, 36-slot, double-layer-wound, three-phase motor has 10 turns per coil and a squirrel-cage rotor with 48 bars. The resistance and reactance of each bar are 20 $\mu\Omega$ and 2 $m\Omega$, respectively. Determine the equivalent rotor impedance as referred to the stator on a per phase basis.

Solution

Since the stator is wound using double-layer winding, there are 36 coils. The coils per pole per phase are $n = 36/(4 \times 3) = 3$. The pole spans $36/4 = 9$ slots. The slot span is $180/9 = 20°$ (electrical). The coil pitch, as can be determined from the developed diagram, is 7 slots or $140°$ (electrical). The pitch factor is $k_{p_1} = \sin(140/2) = 0.94$. The distribution factor is

$$k_{d_1} = \frac{\sin[3 \times (20/2)]}{3 \sin(20/2)} = \frac{\sin 30}{3 \sin 10} = 0.96$$

The winding factor is $k_{\omega_1} = 0.94 \times 0.96 = 0.9$. Total turns per phase are $10 \times 36/3 = 120$. It is assumed that all coils in a phase group are connected in series. The number of phases in the stator is $m_1 = 3$.

For the rotor, $k_{\omega_2} = 1$. The number of phases is $m_2 = Q/P = 48/4 = 12$. Turns per phase are $N_2 = P/2 = 4/2 = 2$. Thus, the a ratio is

$$a = \sqrt{\frac{3}{12}} \left(\frac{0.90 \times 120}{1 \times 2} \right) = 27.0$$

The rotor parameters as referred to the stator are

$$r_2 = a^2 r_b = 27.0^2 \times 20 \times 10^{-6} = 0.0146 \ \Omega$$

and

$$x_2 = a^2 x_b = 27.0^2 \times 2 \times 10^{-3} = 1.458 \ \Omega$$

8.12   SPEED CONTROL OF INDUCTION MOTORS

It was pointed out in the foregoing sections that the speed of the induction motor for the stable operation must be higher than that at which the motor develops maximum torque. For low-resistance motors, the slip at which the maximum torque occurs is usually less than 10%. Thus, for low-resistance motors, the speed regulation is within 5%. We must devise some methods to control its speed.

We already know that the synchronous speed is directly proportional to the frequency and inversely proportional to the number of poles; the motor speed at any slip, s, is

$$N_m = (1 - s)\,\frac{120f}{P}\ \text{(rpm)}$$

This equation highlights the fact that the speed of an induction motor can be controlled by changing the frequency of the applied voltage and the number of poles. We can also change the speed of operation of an induction motor by changing the applied voltage, changing the armature resistance, or introducing external voltage in the rotor circuit.

Frequency Control   Since the synchronous speed is directly proportional to the frequency of the applied voltage, by continuously increasing or decreasing the line frequency a wide variation in motor speed can be obtained. The only requirement is that we must have a variable-frequency supply. To maintain constant flux density and thereby the maximum torque developed, the applied voltage must be varied in direct proportion to the frequency. This is due to the fact that the induced voltage in the stator winding is directly proportional to the frequency. The speed–torque characteristics for an induction motor at four different frequencies are shown in Fig. 8.29. Also shown is a typical load curve. The intersections of the load and speed–torque curves indicate the speed of operation.

FIGURE 8.29
Motor Performance as a Function of Frequency for Adjusted Supply Voltage

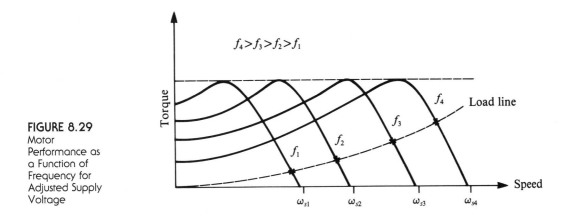

Changing Poles By increasing or decreasing the number of poles in the stator the synchronous speed (and hence the operating speed) of the motor can be decreased or increased, respectively. This method is quite suitable for squirrel-cage induction motors. The stator can be wound with two or more entirely independent windings. Each winding corresponds to a different number of poles. At any time, there is only one winding that is in operation. All other windings are disconnected. A motor wound for 4 and 6 poles can give us two synchronous speeds just by changing connections from one winding to another. This method of speed control, although limited, is very simple and has good speed regulation and high efficiency for each speed setting. However, with the help of some switches, a 4-pole winding can be reconnected to give an equivalent 8-pole winding. Similarly, a 6-pole winding, if reconnected, can perform as a 12-pole winding. Thus, a motor wound with two independent windings can yield four different synchronous speeds.

Rotor Resistance Control The speed of operation can be changed by causing a change in the slip at which the motor operates. One easy way to alter the speed–torque characteristic of a wound-rotor motor is to add an external resistance in the rotor circuit. From that point of view, this method is very similar to the armature resistance control method for dc shunt motors. However, with the increase in the rotor resistance the motor becomes inefficient. The speed–torque characteristics for three different values of the rotor resistance are shown in Fig. 8.22.

Stator Voltage Control Since the torque developed by the motor is proportional to the square of the applied voltage, the reduction in speed can be achieved by reducing the terminal voltage. Figure 8.30 shows typical speed–torque curves for various values of the terminal voltage. This method is convenient to use but is limited, because to achieve an appreciable change in speed a relatively large change in the applied voltage is required.

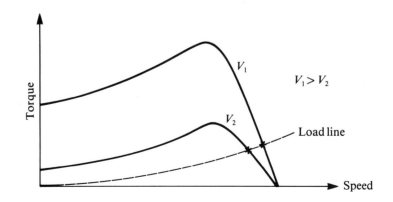

FIGURE 8.30
Supply Voltage
Effect on Motor
Performance

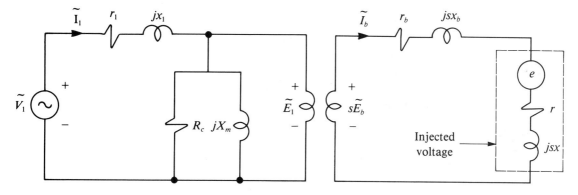

FIGURE 8.31
Equivalent Circuit of an Induction Motor with an External Source in the Rotor Circuit

Introducing Voltage in the Rotor Circuit   We can alter the operation of a three-phase induction motor by introducing a voltage in the secondary circuit consisting of a wound rotor, as shown in Fig. 8.31. The frequency of the injected voltage must be equal to the rotor frequency. However, there is no restriction on the phase of the injected voltage. If the phase of the injected voltage is the same as the phase of the induced voltage in the secondary, the rotor current will increase. The secondary circuit of the motor will appear to have a lower resistance. On the other hand, if the introduced voltage is in phase opposition to the induced voltage, the decrease in the rotor current will amount to an increase in rotor resistance. Therefore, by changing the phase of the injected voltage, we can change the effective resistance of the rotor circuit and hence the speed of operation of the motor. Speed control can also be achieved by varying the magnitude of the inserted voltage.

PROBLEMS

8.1. The field produced by the primary winding of a three-phase induction motor revolves at a speed of 1000 rpm. If the frequency of the applied voltage is 50 Hz, determine the number of poles in the motor.

8.2. The frequency of the induced voltage in the secondary winding of an 8-pole three-phase induction motor is 10 Hz. At what speed does the mmf of the secondary revolve with respect to the secondary winding?

8.3. A 2-pole, 230-V, 50-Hz, three-phase induction motor is running at a speed of 2800 rpm. Determine (a) per unit slip and (b) the frequency of the induced voltage in the secondary circuit.

8.4. A 10-pole, 120-V, 50-Hz, three-phase induction motor is designed to operate at a slip of 5% on full load. Determine (a) the rated speed, (b) the rotor frequency, and (c) the speed of the rotor revolving field with respect to the rotor and (d) with respect to the stator.

8.5. A three-phase induction motor operating at a slip of 3% has a rotor copper loss of 300 W. The rotational loss of the motor is 1500 W. Determine (a) the air-gap power and (b) the power output.

8.6. A 10-hp, 440-V, 60-Hz, 4-pole, three-phase induction motor is designed to run at 3% slip on full load. The rotational loss is 4% of the output. Compute at full load the (a) rotor copper loss, (b) air-gap power, (c) power developed and (d) torque available at the shaft.

8.7. A 4-hp, 230-V, 60-Hz, 6-pole, three-phase induction motor operates at 1050 rpm on full load. The rotational losses are 300 W. Determine the equivalent rotor resistance on a per phase basis if the rotor current is not to exceed 100 A.

8.8. A 4-pole, 230-V, 60-Hz, Y-connected, three-phase induction motor has the following parameters on a per phase basis: $r_1 = 73.12 \ \Omega$, $x_1 = 38.61 \ \Omega$, $r_2 = 31.97 \ \Omega$, $x_2 = 11.56 \ \Omega$, and $X_m = 432.48 \ \Omega$. The core loss is 10.72 W and the friction and windage loss is 1.9 W. Determine the (a) stator current, (b) magnetization current, (c) rotor current, (d) power input, (e) stator copper loss, (f) rotor copper loss, (g) power output, (h) shaft-torque developed, and (i) efficiency of the motor at its rated speed of 1550 rpm using the exact equivalent circuit. (Ignore the equivalent core-loss resistance.)

8.9. A 230-V, 4-pole, 60-Hz, three-phase induction motor draws a line current of 4.2 A at 0.9 pf lagging while operating at 1725 rpm. The friction and windage losses are 25 W, the core loss is 75 W, and the stator copper loss is 100 W. Determine the power input, air-gap power, rotor copper loss, power output, shaft torque, and efficiency of the motor. Assume that the core loss is a part of the input power.

8.10. A 230-V, 60-Hz, 6-pole, Y-connected, three-phase induction motor is rated at 1125 rpm. It has the following parameters in ohms per phase as referred to the stator: $r_1 = 12.5$, $r_2 = 28.6$, $x_1 = 21.3$, $x_2 = 13.6$, and $X_m = 449.1$. The core loss is 5 W and the mechanical losses are 1.75 W. Determine the power input, stator copper loss, rotor copper loss, air-gap power, power output, available torque, and efficiency of the motor using the exact equivalent circuit. Treat core loss as part of the input.

8.11. Determine the breakdown slip and the maximum torque developed by the motor whose parameters are given in Problem 8.8 using the exact expressions.

8.12. Consider the motor whose parameters are given in Problem 8.10. Find the maximum torque developed by the motor using exact expressions.

8.13. A three-phase induction motor has an equivalent rotor impedance of 0.04 + $j0.12$ Ω per phase. What is the per unit slip at maximum torque? What is the rotor resistance if the starting torque of the motor is 75% of its maximum value?

8.14. A 440-V, Y-connected, three-phase induction motor has the following parameters on a per phase basis as referred to the stator: $r_1 = 0.25$ Ω, $x_1 = 0.8$ Ω, $r_2 = 0.5$ Ω, $x_2 = 0.8$ Ω, $R_c = 150$ Ω, $X_m = 50$ Ω, and $N_s = 1800$ rpm. Calculate (a) the breakdown slip, (b) the maximum torque, and (c) the power factor at a slip of 4% using an approximate equivalent circuit.

8.15. A 10-hp, 660-V, 60-Hz, Y-connected, three-phase induction motor runs at 1125 rpm on full load. The unregulated supply voltage fluctuates from 600 to 720 V. What is the torque range of the motor if the speed is held constant? What is the speed range for the motor to develop the same torque?

8.16. A 2-hp, 120-V, 60-Hz, 6-pole, three-phase induction motor operates at a speed of 1050 rpm on full load. Under reduced load the speed increases to 1125 rpm. Determine the torque developed and the power output of the motor when the motor operates at (a) full load and (b) reduced load.

8.17. The following data were obtained on a 230-V, 60-Hz, 4-pole, Y-connected, three-phase induction motor:

No-load test: input power = 130 W, line current = 0.45 A at rated supply voltage

Blocked-rotor test: power input = 65 W, line current = 1.2 A at reduced line voltage of 47 V

The friction and windage losses are 15 W and the resistance between any two lines is 4.1 Ω. Calculate the motor circuit parameters and draw its exact equivalent circuit.

8.18. The following test data were obtained on a 460-V, 60-Hz, 6-pole, delta-connected, three-phase induction motor:

No-load test: power input = 380 W, line current = 1.15 W at its rated voltage

Blocked rotor test: power input = 14.7 W, line current = 2.1 A at a reduced line voltage of 21 V

The mechanical losses are 21 W and the winding resistance between any two lines is 1.2 Ω. Determine the approximate equivalent circuit of the motor. Obtain the motor efficiency at its rated slip of 5%.

8.19. A double-cage, three-phase, 6-pole, Y-connected induction motor has the following impedances on a per phase basis at standstill: inner cage = 0.1 + $j0.6$ Ω; outer cage = 0.4 + $j0.1$ Ω. Determine the ratio of the torques developed by the two cages at (a) standstill and (b) 5% slip. What is the slip at which the torque developed by the two cages is the same?

8.20. If the motor in Problem 8.19 is connected across a line voltage of 230 V, 60 Hz and has a stator impedance of $1.5 + j2.5 \ \Omega$, obtain the torques developed by the motor at (a) standstill and (b) 5% slip.

8.21. In a three-phase wound-rotor induction motor the primary has twice as many turns as the secondary. The winding distribution factor for the primary is 0.85 and that for the secondary is 0.8. The impedance of the secondary under a blocked-rotor condition is $0.18 + j0.25 \ \Omega$. What is its impedance as referred to the primary?

CHAPTER 9

Single-Phase Motors

Single-Phase Induction Motor (Courtesy of
Bodine Electric Company)

In the previous chapter we discussed how the torque is developed in a three-phase induction motor. It is an ingenious scheme of placing the three-phase windings 120° apart and feeding them from a three-phase power supply. It resulted in a flux that was constant in magnitude, but revolved around the periphery of the rotor with the synchronous speed. This revolving flux produced the cutting action and consequently induced the voltages in the rotor conductors. Since the rotor conductors were closed onto themselves, the induced voltage produced a current in them. The interaction of the rotor current and the revolving magnetic field resulted in the rotation of the rotor.

Consider a situation when one of the three phases of the supply is not connected to one of the three-phase windings of the motor. Will the rotor be able to start from its standstill position and take up some mechanical load? The answer to this is definitely yes, as we will soon see. Such a motor can be labeled a two-phase motor because only two windings are being excited by two of the phases. Let us go one step further. What will happen if only one of the three phases of an induction motor is excited? At this time, you will be surprised to notice that if the rotor is already rotating it will continue rotating in the same direction. However, if the rotor was initially stationary and then one of the phases of a three-phase motor is excited by a single-phase supply, the rotor will buzz but not rotate. An induction motor in which only one phase of a three-phase power supply is used is referred to as a single-phase induction motor. In such a motor, we only need one single-phase winding to keep the motor running. However, such a motor is not self-starting. Therefore, we must provide some means to start the motor. To get some ideas on how we might start such a motor let us examine the operation of a two-phase induction motor.

9.1 TWO-PHASE INDUCTION MOTOR

In construction, a two-phase induction motor is very similar to a three-phase induction motor with the exception that it requires only two identical phases for its operation. Since a balanced two-phase power supply has two identical phase voltages that are 90° apart, the two phase windings must be placed in space quadrature.

Let us consider the effect of exciting the two windings of a two-pole, two-phase induction motor from a balanced alternating-current two-phase power source. Figure 9.1 shows the current waveforms of a balanced two-phase power source. The two waveforms can also illustrate the behavior of the fluxes produced by each of the two phases for one complete cycle.

In the usual representation of a two-phase induction motor, the two phase windings are drawn at right angles to each other, as shown in Fig. 9.2. The standard terminal markings are also indicated. S_1 and F_1 are the starting and finishing ends for phase 1, while S_2 and F_2 are for phase 2. For our reference purposes, let us assume that the current flow in the windings will be from S to F when the current

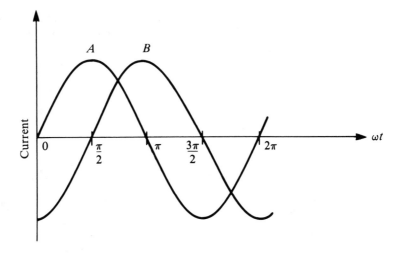

FIGURE 9.1
Two-Phase
Supply
Waveforms

is positive. At the same time, the windings are so wound that a coil produces a south pole when the current direction is marked by an arrow in the clockwise direction.

At time $t = 0$, the current in the first-phase winding is zero. Therefore, no flux is produced by the first-phase winding, which is marked as 1 and 1'. On the other hand, the current in the second-phase winding is at its maximum value in the negative direction. Thus, it is flowing from the terminal F_2 toward S_2, as

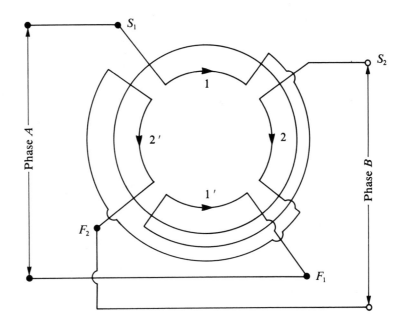

FIGURE 9.2
Stator Windings
of a Two-Phase
Induction Motor

shown in Fig. 9.3a. This is the only winding that is producing any flux in the motor, and the direction of the flux is in accordance with our assumptions. Therefore, the phase-group 2 acts like a north pole, while 2′ is a south pole and the magnetic axis of the flux is horizontal.

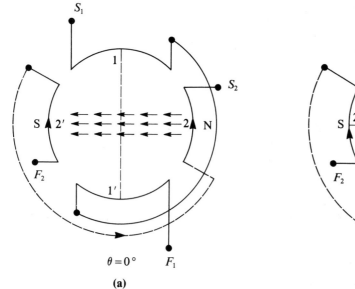

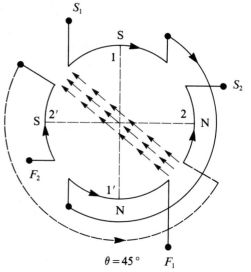

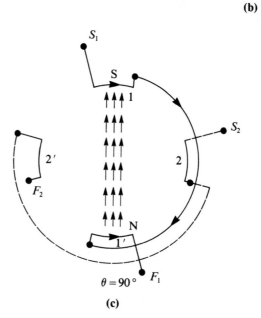

FIGURE 9.3
Production of Revolving Flux in a Two-Phase Motor

One-eighth of a cycle later (i.e. at $\omega t = 45°$), the current in the first-phase winding is now positive, while that in the second-phase winding is still negative, as indicated in Fig. 9.3b. Since phase groups 2 and 1′ are now carrying currents in the same direction, they produce a flux in the same direction. It can therefore be said that these two groups act as a single group with its magnetic axis 45° from the horizontal axis. The same is true for phase groups 1 and 2′. The direction of flux is now at an angle of 45° with respect to the horizontal axis. In other words, a 45° advancement in the phase currents has resulted in a 45° (electrical) shift in the magnetic flux.

A quarter-cycle later (i.e., at $\omega t = 90°$), the first-phase current is at its positive peak, while the second-phase current has reduced to zero. At this instant only the first-phase winding is carrying current from S_1 to F_1. Coil 1 acts a south pole, while coil 1′ behaves like a north pole. The direction of flux is therefore vertically upward, as illustrated in Fig. 9.3c. A time phase increment of 90° in the phase currents has rotated the magnetic flux in the motor by 90° (electrical).

If this process is continued, after one complete cycle, the orientation of the magnetic axis would be same as shown in Fig. 9.3a. Therefore, one complete cycle of currents in two phases of a two-phase power source corresponds to a complete cycle of rotation of the flux in a two-phase, two-pole motor. This situation is similar to what we studied earlier for three-phase induction motors.

The flux produced by each phase of a multiphase motor pulsates from zero to a maximum value in both the positive and negative directions, as shown in Fig. 9.1. The combination of the fluxes produces the resultant field revolving at its synchronous speed. The revolving field is responsible for the production of torque in a multiphase motor. Since the field is revolving in the clockwise direction, the rotor will also rotate in that direction.

To recapitulate, a true revolving field will be set up in a two-phase motor if the following conditions are satisfied:

1. The two windings are symmetric and are distributed in space quadrature.

2. The power source is a balanced type with quadrature time–phase difference between the phases.

It can now be construed that the minimum number of phases in a multiphase motor must be at least 2 in order to produce a revolving field and consequently make the motor self-starting.

The strength of the rotating magnetic field is constant. Let us illustrate this point by considering a few time intervals. If we assume that the fluxes in the motor have similar waveforms, as shown in Fig. 9.1, the flux created by the first phase windings leads the flux produced by the second phase winding by 90° (electrical). Let

$$\phi_a = \phi_m \sin \omega t \qquad (9.1)$$

and

$$\phi_b = \phi_m \sin(\omega t - 90°) \qquad (9.2)$$

where
ϕ_a = flux produced by the first phase winding
ϕ_b = flux produced by the second phase winding
ϕ_m = maximum value of the flux

At $\omega t = 0$,

$$\phi_a = 0 \quad \text{while} \quad \phi_b = \phi_m$$

The resultant flux is

$$\phi_r = \phi_b = \phi_m \tag{9.3}$$

When $\omega t = 30°$, $\phi_a = \phi_m \sin 30°$, $\phi_b = -\phi_m \sin 60°$, and

$$\phi_r = \sqrt{\phi_a^2 + \phi_b^2} = \phi_m\sqrt{\sin^2 30° + \sin^2 60°} = \phi_m \tag{9.4}$$

Once again the resultant flux is equal to ϕ_m. It can therefore be said that at any angle θ the resultant flux is not only constant but its magnitude is equal to the maximum value of the flux produced by either winding.

9.2 SINGLE-PHASE MOTOR

The fact that there must be at least two phase windings in a motor connected to a two-phase power supply in order to produce a revolving field plays an important role in the design of a single-phase induction motor. Strictly speaking, a single-phase induction motor will have only a one-phase winding, which is excited by a single-phase power supply. Since most single-phase induction motors are in the fractional-horsepower range, each phase group may have more than one coil, which are wound concentric to each other. A cross-sectional view of a single-phase two-pole motor with a squirrel-cage rotor is shown in Fig. 9.4. For clarity, we have shown rotor conductors on the outer periphery of the rotor. In an actual motor, the rotor conductors are usually die-cast into the slots in the rotor.

Let us suppose that the supply voltage is increasing in the positive direction and causing a current flow, as indicated in Fig. 9.4. The current flowing through each winding produces a magnetic flux, which is also increasing in the upward direction. A rotor conductor forms a full-pitch closed loop through another rotor conductor, which is 180° (electrical) apart. The conductors can then be paired as shown. Let us consider one of these pairs, say the loop formed by conductors 2 and 2'. The flux is passing through this loop and is increasing in the upward direction. A voltage is induced by this flux in the loop. Since the loop forms a closed circuit with some resistance, current flows in that loop. The direction of the current in the loop is such that it produces a magnetic flux, which tends to oppose the increase in the magnetic flux. For that to happen, the current must flow out of conductor 2 and into the conductor 2′, as shown by the dots and crosses, respectively. The same would be true for conductors 3–3′, 4–4′, 5′–5, 6′–6, and 7′–7. However, there must be no current in the loop 1–1′. Now we have conductors carrying currents at right angles to the magnetic field in the motor. Each current-carrying conductor must experience a force, as indicated in the Fig. 9.4. As is evident, the net torque developed by the motor is zero. Hence the rotor will not rotate. However, if the rotor is made to rotate in any direction while the single-phase winding is excited, the motor will develop torque in that direction. There are two basic theories to explain why a torque is produced in the rotor once it is rotating: the *double revolving field theory* and the *cross-field theory*. In this book, we will restrict our discussion to the double revolving field theory.

Double Revolving Field Theory   According to this theory, a magnetic field that pulsates in time but is stationary in space can be resolved into two magnetic fields that are equal in magnitude but rotate in opposite directions. Let us consider the standstill condition again. The magnetic field produced by the motor shown in

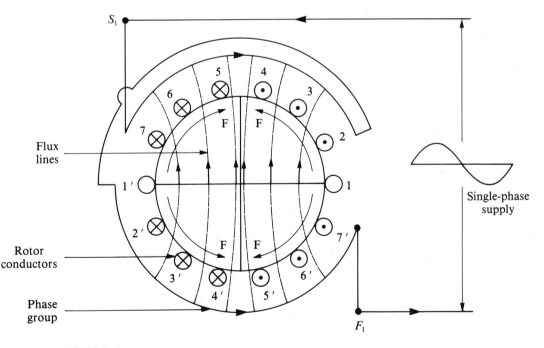

FIGURE 9.4
Field Current Interactions in a Two-Pole, Single-Phase Motor

Fig. 9.4 pulsates up and down with time, and at any instant its magnitude can be given as

$$B = B_m \cos \omega t \tag{9.5}$$

where B_m is the maximum flux density in the motor.

The flux density, B, can be resolved into two components, B_1 and B_2, such that the magnitude of B_1 is equal to the magnitude of B_2; but B_1 is rotating in the clockwise direction, while B_2 is rotating in the counterclockwise direction, as shown in Fig. 9.5.

Now we have two revolving fields of constant magnitude rotating in opposite directions. Each magnetic field tends to produce torque in its direction of rotation. At standstill, both magnetic fields are of equal strength and each tends to develop the same amount of torque. Since the torques thus produced are equal in magnitude but opposite in direction, the net torque developed by the motor is zero. This is the same conclusion we have arrived at before. However, we have gained some insight into the working of a single-phase motor according to the double revolving field theory. We have come to the understanding that a single-phase motor in fact behaves exactly as two three-phase motors that are producing torques in opposite directions. The magnetic-field strength of each motor is exactly half the maximum strength of the field produced by one phase. On this basis, we can develop an equivalent circuit for a single-phase induction motor as shown in Fig. 9.6. Note that the magnetization reactance and the rotor circuit impedance are decomposed into two sections to highlight the effect of two equal and opposite rotating fields. One branch is usually referred to as the forward branch, while the other is called the backward branch. If the rotor is to rotate in the clockwise direction, the forward branch will be the branch that represents the effect of the

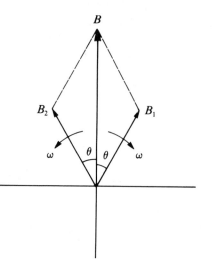

FIGURE 9.5
Resolution of a Pulsating Vector into Two Equal and Oppositely Revolving Vectors

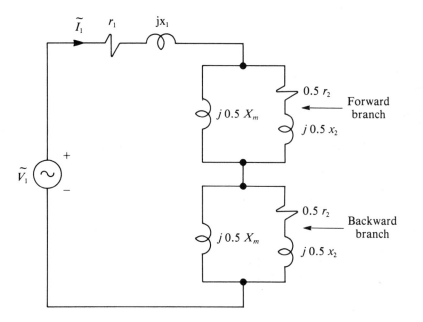

FIGURE 9.6
Equivalent
Circuit of a
Single-Phase
Motor under a
Standstill
Condition

rotating field in the clockwise direction. The backward branch will correspond to the counterclockwise rotating field.

In three-phase motors we included the effect of speed in the form of per unit slip and incorporated it in representing the effective rotor resistance. Let us now assume that the rotor is rotating in the clockwise direction with an angular velocity of ω_m. The magnetic field rotating in the clockwise direction has a synchronous angular velocity of ω_s, while the synchronous angular velocity of the counter-clockwise rotating field is $-\omega_s$. The per unit slip for the motor producing torque in the forward direction is

$$s = \frac{\omega_s - \omega_m}{\omega_s} = 1 - \frac{\omega_m}{\omega_s} \tag{9.6}$$

The per-unit slip of the motor producing torque in the counterclockwise or backward direction is

$$s_b = \frac{-\omega_s - \omega_m}{-\omega_s} = 1 + \frac{\omega_m}{\omega_s} = 2 - s \tag{9.7}$$

Note that at standstill $s = s_b = 1$. We can now incorporate the effect of slips in the forward and backward branches of the single-phase induction motor as we did for the three-phase motor. The modified equivalent circuit is shown in Fig. 9.7. In our discussion of three-phase induction motors, we found that the torque

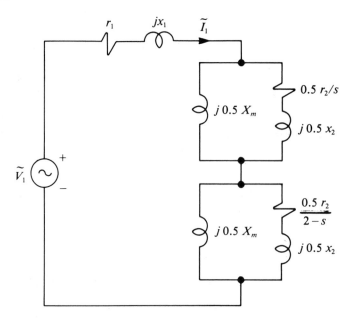

FIGURE 9.7
Equivalent
Circuit of a
Single-Phase
Motor at any
Slip s in the
Forward
Direction

developed by the motor is proportional to its effective resistance in the rotor branch. At standstill, $s = 1$, the effective resistance in both the branches is the same. Therefore, the torques developed by the forward and backward equivalent motors are equal and opposite. On the other hand, for any slip s less than unity, the effective rotor resistance in the forward branch is higher than that in the backward branch. Thus, the torque developed by the forward revolving field is more than that developed by the backward revolving field. The resultant torque is in the forward direction, which will tend to maintain the rotation in that direction. This is why the motor will continue developing torque in one direction as long as it is rotating in that direction. The speed–torque curves for the two revolving fields are shown in Fig. 9.8 for slips varying from 0 to 2.

■ EXAMPLE 9.1

A 115-V, 60-Hz, 4-pole, single-phase induction motor is rotating in the clockwise direction at a speed of 1710 rpm. Determine the per unit slip of the motor in (a) the direction of rotation and (b) the backward direction.

Solution

$$N_s = \frac{120f}{p} = \frac{120 \times 60}{4} = 1800 \text{ rpm}$$

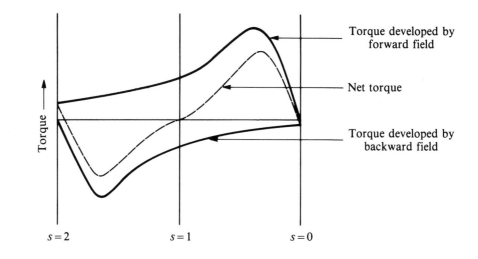

FIGURE 9.8
Speed–Torque
Characteristics of
a Single-Phase
Induction Motor

$s = 2$ $\qquad\qquad$ $s = 1$ $\qquad\qquad$ $s = 0$

(a) The slip in the forward direction is

$$s = \frac{N_s - N_m}{N_s} = \frac{1800 - 1710}{1800} = 0.05 \quad \text{or} \quad 5\%$$

(b) The slip in the backward direction is

$$s_b = 2 - s = 2 - 0.05 = 1.95$$

■ EXAMPLE 9.2

If the actual rotor resistance of the motor discussed in Example 9.1 is 12.5 Ω referred to the stator side, determine the effective rotor resistance in the forward and backward branches.

Solution

Since the per unit slip of the motor is 0.05, the effective rotor resistance in the forward branch is

$$\frac{r_2}{s} = \frac{12.5}{0.05} = 250 \ \Omega$$

The effective resistance in the backward branch is

$$\frac{r_2}{s_b} = \frac{12.5}{1.95} = 6.41 \ \Omega$$

9.3   ANALYSIS OF A SINGLE-PHASE MOTOR ─────────────────────

As the equivalent circuit of a single-phase induction motor was developed on the basis of the equivalent circuit for a three-phase motor, we must be able to analyze a single-phase motor by using the relations developed for the three-phase motor. However, we have to treat the single-phase motor as a pair of three-phase motors, one developing the power and torque in one direction while the other is doing the same in the opposite direction. The net power developed would then be the difference of the powers developed in the forward and backward directions. The equivalent circuit of Fig. 9.7 can be simplified and redrawn as shown in Fig. 9.9, where

$$\hat{Z}_f = R_f + jX_f = 0.5 \frac{jX_m[(r_2/s) + jx_2]}{(r_2/s) + j(x_2 + X_m)} \tag{9.8}$$

is the effective impedance in the forward branch and

$$\hat{Z}_b = R_b + jX_b = 0.5 \frac{jX_m\left(\dfrac{r_2}{2-s} + jx_2\right)}{[r_2/(2-s)] + j(x_2 + X_m)} \tag{9.9}$$

is the effective impedance in the backward branch. The current is

$$\tilde{I}_1 = \frac{\tilde{V}_1}{r_1 + jx_1 + \hat{Z}_f + \hat{Z}_b} \tag{9.10}$$

The power input is

$$P_{in} = V_1 I_1 \cos\theta \tag{9.11}$$

where θ is the power factor angle by which the current $\tilde{I}_1$ lags the applied voltage $\tilde{V}_1$. The stator copper loss is

$$SCL = I_1^2 r_1 \tag{9.12}$$

If we subtract the stator copper loss from the total power input, we are left with the total power in the air-gap region. However, the air-gap power is distributed between the air-gap powers due to the forward and backward revolving fields. In the three-phase induction motor, the air-gap power was found to be equal to $I_2^2 r_2/s$ on a per phase basis, where $\tilde{I}_2$ was the current in the rotor circuit flowing through the effective rotor resistance r_2/s. If we have to determine the air-gap powers due to the forward and backward revolving fields, we should first determine the effective currents in the equivalent rotor circuits. This appears to be a

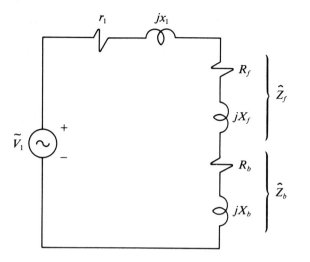

FIGURE 9.9
Simplified
Equivalent
Circuit of a
Single-Phase
Induction Motor

tedious task, which can simply be avoided. The only resistance that consumes the total air-gap power due to each revolving field is r_2 divided by the per unit slip for that revolving field. If $\hat{Z}_f$ and $\hat{Z}_b$ are the equivalent impedances for the forward and backward branches, then the power consumed in R_f and R_b must be the same as would have been consumed by r_2/s and $r_2/(2 - s)$, respectively. Therefore, the air-gap power in the forward revolving field is

$$P_{gf} = I_1^2 R_f \qquad (9.13)$$

and in the backward revolving field is

$$P_{gb} = I_1^2 R_b \qquad (9.14)$$

The net air-gap power is

$$P_g = P_{gf} - P_{gb} = I_1^2(R_f - R_b) \qquad (9.15)$$

From the analysis of the three-phase motor, the torque developed by the motor is the ratio of air-gap power to the synchronous angular velocity of the revolving field. In a single-phase motor, the net air-gap power is in the forward direction. Thus, the torque developed by the single-phase motor is

$$T_d = \frac{P_g}{\omega_s} \qquad (9.16)$$

If we subtract the rotor copper losses in the rotor circuits of the single-phase motor, we can obtain the mechanical power developed by the motor. However, in a three-phase induction motor we found an easy way of calculating the mechanical power developed in terms of air-gap power and the per unit slip. Using the same analogy, the net mechanical power developed by the single-phase motor is

$$P_d = (1 - s)P_g \tag{9.17}$$

The power available at the shaft is

$$P_o = P_d - P_{rot} \tag{9.18}$$

The load torque of the motor is

$$T_s = \frac{P_o}{\omega_m} \tag{9.19}$$

Finally, the motor efficiency is the ratio of power available at the shaft to the total power input.

■ EXAMPLE 9.3

A 120-V, 60-Hz, $\frac{1}{3}$-hp, 4-pole, single-phase induction motor has the following circuit parameters: $r_1 = 2.5\ \Omega$, $x_1 = 1.25\ \Omega$, $r_2 = 3.75\ \Omega$, $x_2 = 1.25\ \Omega$, and $X_m = 65.0\ \Omega$. The motor is running at a speed of 1710 rpm and has a core loss of 25 W. The friction and windage loss is 2 W. Determine the motor performance assuming that the stray load loss is zero.

Solution

The synchronous speed is

$$N_s = \frac{120f}{P} = \frac{120 \times 60}{4} = 1800 \text{ rpm}$$

The per unit slip is

$$s = \frac{N_s - N_m}{N_s} = \frac{1800 - 1710}{1800} = 0.05$$

$$\hat{Z}_f = 0.5 \times \frac{j65[(3.75/0.05) + j1.25]}{(3.75/0.05) + j(1.25 + 65)} = 15.82 + j18.52\ \Omega$$

$$\hat{Z}_b = 0.5 \times \frac{j65[(3.75/1.95) + j1.25]}{(3.75/1.95) + j(1.25 + 65)} = 0.92 + j0.64\ \Omega$$

The total impedance is

$$\hat{Z} = r_1 + jx_1 + \hat{Z}_f + \hat{Z}_b$$
$$= 2.5 + j1.25 + 15.82 + j18.52 + 0.92 + j0.64$$
$$= 19.24 + j20.41 = 28.05\underline{/46.69°}\ \Omega$$

Thus, the primary current is

$$\tilde{I}_1 = \frac{120\underline{/0°}}{28.05\underline{/46.69°}} = 4.28\underline{/-46.69°}\ A$$

Power input:

$$P_{in} = 120 \times 4.28 \times \cos(46.69) = 352.3\ W$$

Power in the forward branch:

$$P_{gf} = I_1^2 R_f = 4.28^2 \times 15.82 = 289.8\ W$$

Power in the backward branch:

$$P_{gb} = I_1^2 R_b = 4.28^2 \times 0.92 = 16.85\ W$$

Air-gap power:

$$P_g = P_{gf} - P_{gb} = 289.8 - 16.85 = 272.95\ W$$

Gross power developed:

$$P_d = (1 - s)P_g = 0.95 \times 272.95 = 259.3\ W$$

Net power output:

$$P_o = 259.3 - 25 - 2 = 232.3\ W$$

Efficiency:

$$\eta = \frac{P_o}{P_{in}} = \frac{232.3}{352.3} = 0.659 \quad \text{or} \quad 65.9\%$$

Angular velocity:

$$\omega_m = \frac{2\pi N_m}{60} = \frac{2\pi \times 1710}{60} = 179.07\ rad/s$$

Load torque:

$$T_s = \frac{P_o}{\omega_m} = \frac{232.3}{179.07} = 1.297\ \text{N-m}$$

9.4 SPLIT-PHASE MOTOR

A single-phase induction motor usually has two windings that are wound in quadrature with respect to each other. One winding, known as the main winding, is active all the time while the other winding, known as the auxiliary winding, may be opened at a predetermined speed. The main winding is very inductive in nature

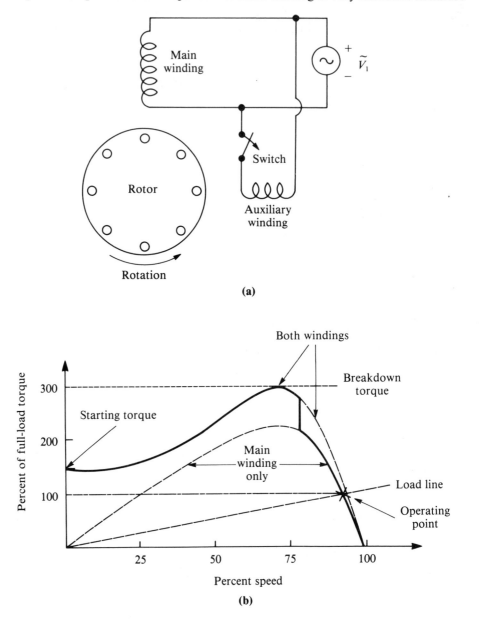

FIGURE 9.10
(a) Schematic Representation of a Split-Phase Motor; and (b) Its Typical Speed–Torque Characteristic

and is wound with a large wire to keep the winding resistance and consequently the copper losses to a minimum. If the auxiliary winding is made highly resistive by using comparatively thinner wire and has very few turns for low inductance, the single-phase induction motor is referred to as a *split-phase* motor. The currents in the two windings will have a phase difference varying from 30° to 45°. The rotor is of conventional squirrel-cage type. Since the windings are wound in quadrature, the motor behaves as an unbalanced two-phase motor and will establish a revolving field. The revolving field will have unequal strength around the rotor periphery. However, the motor develops a starting torque that is usually lower than its maximum torque. The starting torque is typically 150 to 200 percent of the full-load torque and the starting current is 6 to 8 times the full-load current. When the motor has reached a certain speed, normally 75% of its synchronous speed, the auxiliary winding is "cut out" from the circuit. If the auxiliary winding is left in the circuit, it will soon burn out due to excessive current. The auxiliary winding is opened with the help of a centrifugal switch that is rugged in construction, more reliable in operation, and costs less than other electronic switching schemes used for this purpose.

Split-phase induction motors are commonly designed for mechanical duty applications and are in the fractional-horsepower range. A schematic representation of a split-phase induction motor is shown in Fig. 9.10 along with a typical speed–torque curve. Notice the drop in torque at the time the auxiliary winding is opened.

9.5 CAPACITOR MOTORS

In a capacitor motor, an external capacitor is added in series with the auxiliary winding to provide the needed phase difference between the currents in two windings. If the capacitor value is properly chosen, the performance of the capacitor motor comes very close to that of a two-phase motor as far as the development of the revolving field and production of the torque are concerned. There are three types of capacitor motors.

Capacitor-start Motors   In a capacitor-start motor, the capacitor selection depends on the starting torque to be developed by the motor. The auxiliary winding and the capacitor are removed from the circuit as soon as the motor attains a speed of 75% or so of its synchronous speed. As is evident, a capacitor-start motor is more expensive than a split-phase motor. However, capacitor-start motors are needed when the starting torque requirements are four to five times the rated running torque. Such a high starting torque is not within the realm of a split-phase induction motor. On the other hand, the capacitor is used only during the starting phase; its duty cycle is very intermittent. Thus, an inexpensive and relatively small ac electrolytic-type capacitor can be used for all capacitor-start induction-run motors. Figure 9.11 shows a schematic representation of a capacitor-

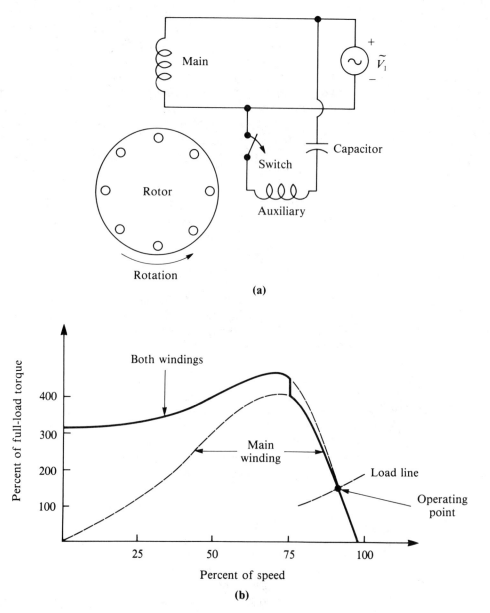

FIGURE 9.11
Capacitor-Start
Motor: (a)
Schematic
Representation;
(b) Typical
Speed–Torque
Characteristic

start induction-run motor. A typical speed–torque curve is also shown in the figure.

Capacitor-start, Capacitor-run Motors   Both the capacitor-start and split-phase induction-run motors satisfy the rated torque requirements, but the only drawback is the low power factor at the rated speed. Since the main winding is mostly inductive, the power-factor angle is low compared to a polyphase induction motor. The lower the power factor angle, the higher is the power input to the motor for the same power output. Therefore, the efficiency of the motor is low. For fractional-horsepower motors, the efficiency is usually 50% to 60% for split-phase and capacitor-start motors. For the same applications a three-phase induction motor may have an efficiency of 75% to 85%.

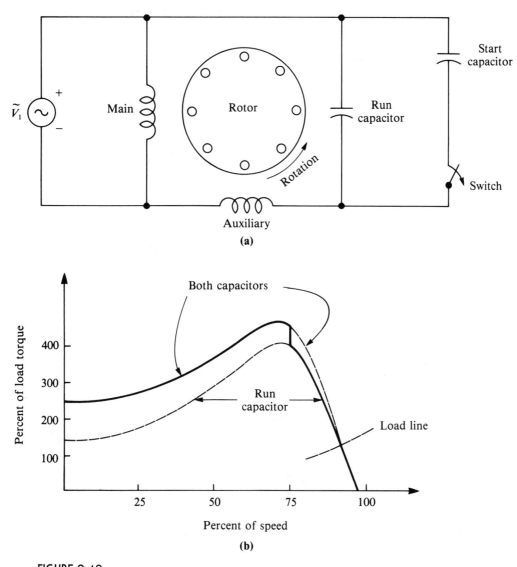

(a)

(b)

FIGURE 9.12
Capacitor-Start, Capacitor-Run Motor: (a) Schematic Representation; (b) Typical Speed–Torque Characteristic

To make a single-phase induction motor more efficient, further techniques must be devised. A capacitor-start, capacitor-run motor is one which is more efficient. The start capacitor is selected on the basis of starting torque requirements, while the run capacitor is designed for running performance. Hence, the auxiliary winding stays in the circuit at all times the motor is in operation. However, the centrifugal switch helps in selecting the proper size of capacitor. The starting capacitor is of the ac electrolytic type. The running capacitor is usually an ac oil-type capacitor that is rated for continuous duty. These motors are generally employed for mechanical duty, direct drive applications. A schematic re-

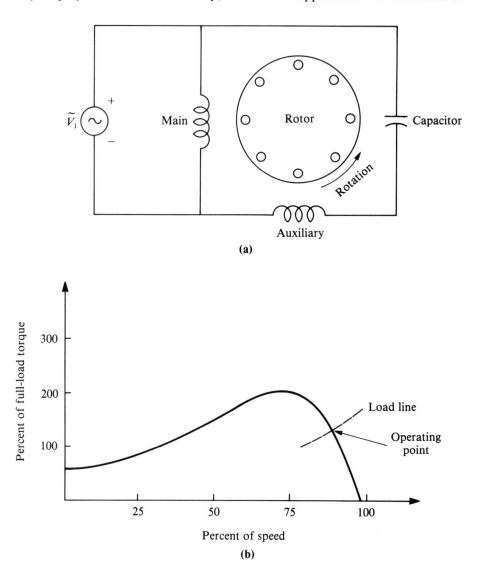

FIGURE 9.13
Permanent Split-Capacitor Motor:
(a) Schematic Representation;
(b) Typical Speed–Torque Characteristic

presentation of a capacitor-start, capacitor-run motor, also known as a two-value capacitor motor, is shown in Fig. 9.12 along with its typical speed–torque curve.

Permanent Split Capacitor Motors   A less expensive version of a capacitor-start, capacitor-run motor is a permanent split capacitor (PSC) motor. The same capacitor is used for starting and running without any switching. Due to the absence of the centrifugal switch, the overall length of the motor is smaller compared to other motors of the single-phase type, and the motor is less expensive. However, the starting torque is low for permanent split capacitor motors. Such motors are therefore suitable for such applications as fans where the starting torque requirements are minimal. These motors are also good candidates for applications that require frequent starts. Other types of single-phase induction motors tend to overheat when started frequently and this badly affects the reliability of the system. With fewer rotating parts, the motor is usually quieter and has a high efficiency in the range of 60% to 70% at full load. The equivalent circuit of a permanent split capacitor motor with its speed–torque characteristic is shown in Fig. 9.13.

9.6   ANALYSIS OF SINGLE-PHASE INDUCTION MOTORS USING BOTH WINDINGS _____

In the foregoing sections we directed our efforts at determining the performance of a single-phase induction motor running on the main winding only. The auxiliary winding, although physically present, was not included in sharing the load at the rated speed. As pointed out earlier, the auxiliary winding plays a significant role in self-starting a single-phase induction motor. The design and/or analysis of a single-phase motor is not complete without the knowledge of its starting performance. The main question would be: Is the auxiliary winding strong enough to provide the required starting torque? In the case of capacitor-start motors, it would be wise to know if the voltage across the capacitor is within its maximum allowable limits. Furthermore, how can we determine the performance of a permanent-split capacitor motor that uses both windings at all times? It is evident that the performance under these requirements cannot be predicted based on a single-winding analysis. Therefore, in our study we must include both windings.

Prior to proceeding further, let us make certain assumptions that are commonly accepted in this area. We will assume that the auxiliary winding is displaced 90° (electrical) from the main winding. Therefore, the flux produced by one winding will not induce a voltage in the other winding. The main winding and the auxiliary winding resistances are known. Also known are the leakage and magnetization reactances of the main winding. The leakage and magnetization reactances of the auxiliary winding can be expressed in terms of the parameters of the main winding and the transformation ratio. The transformation or a ratio is defined as the rate of the effective number of turns in the auxiliary winding to the effective number of turns in the main winding. The rotor winding is broken into two components, a main winding component and an auxiliary winding component.

It has already been explained in length how a stationary yet pulsating field can be represented by two oppositely revolving fields. When both the main and auxiliary windings are excited, both windings will produce a pair of forward and backward revolving fields. Thus each winding can be represented by an equivalent circuit with two parallel branches, one for the forward field and the other for the backward field. There are, in fact, four revolving fields in the motor. A revolving field, regardless of which winding has produced it, will generate voltages in both windings. Let us assume that the main winding is displaced forward in space by 90° (electrical) with respect to the auxiliary winding. Then the forward field produced by the auxiliary winding will induce a voltage in the main winding, which will lag by 90° from the voltage produced by the same field in the auxiliary winding. This is a very important concept and must be clearly understood in order to properly account for the voltages induced in one winding by the fields of the other winding. The equivalent circuit of a single-phase motor with both windings is shown in Fig. 9.14, where

r_a = resistance of the auxiliary winding

a = ratio of effective turns in the auxiliary to the effective turns in the main winding

$\tilde{E}_1$ = voltage induced in the forward branch of the main winding by the forward revolving field of the auxiliary winding

$\tilde{E}_2$ = voltage induced in the backward branch of the main winding by the backward revolving field of the auxiliary winding

$\tilde{E}_3$ = voltage induced in the forward branch of the auxiliary by the forward revolving field of the main winding

$\tilde{E}_4$ = voltage induced in the backward branch of the auxiliary by the backward revolving field of the main winding

Note that the corresponding parameters of the auxiliary winding in Fig. 9.14 have been defined in terms of the a ratio and the main winding parameters. The forward impedance of the main winding is

$$\hat{Z}_f = R_f + jX_f = 0.5 \frac{jX_m[(r_2/s) + jx_2]}{(r_2/s) + j(x_2 + X_m)} \tag{9.20}$$

The backward impedance of the main winding is

$$\hat{Z}_b = R_b + jX_b = 0.5 \frac{jX_m[(r_2/2 - s) + jx_2]}{(r_2/2 - s) + j(x_2 + X_m)} \tag{9.21}$$

Figure 9.15 shows the equivalent circuit in terms of $\hat{Z}_f$ and $\hat{Z}_b$. The voltage

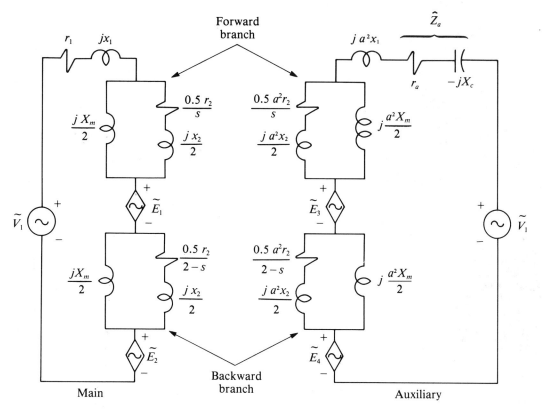

FIGURE 9.14
Equivalent Circuit of a Single-Phase Two-Winding Capacitor Motor

induced in the main winding by its forward revolving field is

$$\tilde{E}_{fm} = \tilde{I}_1 \hat{Z}_f \tag{9.22}$$

The voltage induced in the main winding by its backward revolving field is

$$\tilde{E}_{bm} = \tilde{I}_1 \hat{Z}_b \tag{9.23}$$

The voltage induced in the auxiliary winding by its forward revolving field is

$$\tilde{E}_{fa} = \tilde{I}_2 \, a^2 \hat{Z}_f \tag{9.24}$$

The voltage induced in the auxiliary winding by its backward revolving field is

$$\tilde{E}_{ba} = \tilde{I}_2 \, a^2 \hat{Z}_b \tag{9.25}$$

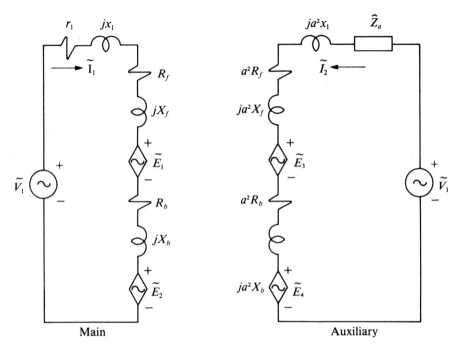

FIGURE 9.15
Simplified
Version of
Figure 9.14

Main   Auxiliary

Since the main winding is displaced ahead by 90° (electrical) from the auxiliary winding, the voltage induced in the main winding by the forward revolving field of the auxiliary winding must lag by 90° from the voltage induced by the same field in the auxiliary. Furthermore, the voltage induced in the main must be $1/a$ times the voltage induced in the auxiliary. That is,

$$\tilde{E}_1 = -j\frac{1}{a}\tilde{E}_{fa} = -ja\tilde{I}_2\hat{Z}_f \tag{9.26}$$

Similarly, the voltage induced in the main winding by the backward revolving field produced by the auxiliary winding must lead by 90° from the voltage it induces in the auxiliary winding. That is,

$$\tilde{E}_2 = j\frac{1}{a}\tilde{E}_{ba} = ja\tilde{I}_2\hat{Z}_b \tag{9.27}$$

By the same token, the voltage induced in the forward branch of the auxiliary winding by the forward revolving field of the main winding is

$$\tilde{E}_3 = ja\tilde{I}_1\hat{Z}_f \tag{9.28}$$

Finally, the voltage induced in the auxiliary winding by the backward revolving field of the main winding is

$$\tilde{E}_4 = -ja\tilde{I}_1\hat{Z}_b \tag{9.29}$$

Since all the voltages are now known in terms of the unknown currents and circuit parameters, the simultaneous equations for both branches are

$$\tilde{I}_1(r_1 + jx_1) + \tilde{E}_{fm} + \tilde{E}_{bm} + \tilde{E}_1 + \tilde{E}_2 = \tilde{V}_1 \tag{9.30}$$

and $$\tilde{I}_2(\hat{Z}_a + ja^2x_1) + \tilde{E}_{fa} + \tilde{E}_3 + \tilde{E}_{ba} + \tilde{E}_4 = \tilde{V}_1 \tag{9.31}$$

The equations can be rewritten as

$$\tilde{I}_1\hat{Z}_{11} + \tilde{I}_2\hat{Z}_{12} = \tilde{V}_1 \tag{9.32}$$

$$\tilde{I}_1\hat{Z}_{21} + \tilde{I}_2\hat{Z}_{22} = \tilde{V}_1 \tag{9.33}$$

where

$$\hat{Z}_{11} = r_1 + \hat{Z}_f + \hat{Z}_b + jx_1 \tag{9.34a}$$

$$\hat{Z}_{12} = -ja(\hat{Z}_f - \hat{Z}_b) \tag{9.34b}$$

$$\hat{Z}_{21} = ja(\hat{Z}_f - \hat{Z}_b) \tag{9.34c}$$

$$\hat{Z}_{22} = \hat{Z}_a + a^2(\hat{Z}_f + \hat{Z}_b + jx_1) \tag{9.34d}$$

where $\hat{Z}_a$ is either the resistance of the auxiliary winding alone in the case of a split-phase induction motor or it also includes the equivalent reactance of the capacitor for capacitor motors.

From the preceding simultaneous equations, we can determine the currents in both the branches as

$$\tilde{I}_1 = \frac{\tilde{V}_1(\hat{Z}_{22} - \hat{Z}_{12})}{\hat{Z}_{11}\hat{Z}_{22} - \hat{Z}_{12}\hat{Z}_{21}} \tag{9.35}$$

$$\tilde{I}_2 = \frac{\tilde{V}_1(\hat{Z}_{11} - \hat{Z}_{21})}{\hat{Z}_{11}\hat{Z}_{22} - \hat{Z}_{12}\hat{Z}_{21}} \tag{9.36}$$

The line current is

$$\tilde{I}_L = \tilde{I}_1 + \tilde{I}_2 \tag{9.37}$$

The power input is

$$P_{\text{in}} = V_1 I_L \cos \theta \tag{9.38}$$

where θ is the power factor angle by which the line current lags the applied voltage.

The stator copper losses are

$$\text{SCL} = I_1^2 r_1 + I_2^2 r_a \qquad (9.39)$$

If the stator copper losses are subtracted from the input power, the air-gap power is obtained, which is distributed among the forward and backward revolving fields. However, we can write an expression for the air-gap power just as we did for the single-phase motor operating on a single winding. Also, we have to take into account the presence of the induced voltages in one winding due to the fields produced by the other winding. On that basis, the air-gap power developed due to the forward revolving field of the main winding is

$$P_{gfm} = Re[(\tilde{E}_{fm} + \tilde{E}_1)\tilde{I}_1^*] \qquad (9.40)$$

Similarly, the forward air-gap power due to the auxiliary winding is

$$P_{gfa} = Re[(\tilde{E}_{fa} + \tilde{E}_3)\tilde{I}_2^*] \qquad (9.41)$$

Thus, the total air-gap power due to the forward fields of both the main and auxiliary windings is

$$P_{gf} = Re[(\tilde{E}_{fm} + \tilde{E}_1)\tilde{I}_1^* + (\tilde{E}_{fa} + \tilde{E}_3)\tilde{I}_2^*] \qquad (9.42)$$

Similarly, the total air-gap power due to the backward revolving fields of both the main and the auxiliary windings is

$$P_{gb} = Re[(\tilde{E}_{bm} + \tilde{E}_2)\tilde{I}_1^* + (\tilde{E}_{ba} + \tilde{E}_4)\tilde{I}_2^*] \qquad (9.43)$$

A very useful relation is obtained if we evaluate the expression for the air-gap power by expressing the main and auxiliary winding currents as

$$\tilde{I}_1 = I_1\underline{/\theta_1} \qquad (9.44)$$

and

$$\tilde{I}_2 = I_2\underline{/\theta_2} \qquad (9.45)$$

where θ_1 and θ_2 are the phase angles of the main and auxiliary winding currents with respect to the applied voltage. The air-gap power due to the forward fields can be rewritten as

$$
\begin{aligned}
P_{gf} &= Re[(\tilde{I}_1\hat{Z}_f - ja\tilde{I}_2\hat{Z}_f)\tilde{I}_1^* + (\tilde{I}_2 a^2\hat{Z}_f + ja\tilde{I}_1\hat{Z}_f)\tilde{I}_2^*] \\
&= Re[(I_1^2 + a^2 I_2^2)\hat{Z}_f - ja\hat{Z}_f(\tilde{I}_2\tilde{I}_1^* - \tilde{I}_1\tilde{I}_2^*)] \qquad (9.46) \\
&= (I_1^2 + a^2 I_2^2)R_f + 2aI_1 I_2 R_f \sin\theta_{21}
\end{aligned}
$$

where $\theta_{21} = \theta_2 - \theta_1$.

Similarly, we can rewrite the expression for the air-gap power due to the backward fields as

$$P_{gb} = (I_1^2 + a^2 I_2^2) R_b - 2a I_1 I_2 R_b \sin \theta_{21} \tag{9.47}$$

The net air-gap power is then

$$P_g = (I_1^2 + a^2 I_2^2)(R_f - R_b) + 2a(R_f + R_b) I_1 I_2 \sin \theta_{21} \tag{9.48}$$

At standstill (i.e., blocked-rotor condition), the per unit slip of the motor is unity and the rotor impedances in the forward and backward branches are the same. The net air-gap power developed by the motor is

$$P_{gs} = 4a I_1 I_2 R_f \sin \theta_{21} \tag{9.49}$$

Note that the net power developed is proportional to the sine of the angle between the currents in the two windings. The power developed is maximum when that angle is equal to 90°. However, in a split-phase motor, the angle is between 30° to 45°, whereas it almost approaches 90° for capacitor motors. This is why a capacitor motor of the same size can develop more starting torque than a split-phase motor.

■ EXAMPLE 9.4

A $\frac{1}{4}$-hp, 120-V, 60-Hz, single-phase capacitor-start, induction-run motor has a main winding impedance of $10 + j100$ Ω and an auxiliary winding impedance of $15 + j70$ Ω. What value of the starting capacitor will result in an auxiliary current in quadrature with the main current at starting?

Solution

The main winding impedance is

$$\hat{Z}_m = 10 + j100 \ \Omega$$

The main winding current is

$$\tilde{I}_1 = \frac{120\underline{/0°}}{10 + j100} = 1.194\underline{/-84.29°} \ A$$

The auxiliary winding current, $\tilde{I}_2$, must lead the main winding current by 90°.

Thus,

$$\theta_2 = 90° - 84.29° = 5.71°$$

Let the impedance of the auxiliary winding be $\hat{Z}_a = r_a - jx_a$. The auxiliary winding current is

$$\tilde{I}_2 = \frac{120\underline{/0°}}{\hat{Z}_a} = \frac{120}{\sqrt{r_a^2 + x_a^2}}\underline{/\tan^{-1}(x_a/r_a)}$$

but

$$\tan^{-1}(x_a/r_a) = 5.71°$$

Hence

$$x_a = 0.1r_a$$

$$= 0.1 \times 15 = 1.5 \ \Omega$$

The capacitive reactance is $X_c = 70 + 1.5 = 71.5 \ \Omega$. Therefore,

$$C = \frac{1}{2\pi \times 60 \times 71.5} = 37 \ \mu F$$

■ EXAMPLE 9.5

A 230-V, 50-Hz, 6-pole single-phase permanent split capacitor motor is rated at 940 rpm. The motor has the following parameters:

$r_1 = 34.14 \ \Omega, \quad x_1 = 35.9 \ \Omega, \quad r_a = 149.78 \ \Omega, \quad x_2 = 29.32 \ \Omega, \quad X_m = 248.59 \ \Omega$

$r_2 = 23.25 \ \Omega, \quad a = 1.73, \quad C = 4 \ \mu F,$ rated at 440 V

The core loss is 19.88 W and the friction and windage loss is 1.9 W. Determine the (a) line current, (b) power input, (c) efficiency, (d) net torque developed, (e) voltage across the capacitor, and (f) starting torque of the motor.

Solution

Synchronous speed:
$$N_s = \frac{120f}{P} = \frac{120 \times 50}{6} = 1000 \ \text{rpm}$$

Per unit slip:
$$\frac{N_s - N_m}{N_s} = \frac{1000 - 940}{1000} = 0.06$$

Angular velocity:
$$\omega_m = \frac{2\pi N_m}{60} = \frac{2\pi \times 940}{60} = 98.44 \ \text{rad/s}$$

Capacitive reactance:
$$X_c = \frac{1}{2\pi f C} = \frac{10^6}{2\pi \times 50 \times 4} = 795.78 \ \Omega$$

$$\hat{Z}_f = R_f + jX_f = 0.5 \times \left[\frac{j248.59 \left(\dfrac{23.25}{0.06} + j29.32 \right)}{\dfrac{23.25}{0.06} + j(29.32 + 248.59)} \right] = 52.64 + j86.41 \ \Omega$$

$$\hat{Z}_b = R_b + jX_b = 0.5 \times \left[\frac{j248.59 \left(\dfrac{23.25}{1.94} + j29.32 \right)}{\dfrac{23.25}{1.94} + j(29.32 + 248.59)} \right] = 4.76 + j13.31 \ \Omega$$

$$\hat{Z}_{11} = 34.14 + 52.64 + j86.41 + 4.76 + j13.31 + j35.9$$

$$= 91.54 + j135.62 = 163.62\underline{/55.98°} \ \Omega$$

$$\hat{Z}_{12} = -j1.73 \times (52.64 + j86.41 - 4.76 - j13.31)$$

$$= 126.46 - j82.82 = 151.17\underline{/-33.22°} \ \Omega$$

$$\hat{Z}_{21} = j1.73 \times (52.64 + j86.41 - 4.76 - j13.31)$$

$$= -126.46 + j82.82 = 151.71\underline{/146.78°} \ \Omega$$

$$\hat{Z}_{22} = r_a - jX_c + a^2(Z_f + Z_b + jx_1)$$

$$= 149.78 - j795.78 + 1.73^2 \times (52.64 + j86.41 + 4.76 + j13.31 + j35.9)$$

$$= 321.57 - j389.88 = 505.38\underline{/-50.48°} \ \Omega$$

The main winding current is

$$\tilde{I}_1 = \frac{230(321.57 - j389.88 - 126.46 + j82.82)}{(163.62\underline{/55.98°})(505.38\underline{/-50.48°}) - (151.17\underline{/-33.22°})(151.17\underline{/146.78°})}$$

$$= 0.59 - j0.69 = 0.91\underline{/-49.47°} \ A$$

The auxiliary winding current is

$$\tilde{I}_2 = \frac{230(91.54 + j135.62 + 126.46 - j82.82)}{(163.62\underline{/55.98°})(505.38\underline{/-50.48°}) - (151.17\underline{/-33.22°})(151.17\underline{/146.78°})}$$

$$= 0.52 + j0.21 = 0.56\underline{/21.72°} \ A$$

(a) The line current is

$$\tilde{I}_L = \tilde{I}_1 + \tilde{I}_2 = 0.59 - j0.69 + 0.52 + j0.21$$
$$= 1.11 - j0.48 = 1.21\underline{/-23.39°} \text{ A}$$

(b) The power input is

$$P_{in} = 230 \times 1.21 \times \cos(23.39) = 255.44 \text{ W}$$

(c) The air-gap power from Eq. (9.48) is

$$P_g = (0.91^2 + 1.73^2 \times 0.56^2) \times (52.64 - 4.76)$$
$$+ 2 \times 1.73 \times 0.91 \times 0.56 \times (52.64 + 4.76) \times \sin(21.72° + 49.47°)$$
$$= 180.39 \text{ W}$$

Power developed: $P_d = (1 - s)P_g = (1 - 0.06) \times 180.39 = 169.57$ W

Power output: $P_o = 169.57 - 19.88 - 1.9 = 147.79$ W

Efficiency: $\eta = \dfrac{147.79}{255.44} = 0.579$ or 57.9%

(d) The net torque developed is

$$T_s = \frac{P_o}{\omega_m} = \frac{147.79}{98.44} = 1.5 \text{ N-m}$$

(e) The voltage across the capacitor is

$$\tilde{V}_c = -j\tilde{I}_2 X_c = -j(0.56\underline{/21.72°})795.78$$
$$= 445.6\underline{/-68.28°}$$

Since the voltage across the capacitor is higher than its nominal rating of 440 V, for continuous duty applications either the capacitor with a higher voltage rating must be used or the windings must be redesigned.

(f) For the starting torque $s = 1$. Therefore,

$$\hat{Z}_f = R_f + jX_f = 0.5 \times \left[\frac{j248.59(23.25 + j29.32)}{23.25 + j(29.32 + 248.59)}\right] = 9.24 + j13.89 \text{ } \Omega$$

$$\hat{Z}_b = R_b + jX_b = 0.5 \times \left[\frac{j248.59(23.25 + j29.32)}{23.25 + j(29.32 + 248.59)}\right] = 9.24 + j13.89 \text{ } \Omega$$

$$\hat{Z}_{11} = 34.14 + j35.9 + 9.24 + j13.89 + 9.24 + j13.89 = 82.61\underline{/50.43°}\ \Omega$$

$$\hat{Z}_{12} = -j1.73(9.24 + j13.89 - 9.24 - j13.89) = 0\ \Omega$$

$$\hat{Z}_{21} = j1.73(9.24 + j13.89 - 9.24 - j13.89) = 0\ \Omega$$

$$\hat{Z}_{22} = 149.78 - j795.78 + 1.73^2(9.24 + j13.89 + 9.24 + j13.89 + j35.9)$$

$$= 639\underline{/-71.28°}\ \Omega$$

$$\tilde{I}_1 = \frac{230(639\underline{/-71.28°})}{(82.61\underline{/50.43°})(639\underline{/-71.28°})} = 2.78\underline{/-50.43°}\ A$$

$$\tilde{I}_2 = \frac{230(82.61\underline{/50.43°})}{(82.61\underline{/50.43°})(639\underline{/-71.28°})} = 0.36\underline{/71.28°}\ A$$

The air-gap power is

$$P_{gs} = 4 \times 1.73 \times 2.78 \times 0.36 \times 9.24 \times \sin(71.28° + 50.43°)$$

$$= 54.44\ W$$

The torque induced in the motor is

$$T_{ds} = \frac{P_{gs}}{\omega_s} = \frac{54.44}{104.72} = 0.52\ N\text{-}m$$

which is only one-third of its full-load torque. The starting torque developed by a relatively high efficiency motor is small in comparison with the split-phase and other capacitor motors. This is why a permanent-split capacitor motor is usually designed for fan applications.

■ EXAMPLE 9.6

If the permanent-split capacitor motor in Example 9.5 is designed as a two-value capacitor motor with a starting capacitor of 16 μF, calculate the starting torque developed by the motor.

Solution

At starting

$$X_c = \frac{10^6}{2\pi \times 50 \times 16} = 198.94\ \Omega$$

Thus

$$\hat{Z}_{11} = 82.61\underline{/50.43°} \ \Omega$$

$$\hat{Z}_{12} = \hat{Z}_{21} = 0 \ \Omega$$

$$\hat{Z}_{22} = 149.78 - j198.94 + 1.73^2(9.24 + j13.89 + 9.24 + j13.89 + j35.9)$$

$$= 205.26\underline{/-2.33°}$$

$$\tilde{I}_1 = \frac{230}{82.61\underline{/50.43°}} = 2.78\underline{/-50.43°} \ A$$

$$\tilde{I}_2 = \frac{230}{205.26\underline{/-2.33°}} = 1.12\underline{/2.33°} \ A$$

The air-gap power is

$$P_{gs} = 4 \times 1.73 \times 2.78 \times 1.12 \times 9.24 \times \sin(2.33° + 50.43°) = 158.5 \ W$$

The torque developed is

$$T_{ds} = \frac{158.5}{104.72} = 1.51 \ \text{N-m}$$

By changing the size of the capacitor, a threefold increase in the starting torque has been achieved. However, the initial cost of the motor is higher due to an extra capacitor and the centrifugal switch.

9.7 SINGLE-PHASE MOTOR TESTS

The approximate equivalent-circuit parameters of a single-phase induction motor can be determined from the blocked-rotor and no-load tests. Since it is a common practice to express the parameters of the auxiliary winding in terms of the parameters of the main winding and the winding ratio, these tests can be performed by energizing only the main winding.

Blocked-rotor Test   The circuit arrangement for the blocked-rotor test is shown in Fig. 9.16. The auxiliary winding is left open. The test is performed at the rated frequency and rated voltage. If we assume that the magnetizing reactance, X_m, is very large compared to the rotor impedance, the equivalent circuit can be approximated as shown in Fig. 9.17.

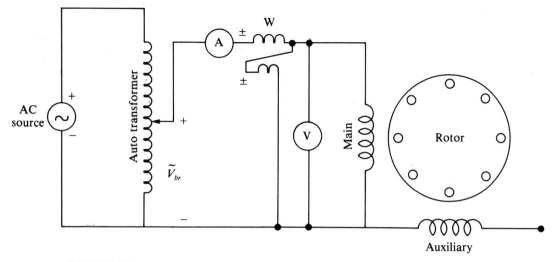

FIGURE 9.16
Blocked-Rotor Test with Auxiliary Winding Open

Let V_{br}, I_{br}, and P_{br} be the measured values of the applied voltage, the current, and the power loss in the motor under blocker-rotor conditions. The blocked-rotor impedance can be computed as

$$Z_{br} = \frac{V_{br}}{I_{br}} \tag{9.50}$$

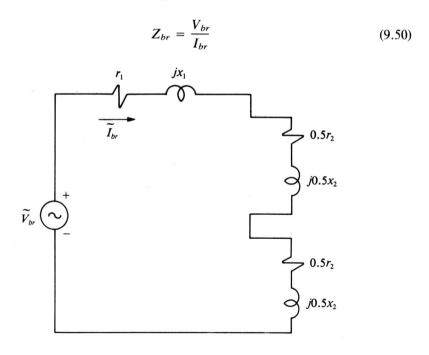

FIGURE 9.17
Approximate
Equivalent
Circuit under
Blocked-Rotor
Condition

The blocked-rotor resistance is

$$R_{br} = \frac{P_{br}}{I_{br}^2} \qquad (9.51)$$

Thus, the blocked-rotor reactance is

$$X_{br} = \sqrt{Z_{br}^2 - R_{br}^2} \qquad (9.52)$$

From the equivalent circuit given in Fig. 9.17,

$$R_{br} = r_1 + r_2 \qquad (9.53)$$

and $\qquad\qquad X_{br} = x_1 + x_2 \qquad\qquad (9.54)$

If the winding resistance is known, the equivalent rotor resistance can be determined from Eq. (9.53).

However, there is no way to isolate the leakage reactance of the main winding and the rotor. For all practical purposes, let us assume that these leakage reactances are equal. That is,

$$x_1 = x_2 = 0.5X_{br} \qquad (9.55)$$

No-load test In a three-phase induction motor operating under no load, the no-load copper loss in the rotor circuit was assumed to be so small that it was neglected. In fact, we considered the rotor circuit branch as an open circuit because of very low slip. In a single-phase induction motor, the no-load slip is not as small as for the three-phase motor. However, if we consider the rotor circuit as an open circuit for the forward branch under a no-load condition, the error introduced in the calculations of motor-circuit parameters will be somewhat greater than that for the three-phase induction motor. But such an approximation will reduce the complexity of circuit manipulations. Therefore, under this approximation, the equivalent circuit of a single-phase motor is shown in Fig. 9.18.

Let V_{nL}, I_{nL}, and P_{nL} be the measured values of the rated voltage, current, and power intake by the motor under the no-load condition. The no-load impedance can be computed as

$$Z_{nL} = \frac{V_{nL}}{I_{nL}} \qquad (9.56)$$

The equivalent no-load resistance can be calculated as

$$R_{nL} = \frac{P_{nL}}{I_{nL}^2} \qquad (9.57)$$

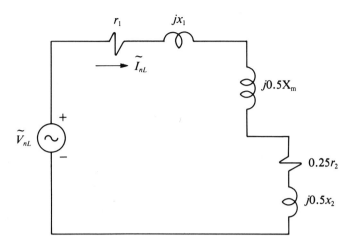

FIGURE 9.18
Approximate
Equivalent
Circuit under No
Load with $s \cong 0$

Thus, the no-load reactance is

$$X_{nL} = \sqrt{Z_{nL}^2 - R_{nL}^2} \qquad (9.58)$$

But $\qquad X_{nL} = x_1 + 0.5X_m + 0.5x_2$

and $\qquad x_1 = x_2 = 0.5X_{br}$

Therefore, $\qquad x_1 + 0.5x_2 = 0.75X_{br}$

Thus, $\qquad X_m = 2X_{mL} - 1.5X_{br} \qquad (9.59)$

The rotational loss is

$$P_{\text{rot}} = P_{nL} - I_{nL}^2(r_1 + 0.25r_2) \qquad (9.60)$$

The rotational loss is the summation of the core loss, friction and windage loss, and stray load loss in single-phase induction motors.

■ EXAMPLE 9.7

A 115-V, 60-Hz, single-phase capacitor-start induction motor is tested as follows with the auxiliary winding left open:

> Blocked rotor test: voltage = 115 V, current = 17.1 A, power = 1825 W
>
> No-load test: voltage = 115 V, current = 3.2 A, power = 72 W

The resistance of the main winding is 2.5 Ω. Determine the equivalent-circuit parameters of the motor.

Solution

From the blocked-rotor test,

$$Z_{br} = \frac{115}{17.1} = 6.725 \ \Omega$$

$$R_{br} = \frac{1825}{17.1^2} = 6.24 \ \Omega$$

$$r_2 = 6.24 - 2.5 = 3.74 \ \Omega$$

$$X_{br} = \sqrt{6.725^2 - 6.24^2} = 2.508 \ \Omega$$

$$x_1 = x_2 = 0.5 \times 2.508 = 1.254 \ \Omega$$

From the no-load test,

$$Z_{nL} = \frac{115}{3.2} = 35.94 \ \Omega$$

$$R_{nL} = \frac{72}{3.2^2} = 7.03 \ \Omega$$

$$X_{nL} = \sqrt{35.94^2 - 7.03^2} = 35.24 \ \Omega$$

$$X_m = 2 \times 35.24 - 1.5 \times 2.508 = 66.718 \ \Omega$$

$$P_{rot} = 72 - 3.2^2(2.5 + 0.25 \times 3.74) = 36.8 \ \text{W}$$

9.8 SHADED-POLE MOTOR

To this point we know that in single-phase induction motors there must be two windings to develop the necessary starting torque. The two windings (i.e. main and auxiliary) must be displaced in space and must carry currents out of phase to set up a revolving field in the motor. A special type of auxiliary winding known as a *shading coil* or a *shading ring* is used to set up the required revolving field in a shaded-pole motor. A shaded-pole motor is simple in construction, very reliable in operation, and the least expensive for fractional-horsepower applications. Most often, the stator of the shaded-pole motor has a salient-pole construction and is wound exactly like the stators for the dc machines. However, the stator pole is physically divided into two sections. A copper ring, which is known as a shading coil, is mounted over the smaller section. Figure 9.19 shows the construction of a two-pole shaded motor. The rotor is of die-cast type, just like the rotor of any induction motor.

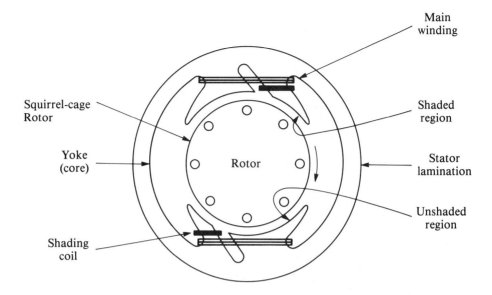

FIGURE 9.19
Two-Pole
Shaded-Pole
Motor

Principle of Operation   Let us investigate how the shading coil helps to set up the revolving field in the motor. In the absence of the shading coil, the flux produced by the main winding is uniformly distributed in the entire pole section and pulsates sinusoidally. Let us view the flux distribution in the presence of the shading coil. Figure 9.20 shows the flux produced by the main winding. As the magnetic flux is increasing in the pole of the motor, a voltage is induced in the shading coil. Since the shading coil is a short circuit with very small resistance, a high current flows in it. The direction of the current is such that it opposes the increase in the flux in the shading region. Therefore, during the period the flux is increasing the bulk of the flux passes through the unshaded part of the pole. The main axis of the magnetic flux would be the center of the unshaded side of the pole. As the flux reaches its maximum value and the rate of change of the flux is zero, there is no induced voltage and thereby no current in the shading coil. The flux distributes uniformly as if there were no shading coil present. The magnetic axis of the flux is then the middle of the entire pole. As we can see, the flux has moved from the center of the unshaded region to the middle of the pole.

After reaching the maximum value, the flux produced by the main winding begins to decrease. The current flow in the shading coil is then such that it opposes the decrease in the flux. In other words, the shading coil wants to sustain the flux at high levels while the flux is decreasing. This is just the opposite to what happened when the flux in the motor was increasing. In this case the bulk of the flux is passing through the shaded region of the pole. The magnetic axis of the flux is almost the same as the center of the shaded region of the pole. Again the flux has moved from the middle of the pole to the center of the shaded region. As we

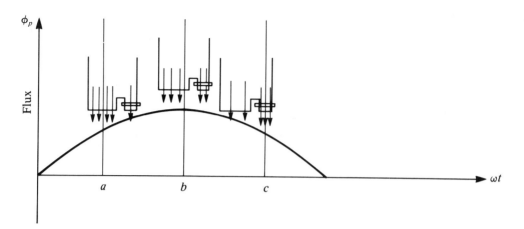

FIGURE 9.20
Shading-Pole Action During the Positive Half-Cycle of a Flux Waveform. *a*, $\omega t < \pi/2$: Almost all the Flux Passing Through Unshaded Region; *b*, $\omega t = \pi/2$: No Shading Action, Flux is Uniformly Distributed over the Entire Pole; *c*, $\omega t > \pi/2$: Most of the Flux is Passing Through the Shaded Region

can see, the shading coil produces a flux that appears to be revolving from the unshaded region to the shaded region. It is apparent that the revolving field is neither continuous nor uniform. However, the mere presence of the revolving field is sufficient to develop a small starting torque and, therefore, start the motor and bring it up to speed.

The starting torque of a shaded-pole motor is low, and the motor is used for low torque applications such as fans. The overall efficiency of this motor ranges from 20% to 40%. Shaded-pole motors are being built from the subfractional-horsepower range to almost $\frac{1}{3}$-horsepower applications. The direction in which

FIGURE 9.21
Typical Speed–
Torque
Characteristic of
a Shaded-Pole
Motor

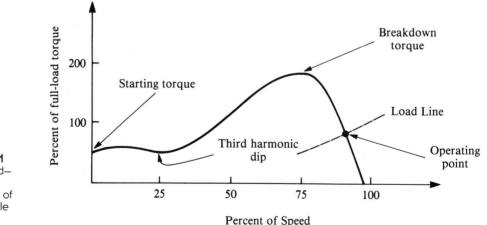

the motor starts depends on the placement of the shading coil. Once the motor is built, its direction of rotation cannot be reversed. To have a reversible motor, we must have two shading coils on each pole face and selectively short one of them.

To increase the starting torque, the leading pole face of the shaded-pole motor may have a wider gap than the rest of the pole. It has been found that if part of the pole face has a wider gap than the other, the motor will need no other starting mechanism and will start. The starting torque will be quite low. Such a motor is known as a *reluctance start motor*. Adding this reluctance feature to the shading-coil motor increases its starting torque. This feature is commonly employed in the design of shaded-pole motors. A typical speed–torque characteristic of a shaded-pole motor is shown in Fig. 9.21.

9.9 UNIVERSAL MOTOR

It would be natural to expect that if the polarities of the armature current and the field-winding current are reversed instantaneously in a dc motor, it would continue developing torque in the same direction. In the case of a dc shunt motor, the field winding is connected in parallel with the armature winding. The field winding has high resistance, which limits the current flow in that winding under dc operation. However, the field winding has a large number of turns and thereby exhibits high inductive reactance when connected to an ac source. The high inductive reactance in addition to high resistance further decreases the current flow in that winding. Therefore, the flux created by the field winding would be quite weak. On the other hand, the highly inductive nature of the field winding tends to delay the reversal of the field winding current. Thus, there exists a considerable phase lag between the armature current and the flux produced by the field winding. This results in reduced average torque, which may not be acceptable for the size of the motor. Therefore, a dc shunt motor is never designed for both ac and dc applications.

Since the field winding and the armature winding are connected in series in a series motor, the current flow through each winding is the same. Thus, the flux produced by the series field winding is in phase with the armature current even when it is connected to an ac source. Since the series winding usually has very few turns, its inductive reactance is very low for ac operation. Therefore, we would expect almost the same performance from a series motor under both ac and dc operations. A motor designed to operate equally well irrespective of the nature of the supply is known as a *universal motor*.

When a series motor is specifically designed for ac operation, every effort is made to improve its performance. To reduce the core loss, the stator yoke and field pole are laminated. For fractional-horsepower applications where universal motors are used in large volumes, the yoke and field pole are punched as one piece. To reduce the overall flux in the pole, a fewer number of turns are used for ac applications. To make up for the loss in performance, the effective conductors in the armature are increased.

Under ac operation, we would expect an induced voltage by transformer action even in the coil that is undergoing commutation. This induced voltage causes extra sparking at the brushes, which reduces brush life and results in more wear and tear on the commutator. To reduce the harmful effects of poor commutation, high-resistance brushes are commonly used. Because a series motor can operate at high speeds and provide large amounts of torque for its size, it has found use in such applications as vacuum cleaners, food blenders, portable electric tools such as saws and routers, and sewing machines, to name a few.

The typical speed–torque characteristics of a universal motor for ac and dc operation are shown in Fig. 9.22. A universal motor can be analyzed using the basic equations developed for the series motor, as demonstrated by the following example.

■ EXAMPLE 9.8

A 120-V, 60-Hz, 2-pole universal motor operates at a speed of 8000 rpm on full load and draws a current of 17.58 A at a lagging power factor of 0.912. The characteristics of the windings are:

Resistance of series field winding:	$R_s = 0.65 \ \Omega$
Reactance of series field winding:	$X_s = 1.2 \ \Omega$
Resistance of armature winding:	$R_a = 1.36 \ \Omega$
Reactance of armature winding:	$X_a = 1.6 \ \Omega$

Determine (a) the generated voltage, (b) the power output, (c) the shaft torque, and (d) the efficiency of the motor if the rotational loss is 80 W.

Solution

From the equivalent circuit of the motor as shown in Fig. 9.22a,

(a) $\qquad \tilde{E}_g = \tilde{V}_s - \tilde{I}_a(R_s + R_a + jX_s + jX_a)$

$\qquad\qquad = 120 - (17.58\underline{/-24.22°})(0.65 + 1.36 + j1.2 + j1.6)$

$\qquad\qquad = 74.1\underline{/-24.22°} \ \text{V}$

As expected, the generated voltage is in phase with the current. The corresponding phasor diagram is shown in Fig. 9.22c.

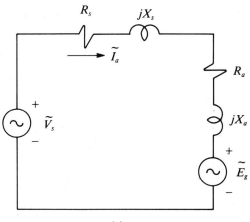

(a)

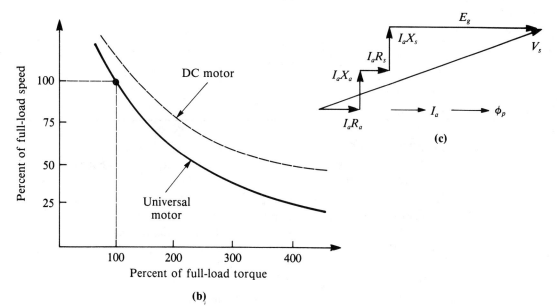

(b)

(c)

FIGURE 9.22
(a) Equivalent Circuit of a Universal Motor; (b) Typical Speed–Torque Characteristic; (c) Phasor Diagram

(b) The power developed is

$$P_d = Re[\tilde{E}_g \tilde{I}_a^*] = 74.1 \times 17.58 = 1302.68 \text{ W}$$

The power output is

$$P_o = P_d - P_{\text{rot}} = 1302.68 - 80 = 1222.68 \text{ W}$$

(c) The rated speed of the motor is given as 8000 rpm. That is,

$$\omega_m = \frac{2\pi \times 8000}{60} = 837.76 \text{ rad/s}$$

The torque output of the motor is

$$T_s = \frac{1222.68}{837.76} = 1.46 \text{ N-m}$$

(d) The power input is

$$P_{in} = 120 \times 17.58 \times 0.912 = 1923.96 \text{ W}$$

The efficiency is

$$\eta = \frac{1222.68}{1923.96} = 0.6355 \quad \text{or} \quad 63.55\%$$

PROBLEMS

9.1. A 230-V, 50-Hz, 2-pole, single-phase induction motor is designed to run at 3% slip. Determine the slip in the opposite direction. What is the speed of the motor in the normal direction of rotation?

9.2. The main winding of a 6-pole, 60-Hz, 120-V, single-phase induction motor has the following parameters: $r_1 = 2.4$ Ω, $x_1 = 3.6$ Ω, $r_2 = 1.6$ Ω, $x_2 = 3.6$ Ω, $X_m = 75$ Ω, and $s = 0.05$. Determine the (a) speed of the motor, (b) effective rotor resistance in the forward branch, (c) effective rotor resistance in the backward branch, (d) forward impedance, $\hat{Z}_f$, and (e) backward impedance, $\hat{Z}_b$.

9.3. A 110-V, 50-Hz, split-phase induction motor has the following parameters: main winding impedance = 1.2 + j25 Ω and auxiliary winding impedance = 12 + j5 Ω. Determine the current in each winding under the standstill condition. Show the phase relation between the currents.

9.4. A 120-V, 50-Hz, capacitor-start induction motor has a main winding impedance of 1.2 + j2.4 Ω and auxiliary winding impedance of 4.8 + j1.2 Ω at standstill. Determine the value of the capacitor that must be connected in series with the auxiliary winding so that the phase difference between the two currents is 80°. What are the currents in the windings?

9.5. A 4-pole, 115-V, 60-Hz, split-phase motor runs at a 4% slip on full load. The motor parameters are $r_1 = 1.9$ Ω, $x_1 = 2.62$ Ω, $r_2 = 0.85$ Ω, $x_2 = $

0.72 Ω, and X_m = 54.6 Ω. Friction, windage, and the core losses are 37 W. Determine the (a) power input, (b) power developed, (c) power output, (d) torque output, and (e) efficiency of the motor.

9.6. A 220-V, 60-Hz, 6-pole, capacitor-start motor has r_1 = 0.15 Ω, x_1 = 0.3 Ω, r_2 = 0.12 Ω, x_2 = 0.3 Ω, and X_m = 16 Ω. If the core loss is 200 W, friction and windage loss is 300 W, and stray load loss is 50 W, determine the efficiency and the net torque at 5% slip.

9.7. A $\frac{1}{4}$-hp, 6-pole, 115-V, 60-Hz, permanent-split capacitor motor has the following parameters: r_1 = 5.9 Ω, x_1 = 4.78 Ω, r_a = 38 Ω, r_2 = 5.1 Ω, x_2 = 3.26 Ω, and X_m = 50.78 Ω. a = 2.548, C = 10 μF, 370 V; full-load speed = 1120 rpm. The friction and windage loss is 1.79 W; core loss is 27.06 W. Determine the (a) line current, (b) power input, (c) efficiency, (d) net torque developed, (e) voltage across the capacitor, and (f) starting torque of the motor.

9.8. A $\frac{1}{6}$-hp, 115-V, 60-Hz, 2-pole, permanent-split capacitor motor is designed for reversible operation. Consequently, the auxiliary winding is identical to the main winding. The motor parameters are r_1 = r_a = 15.5 Ω, x_1 = 11.52 Ω, r_2 = 10.6 Ω, x_2 = 6.75 Ω, X_m = 235.04 Ω, a = 1, rated speed = 3325 rpm; C = 15 μF, 370 V; the core loss is 9.62 W. Friction and windage loss is 1.31 W. Determine the efficiency and the torque output of the motor.

9.9. A 230-V, 60-Hz, permanent-split capacitor motor is tested as follows with the auxiliary winding open:

Blocked rotor test: voltage = 230 V, current = 2.16 A, power = 406.84 W

No-load test: voltage = 230 V, current = 0.8 A, power = 138.52 W

The main winding resistance is 42.4 Ω. Determine the equivalent-circuit parameters of the motor.

9.10. A $\frac{1}{4}$-hp, 4-pole, 120-V, 60-Hz, capacitor-start motor is tested with the auxiliary winding open to obtain the following data.

No load-test: voltage = 120 V, current = 2.7 A, power = 67 W

Blocked-rotor test: voltage = 120 V, current = 15 A, power = 1175 W

The main winding resistance is 2.5 Ω. Determine the equivalent-circuit parameters. Calculate the efficiency and torque output of the motor using these parameters at 1728 rpm.

9.11. A 240-V, 60-Hz, 2-pole, universal motor operates at a speed of 12,000 rpm on full load and draws a current of 6.5 A at 0.94 power factor lagging. The series field-winding impedance is 4.55 + j3.2 Ω and armature circuit impedance is 6.15 + j9.4 Ω. Calculate the (a) back emf of the motor, (b) mechanical power developed by the motor, (c) power output if the rotational loss is 65 W, (d) torque output, and (e) efficiency of the motor.

CHAPTER 10

Special-Purpose Electrical Machines

Brushless dc Motor with PM Rotor and
Electronic Control (Courtesy of Universal
Electric Company)

In the foregoing chapters we focused our attention on various aspects of conventional, wound field machines. Although all machines operate basically on the same principle, special-purpose machines employ some features that distinguish them from conventional machines. With the development of new advanced technologies, new types of machines are continually being conceived and developed. It is beyond the scope of this book to cover all the different types of special-purpose machines. However, an attempt is made to highlight the basic operations of some of special-purpose machines that are being used to a great extent in home, recreational, and industrial applications.

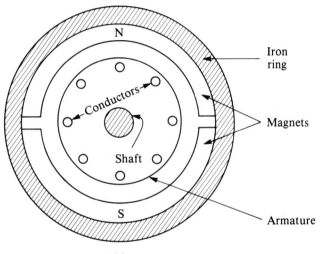

(a)

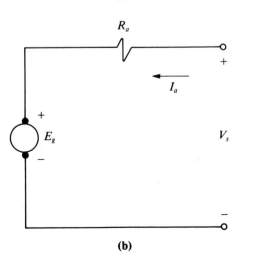

(b)

FIGURE 10.1
(a) Cross-Sectional View of a PM Motor; (b) Equivalent Circuit

10.1 PERMANENT-MAGNET MOTORS

The development of new magnetic materials has led to the replacement of a dc shunt motor with a permanent-magnet (PM) motor. In a PM motor, as shown in Fig. 10.1, the poles are made of permanent magnets. Although dc motors up to 75 hp have been designed with permanent magnets, the major application of permanent magnets is confined to fractional-horsepower motors due to economical reasons. In a wound dc motor the flux per pole depends on the field-winding current and can be controlled, while that in a PM motor is essentially constant and depends on the point of operation as explained in Chapter 1. For the same power output, a PM motor has higher efficiency and requires less material in comparison with wound dc motors. Consequently, it is smaller in size. On the other hand, a PM motor must be so designed that the demagnetization effect of the armature reaction, which is maximum at standstill, is as small as economically possible.

Since the flux of a PM motor is fixed, the speed– and current–torque characteristics are basically straight lines, as shown in Fig. 10.2. These relations can be expressed mathematically as

$$\omega_m = \frac{V_s}{K_a\phi_p} - \frac{R_a}{(K_a\phi_p)^2} T_d \tag{10.1}$$

and

$$I_a = \frac{1}{K_a\phi_p} T_d \tag{10.2}$$

where K_a is the machine constant, V_s is the supply voltage, ϕ_p is the flux per pole, and T_d is the developed torque.

The speed–torque characteristic of a PM motor can be controlled by changing the supply voltage and the effective armature resistance. The change in the supply voltage changes the no-load speed of the motor without affecting the slope

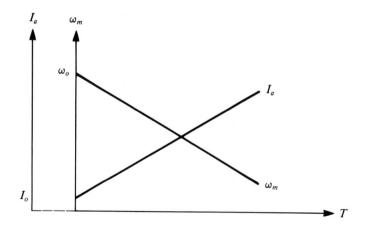

FIGURE 10.2
Speed– and
Current–Torque
Characteristics of
a PM Motor

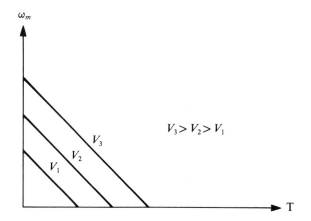

FIGURE 10.3
Operating
Characteristics
for Different
Supply Voltages

of the curves. Thus, for different supply voltages, a set of parallel operating characteristics of the motor can be obtained as shown in Fig. 10.3. On the other hand, the change in the effective resistance in the armature circuit controls the slope of the speed–torque characteristic and has no influence on the no-load speed of the motor, as shown in Fig. 10.4. Using magnets having different flux densities and the same cross-sectional areas, or vice versa, there are almost infinite possibilities for designing a PM motor for given operating conditions, as highlighted in Fig. 10.5. As is evident from the figure, an increase in blocked-rotor torque can be achieved at the expense of lower no-load speed.

■ **EXAMPLE 10.1**

A PM motor operates with a magnetic flux of 3 mWb. The armature resistance is 0.6 Ω and the supply voltage is 25 V. If the motor load is 1 N-m, determine (a)

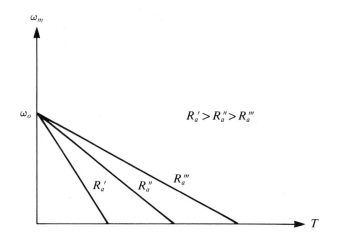

FIGURE 10.4
Operating
Characteristics
for Different
Resistances of
Armature Circuit

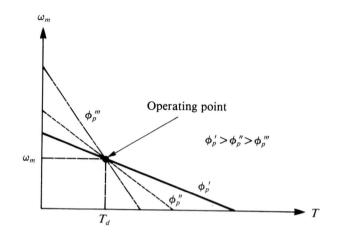

FIGURE 10.5
Operating
Characteristics
for Different
Fluxes in a PM
Motor

the operating speed of the motor, and (b) the torque developed under a blocked-rotor condition. The motor constant is 90.

Solution

(a) From Eq. (10.1),

$$\omega_m = \frac{25}{90 \times 0.003} - \frac{0.6 \times 1}{(90 \times 0.003)^2} = 84.36 \text{ rad/s}$$

or

$$N_m = \frac{84.36 \times 60}{2\pi} = 806 \text{ rpm}$$

(b) For the blocked-rotor condition $\omega_m = 0$. Thus,

$$\frac{25}{90 \times 0.003} = \frac{0.6T_d}{(90 \times 0.003)^2}$$

or

$$T_d = 11.25 \text{ N-m}$$

■ EXAMPLE 10.2

Calculate the magnetic flux in a 150-W, 40-V PM motor operating at 1200 rpm. The motor constant is 78, the armature resistance is 1.44 Ω, and the rotatonal loss is 10 W.

Solution

$$\omega_m = \frac{2\pi \times 1200}{60} = 125.66 \text{ rad/s}$$

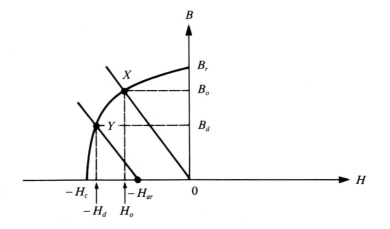

FIGURE 10.6
Demagnetization
Effect in a PM
Motor

The power developed is $P_d = 150 + 10 = 160$ W, and the torque developed is $T_d = 160/125.66 = 1.273$ N-m. From Eq. (10.1),

$$\phi_p^2 - \frac{V_s}{K_a\omega_m}\phi_p + \frac{R_aT_d}{K_a^2\omega_m} = 0$$

$$\phi_p^2 - 0.0041\phi_p + 2.4 \times 10^{-6} = 0$$

or $\qquad\qquad \phi_p \cong 3.4$ mWb

As explained in Chapter 1, the point of operation of a permanent magnet depends on the permeance of the magnetic circuit. The point of intersection of the operating line and the demagnetization curve determines the flux density in the magnetic circuit. The same is true for PM motors as long as we ignore the demagnetization effect of the armature reaction. The point of operation is marked by X in Fig. 10.6. Even though PM motors are designed with relatively larger air gaps to minimize the armature reaction, its demagnetization effect must be included to determine the proper operating point. The operating line will move to

FIGURE 10.7
Intrinsic
Operating Point
of the Trailing
Tip of the
Magnet During
Start or Stall

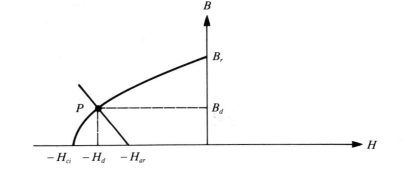

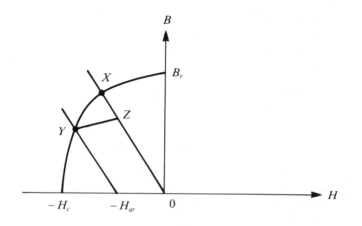

FIGURE 10.8
Recoil Effect on
the Operation
of a PM Motor

the left, as shown in Fig. 10.6, where H_{ar} corresponds to the magnetic field intensity of the armature. The actual operating point, therefore, moves to Y. It is apparent from the figure that the useful magnetic-flux density of the motor decreases with the increase in armature reaction.

The demagnetization effect of the armature reaction is maximum under a blocked-rotor condition. To investigate its effect on the magnets, it is a good practice to consider the intrinsic curve as given in Fig. 10.7. The curve can be extracted from the normal demagnetization curve using $B_i = B_n + \mu_0 H_n$ where B_i is the intrinsic flux density and B_n and H_n are the normal flux density and the corresponding field intensity. It is obvious from Fig. 10.7 that the magnet will be completely demagnetized if H_{ar} is greater than the intrinsic coercive force H_{ci}.

Let us assume that load line intersects the demagnetization curve at point X, as shown in Fig. 10.8, with no armature current. With an increase in armature

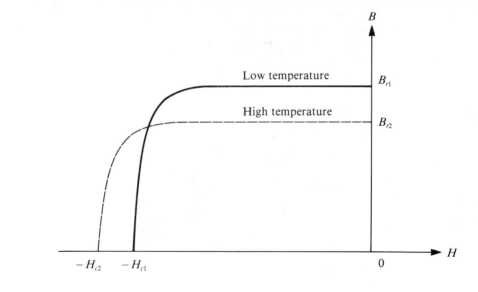

FIGURE 10.9
Temperature
Effect on the
Intrinsic Curve

current, the point of operation shifts to Y due to armature reaction. We would expect the point of operation to move to X again as soon as the armature current is switched off. This, in fact, is not so, and the new operating point will be Z on the original operating line. The line from Y to Z is known as the *recoil line*. It is approximately parallel to the slope of the demagnetization curve at point B_r. The overall effect is a loss in the operating flux density in the motor. However, in the case of ceramic magnets the loss is insignificant, as the demagnetization curve is basically a straight line.

The effect of temperature must also be taken into consideration when designing a PM motor. Figure 10.9 illustrates the changes in the demagnetization characteristic at two different temperatures. As the temperature increases, the residual flux density in the magnet decreases while the intrinsic coercive force increases. On the other hand, the lower the temperature, the more pronounced is the demagnetization effect of the armature reaction.

10.2 STEP MOTORS

Step motors, also referred to as stepping or stepper motors, are essentially incremental motion actuators. A step motor receives a digital input in the form of a pulse and responds by rotating a specified number of degrees. Step motors are excellent devices for precise position controls without any feedback. They are relatively simple in construction and can be made to step in equal increments in either direction. They are good candidates for such devices as printers, electric typewriters, and lathes. They do not offer the flexibility of adjusting the angle of advance. The step response is oscillatory in nature with a considerable overshoot. Step motors can be classified into three main categories: variable reluctance, permanent magnet, and hybrid.

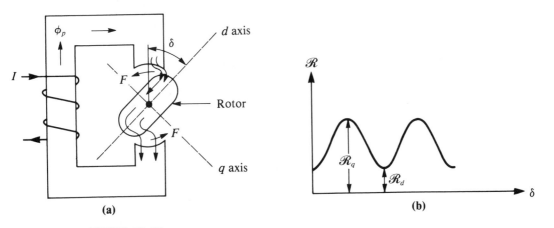

(a)   **(b)**

FIGURE 10.10
(a) Magnetic Circuit with a Freely Rotating Member; (b) Reluctance as a Function of Position

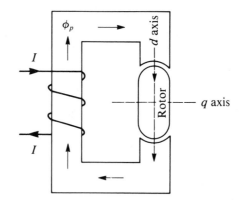

FIGURE 10.11
Minimum
Reluctance,
Equilibrium, or
No-Rotation
Position

10.2.1 Variable-reluctance Step Motors

Variable-reluctance step motors operate on the principle of minimizing the reluctance along the path of the applied magnetic field. Consider a simple magnetic
circuit with a freely rotating rotor as shown in Fig. 10.10. When a direct current
is passed through the coil, it establishes a magnetic field that attracts the rotor
teeth until they line up as shown in Fig. 10.11. This is the minimum reluctance
position and the motor is said to be in equilibrium. The reluctance motor has

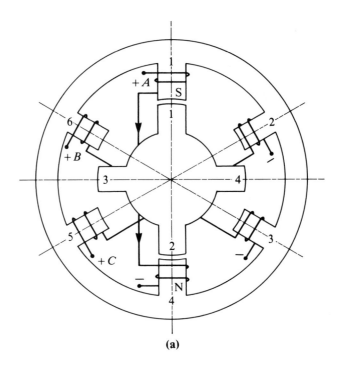

FIGURE 10.12
Variable-
Resistance Step
Motor

(a)

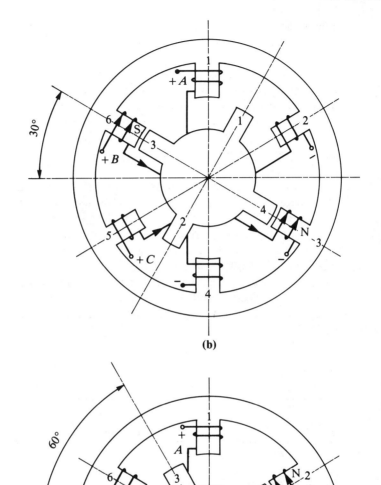

(b)

(c)

FIGURE 10.12
(*continued*)

already been analyzed mathematically in Chapter 2. This principle is used in the design of variable-reluctance step motors.

The stator of a variable-reluctance step motor has a magnetic core constructed with a single stack of steel laminations, and phase windings are placed on the stator teeth as shown in Fig. 10.12. In this case only one set of stator and rotor teeth is made to align by designing the number of teeth in the rotor different than that of the stator. In Fig. 10.12, the motor has six teeth on the stator and four teeth on the rotor. The stator usually has multiphase windings. Figure 10.12 shows three phases A, B, and C with teeth 1 and 4, 2 and 5, and 3 and 6, respectively.

Suppose phase A is energized by passing a constant current through it and rotor teeth 1 and 2 have aligned with stator teeth 1 and 4, as shown in Fig. 10.12a. As long as A is energized, the rotor will be stationary and counteract the torque due to the mechanical drive. It is apparent that the angle between the magnetic axis of phase B or phase C and the rotor axis consisting of teeth 3 and 4 is 30°. If A is switched off and B is energized, rotor teeth 4 and 3 will align under stator teeth 3 and 6, causing a 30° of displacement of the rotor as indicated in Fig. 10.12b. If we now energize C and switch B off, the rotor will rotate another 30° and align with C, as shown in Fig. 10.12c. The rotor can be made to rotate continuously

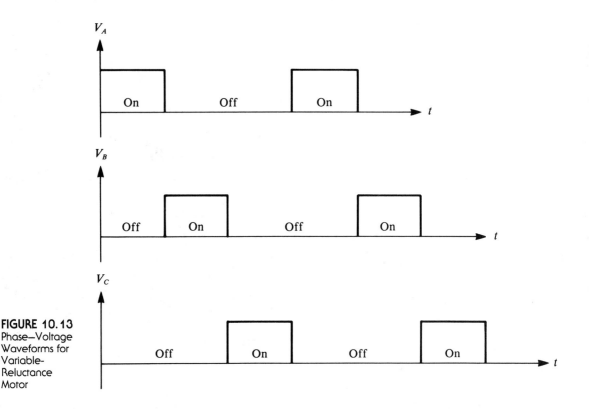

FIGURE 10.13
Phase–Voltage Waveforms for Variable-Reluctance Motor

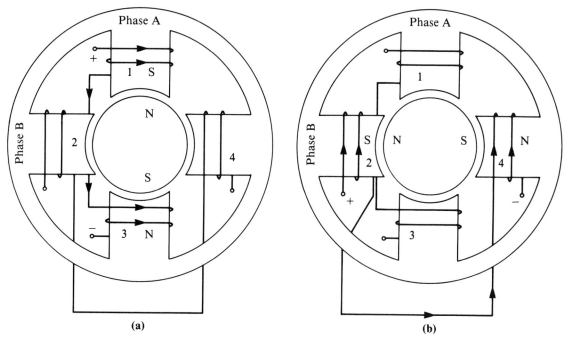

(a)

(b)

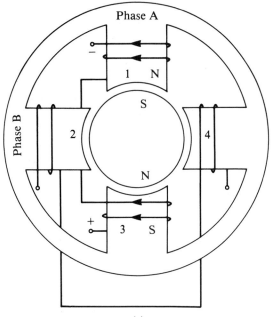

(c)

FIGURE 10.14
Two-Phase Permanent Magnet Step Motor

	Cycle	Phase			Position δ°
		A	B	C	
TABLE 10.1 Switching Sequence for a Variable-reluctance Motor	1	1	0	0	0
		0	1	0	30
		0	0	1	60
	2	1	0	0	90
		0	1	0	120
		0	0	1	150
	3	1	0	0	180
		0	1	0	210
		0	0	1	240
	4	1	0	0	270
		0	1	0	300
		0	0	1	330
	5	1	0	0	360

in the clockwise direction by following the above sequence. The motor can be made to rotate in the counterclockwise direction if A, C, and B are sequentially energized.

The input voltage waveforms of the step motor discussed previously are given in Fig. 10.13. Table 10.1 shows the proper switching sequence for clockwise rotation. For this particular motor, the applied voltage must have at least five cycles for one revolution. In Table 10.1, 1 and 0 correspond to positive and zero current in a phase winding.

10.2.2 Permanent-magnet Step Motors

The stator of a PM step motor is similar to that of a variable-reluctance motor. The only difference is in the rotor, which consists of permanent magnets in a PM step motor. In Fig. 10.14 is shown a two-phase, two-pole rotor PM step motor. The rotor is magnetized in the radial direction so that the rotor poles will line up with appropriate stator teeth.

Let us apply a positive direct current to phase A. For the direction of the current flow in the winding, tooth 1 will act as a south pole, while tooth 3 will be

	Cycle	Phase		Position δ°
		A	B	
TABLE 10-2 Switching Sequence for a Two-phase PM Step Motor	+	1	0	0
		0	1	90
	−	−1	0	180
		0	−1	270
	+	1	0	360

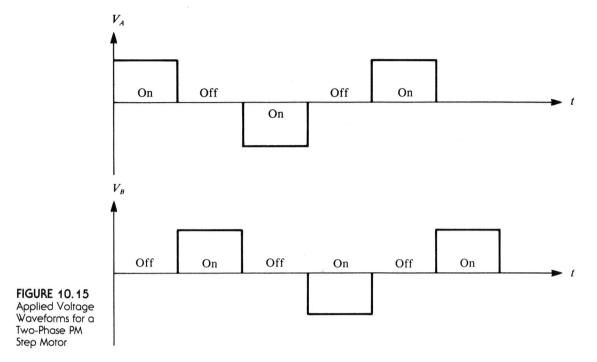

FIGURE 10.15
Applied Voltage
Waveforms for a
Two-Phase PM
Step Motor

a north pole, and the rotor will align as shown in Fig. 10.14a. If the positive current is now injected into phase B while the current in phase A is switched off, the rotor will rotate 90° in the counterclockwise direction, as shown in Fig. 10.14b. If the current is turned off in phase B and a negative current is allowed to flow in phase A, the rotor will rotate one step further in the counterclockwise direction, as shown in Fig. 10.14c. The rotor has gone through half a revolution. The switching sequence for a continuous rotation is given in Table 10.2. Figure 10.15 shows the on–off periods of currents in the two phases, respectively.

10.2.3 Hybrid Step Motors

The hybrid step motor consists of two pieces of soft iron, as well as an axially magnetized round permanent-magnet rotor. The soft-iron pieces are attached to both ends of the permanent-magnet rotor as shown in Fig. 10.16. The rotor teeth are machined on the soft-iron pieces. Consequently, all the rotor teeth at one end are north poles, while those at the other end are south poles. The rotor teeth on one end are offset with respect to the other end for the proper alignment of the rotor pole with that of the stator, as shown. The mode of operation of this motor is very similar to that of a PM step motor.

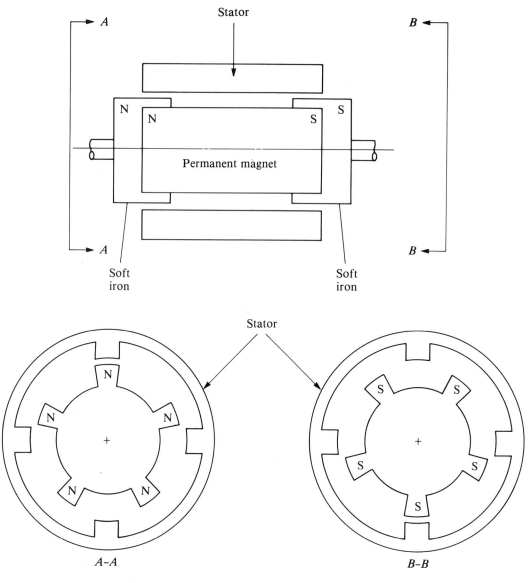

FIGURE 10.16
Various Views of a Hybrid Step Motor

10.3 BRUSHLESS DC MOTORS

Even though a dc motor offers a great variety of controlling schemes compared to other types of motors in electric drive systems, its mechanical commutator–brush arrangement exhibits a major shortcoming. Commutator and brushes are

subject to wear and require maintenance on a regular basis. Furthermore, they
limit the speed of operation and voltage of dc machines. A motor that retains the
features of a dc motor but gets rid off the commutator and brush problems is
recognized in the industry as a brushless dc motor.

A brushless dc motor consists of a multiphase stator winding and a radially
magnetized permanent-magnet rotor. Figure 10.17 shows a schematic diagram of
a brushless dc motor. The multiphase winding may be wound as a single coil or
distributed over the pole span. Direct voltage is applied to individual phase windings by means of a sequential switching operation. For instance, if winding 1 is
energized, the permanent-magnet rotor will line up with the magnetic field produced by winding 1. By turning off the voltage of winding 1 and applying it to
winding 2, the rotor can be made to rotate for alignment with the magnetic flux
of winding 2. As is now obvious, the operation of a brushless dc motor is very
similar to that of a PM step motor. The only difference is that the rotor rotates

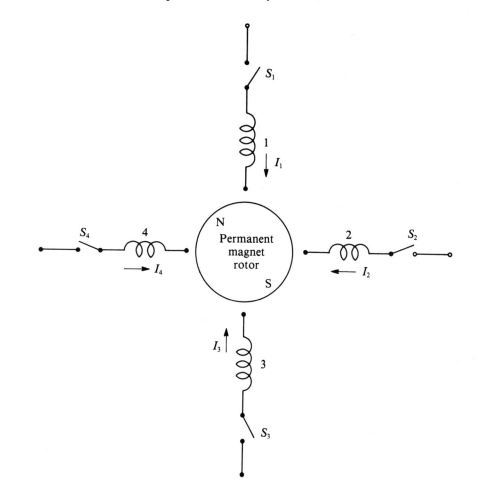

FIGURE 10.17
Schematic of a
Brushless dc
Motor that
Indicates the
Operating
Principle

continuously in a brushless dc motor, whereas it advances in specified steps in a PM step motor. The switching operation in a brushless dc motor is determined by the position of the rotor. The rotor position can be detected by using either Hall-effect or the photoelectric devices.

10.4 LINEAR INDUCTION MOTORS

So far we have examined the fundamental principles of operation of electric motors that produce circular motion. During the last few decades, extensive research in the area of propulsion has led to the development of linear motors. In theory, each type of rotating motor may find a linear counterpart. However, it is the linear induction motor that is being used in a broad spectrum of such industrial applicatons as high-speed ground transportation, sliding door systems, curtain pullers, and conveyors.

If an induction motor is cut and laid flat, a linear induction motor is obtained. The stator and rotor of the rotating motor correspond to the primary and secondary sides, respectively, of the linear induction motor. The primary consists of a magnetic core with a three-phase winding, and the secondary may simply be a metal sheet or a three-phase winding wound over a magnetic core. The basic difference between a linear induction motor and its rotating counterpart is that the latter exhibits endless air-gap and magnetic structure, while the former is open ended due to the finite lengths of the primary and secondary. To maintain a constant force over a considerable distance, one side is shorter than the other. For example, in high-speed ground transportation, a short primary and long secondary are being used. In such a system, the primary is an integral part of the vehicle, while the track manifests as the secondary.

A linear induction motor may be single-sided, as shown in Fig. 10.18a, or double sided, as shown in Fig. 10.18b. To reduce the total reluctance of the magnetic path in a single-sided linear induction motor with a metal sheet as the secondary winding, the metal sheet is backed by a ferromagnetic material such as iron.

When the supply voltage is impressed on the primary winding of a polyphase linear induction motor, the generated magnetic field travels at its synchronous speed in the air-gap region. The interaction of the magnetic field with the induced currents in the secondary exerts a thrust on the secondary to move in the same direction if the primary is held stationary. On the other hand, if the secondary winding is stationary and the primary is free to move, the primary will move in the direction opposite to that of the revolving field.

The speed–torque characteristic of an induction motor becomes the velocity–thrust characteristic for a linear induction motor. A typical velocity–thrust curve is shown in Fig. 10.19, which is similar to its counterpart for a rotating motor. The synchronous velocity of a linear induction motor is

$$v_s = 2\tau f \qquad (10.3)$$

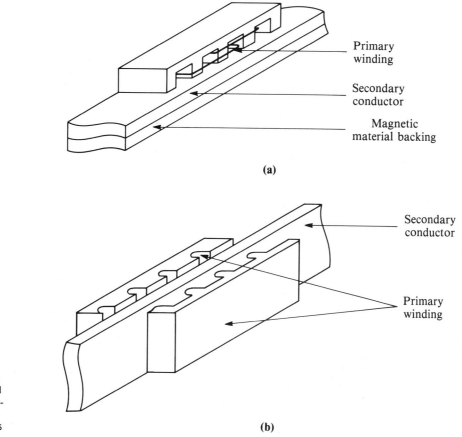

(a)

(b)

FIGURE 10.18
(a) Single-Sided
and (b) Double-
Sided Linear
Induction Motors

where f is the frequency of the applied source and τ is the pole pitch. It is therefore evident that the synchronous velocity is independent of the number of poles in the primary winding. In other words, by changing the pole pitch and/or the frequency, any desired synchronous velocity can be obtained. Furthermore, the synchronous velocity does not depend on the number of poles. Also, the number of poles does not have to be even. However, if v_m is the velocity of the linear induction motor, the slip can be expressed as

$$s = \frac{v_s - v_m}{v_s} \tag{10.4}$$

As is evident from Fig. 10.19, the decrease in velocity is rapid for any increase in the thrust. Thus, a linear induction motor operates at high slip with relatively low efficiency. Owing to the open-ended structure, a linear induction motor displays a phenomenon known as end effects. The end effects can be clas-

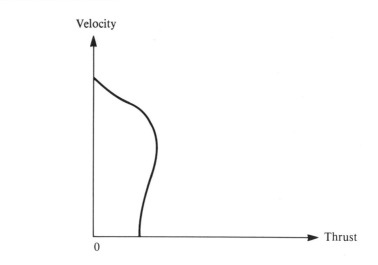

FIGURE 10.19
Typical Velocity
versus Thrust
Characteristics of
a Linear
Induction Motor

sified as static and dynamic. Static end effect occurs solely because of the asymmetric geometry of the primary. In this case, the mutual inductances of the phase windings are not equal to one another. This results in asymmetric flux distribution in the air-gap region and gives rise to unequal induced voltages in the phase windings. The dynamic end effect occurs due to the relative motion of the primary with respect to the secondary. As the primary moves over the secondary, at every instant a new secondary conductor is brought under the leading edge of the primary, while another secondary conductor is leaving the trailing edge of the primary. The conductor coming under the leading edge opposes the magnetic flux in the air gap, while the conductor leaving the trailing edge tries to sustain the flux. Therefore, the flux distribution in the region following the leading edge is weaker than that near the trailing edge. Furthermore, the conductor leaving the trailing edge, although still carrying the current and contributing to the losses, is not contributing to the thrust. Therefore, the increased losses in the secondary lower the efficiency of the motor.

10.5 HYSTERESIS MOTORS

A well-known property of magnetic materials, known as hysteresis, is employed uniquely to produce torque in a hysteresis motor. The stator is wound either as a three phase or single phase, with an auxiliary winding. For a single-phase hysteresis motor, the stator is connected as a permanent split-capacitor motor. The intent is to set up as uniform a revolving field as possible. The rotor is solid magnetic material with no teeth or windings. Figure 10.20 shows a two-phase hysteresis motor.

As soon as the stator windings are excited, a revolving field appears in the air gap of the machine. The rotating magnetic field magnetizes the metal of the

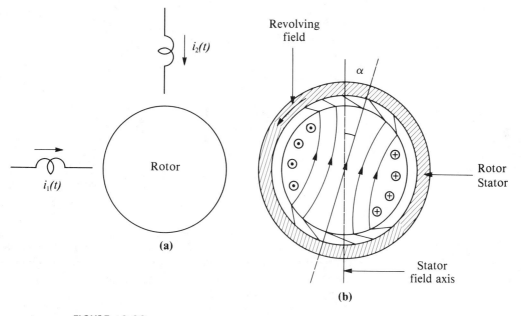

FIGURE 10.20
(a) Schematic of a Two-Phase Hysteresis Motor; (b) Flux Distribution Showing Hysteresis Effect

rotor and induces as many poles on its periphery as there are in the stator. Due to the large hysteresis loss in the rotor, the rotor flux cannot follow the stator flux exactly. Therefore, an angle of lag exists between the magnetic axes of the stator and the rotor magnetic fields. The greater the loss due to hysteresis, the larger is the angle between the magnetic axes of the stator and the rotor fluxes. Owing to the tendency of the rotor to align itself with the stator, a finite torque is produced by the motor. Since the rotor is a solid magnetic material, eddy currents are induced in the rotor by the magnetic field of the stator as long as there is relative motion between the stator magnetic field and the rotor. These eddy currents produce their own magnetic fields, which further increase the torque developed by the motor.

As soon as the rotor starts rotating at the synchronous speed, there is no eddy-current flow in the rotor. In other words, the torque developed by the eddy currents ceases to exist at the synchronous speed of the motor. In this case the rotor behaves as a magnet, and the torque developed is due to hysteresis and is proportional to the angle between the magnetic axes of the stator and rotor magnetic fields. The hysteresis torque developed by the motor depends on the type of magnetic material used and the flux density in the motor. Once the motor is made, the hysteresis torque developed by the motor is almost constant over its entire speed range. On the other hand, the torque developed by the eddy currents is proportional to the slip of the motor. The eddy-current torque is maximum at the time of starting (i.e., when the slip is unity) and goes to zero at synchronous

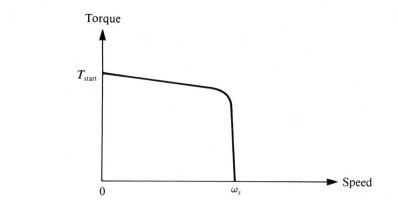

FIGURE 10.21
Speed–Torque
Characteristic of
a Hysteresis
Motor

speed. The speed–torque characteristic of the motor that highlights such a behavior is shown in Fig. 10.21. As is evident from the curve, the torque developed by the motor is higher at any other speed than the synchronous speed. Therefore, the starting torque is never a problem in hysteresis motors.

The principle of the hysteresis motor has also been used with shaded-pole motors for the development of low-cost synchronous motors for running electric clocks.

PROBLEMS

10.1. A 16-V permanent-magnet motor develops a torque of 0.75 N-m at the rated voltage. The magnetic flux in the motor is 2 mWb. The armature resistance is 0.52 Ω and the motor constant is 80. Calculate the operating speed of the motor. Neglect rotational losses.

10.2. Calculate the armature current of the motor given in Problem 10.1 under a blocked-rotor condition.

10.3. The motor described in Problem 10.1 develops a torque of 8 N-m under a blocked-rotor condition. What is the voltage applied to the motor?

10.4. A 20-V, 2-pole, permanent-magnet motor manufactured with ceramic magnets (Fig. P10.4) delivers a load of 100 W with an efficiency of 55%. The ideal no-load speed of the motor is 750 rpm and the armature resistance is 1.2 Ω. The pole length and the average radius are 45 mm and 35 mm, respectively. Determine the operating line of the motor. The motor constant is 68. Consider only the copper losses and neglect the others.

10.5. Determine the performance of the motor given in Problem 10.4 if rare-earth magnets are substituted for the ceramic magnets without changing the dimensions of the motor.

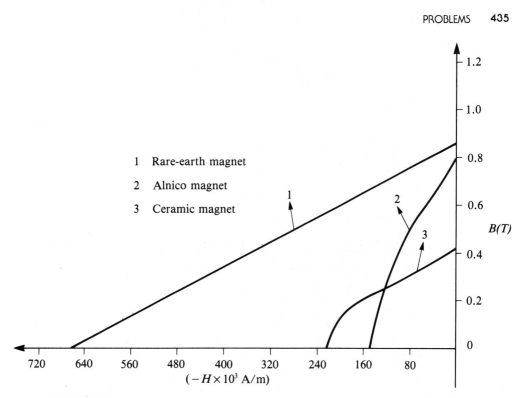

1   Rare-earth magnet

2   Alnico magnet

3   Ceramic magnet

FIGURE P10.4
Illustration for Problem 10.4.

10.6. Determine the magnetic flux in a 100-V, 500-W permanent-magnet motor operating at a speed of 2000 rpm. The motor constant is 92, the armature resistance is 0.5 Ω, and the rotational losses are 50 W.

10.7. A 100-V permanent magnet motor operates at a speed of 300 rad/s under no load. The armature circuit resistance is 1.25 Ω. Determine the speed of the motor when it is required to deliver a torque of 5 N-m at 50 V. Draw the speed–torque characteristics for both 50-V and 100-V operations. Assume that the motor maintains constant flux and there are no rotational losses.

CHAPTER 11

Dynamics of Electric Machines

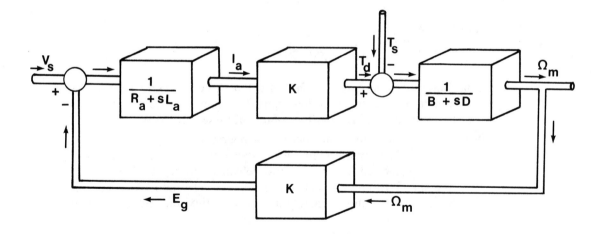

In the previous chapters the electric machines were modeled based on the tacit assumption of steady conditions of operation. However, any changes in the applied voltage and/or torque can cause instantaneous changes in the energy stored in the energy-storing elements, such as equivalent inductances of the machine, and the total mass of the rotating system. From the onset of a change a considerable time may elapse prior to reaching the steady condition. The response during this time is referred to as the *dynamic response* of a machine and consists of a transient response superimposed on the steady-state or normal response. Thus, the steady-state response is one special case of a dynamic response. In this chapter, we will focus our attention on the modeling and analysis of an electric machine to predict its dynamic response.

11.1 ARMATURE-CONTROLLED DC MOTORS

Figure 11.1 shows a separately excited dc motor operating in the linear region. Under dynamic conditions, the torque developed by the motor can be expressed as

$$T_d(t) = T_s(t) + D\omega_m(t) + J\frac{d\omega_m(t)}{dt}$$

or

$$Ki_a(t) = T_s(t) + D\omega_m(t) + J\frac{d\omega_m(t)}{dt} \qquad (11.1)$$

where $K = K_a\phi_p$ and K_a is the machine constant as given in Chapter 4. D is the

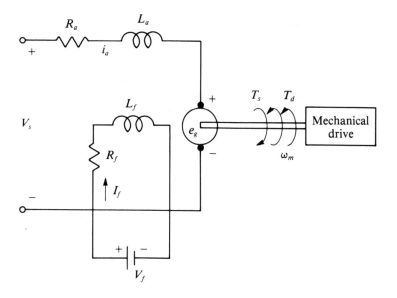

FIGURE 11.1
Separately
Excited dc Motor
with a
Mechanical
Drive

equivalent coefficient of friction and J is the moment of inertia of the rotating members of the motor-load system. The load torque, T_s, depends only on the mechanical drive system. It may be either constant or some function of speed, ω_m.

During the dynamic state the differential equation for the armature circuit of the motor, from Fig. 11.1, can be given as

$$V_s = i_a(t)R_a + L_a \frac{di_a(t)}{dt} + e_g(t)$$

or

$$= i_a(t)R_a + L_a \frac{di_a(t)}{dt} + K\omega_m(t) \tag{11.2}$$

Equations (11.1) and (11.2) describing the behavior of a separately excited dc motor can be expressed in the matrix form as

$$
\begin{bmatrix} \dfrac{d\omega_m(t)}{dt} \\[2em] \dfrac{di_a(t)}{dt} \end{bmatrix}
=
\begin{bmatrix} -\dfrac{D}{J} & \dfrac{K}{J} \\[1.5em] -\dfrac{K}{L_a} & -\dfrac{R_a}{L_a} \end{bmatrix}
\begin{bmatrix} \omega_m(t) \\[2em] i_a(t) \end{bmatrix}
+
\begin{bmatrix} -\dfrac{1}{J} & 0 \\[1.5em] 0 & \dfrac{1}{L_a} \end{bmatrix}
\begin{bmatrix} T_s(t) \\[2em] V_s \end{bmatrix}
\tag{11.3}
$$

Equation (11.3) is the standard form of representing a dynamic system in terms of its state variables ω_m and i_a. These are the variables that are affected by sudden changes in the input variables T_s and V_s. Since there are two energy-storing elements (i.e., inertia of the system, J, and armature winding inductance, L_a), any sudden changes in the operating condition of the system involves the redistribution of the two forms of energy. Let us investigate the dynamic response of the entire system by using the Laplace transform technique and a computer-aided solution.

Laplace Transform Technique   This technique permits the conversion of differential equations into a set of algebraic equations in the complex frequency domain. The Laplace transform of Eq. (11.1) yields

$$KI_a(s) = T_s(s) + D\Omega_m(s) + J[s\Omega_m(s) - \omega_m(0)] \tag{11.4}$$
$$= (D + sJ)\Omega_m(s) + T_s(s) - J\omega_m(0)$$

where $I_a(s)$, $T_s(s)$, and $\Omega_m(s)$ are the Laplace transforms of i_a, T_s, and ω_m, respectively. Similarly, Eq. (11.2) can be expressed in its Laplace transform as

$$V_s(s) = (R_a + sL_a)I_a(s) + K\Omega_m(s) - L_a i_a(0) \tag{11.5}$$

From these equations, after simple algebraic manipulations, we can obtain

$$\Omega_m(s) = \frac{K[V_s(s) + L_a i_a(0)] + [J\omega_m(0) - T_s(s)](R_a + sL_a)}{(D + sJ)(R_a + sL_a) + K^2} \quad (11.6)$$

and
$$I_a(s) = \frac{-K[J\omega_m(0) - T_s(s)] + [L_a i_a(0) + V_s(s)](D + sJ)}{(D + sJ)(R_a + sL_a) + K^2} \quad (11.7)$$

If the motor was at standstill prior to the energization of the armature, the initial speed is zero because of inertia, and the initial armature current is zero because of winding inductance. Therefore, Eqs. (11.6) and (11.7) can be simpified as

$$\Omega_m(s) = \frac{KV_s(s) - T_s(s)(R_a + sL_a)}{(D + sJ)(R_a + sL_a) + K^2} \quad (11.8)$$

and
$$I_a(s) = \frac{(D + sJ)V_s(s) + KT_s(s)}{(D + sJ)(R_a + sL_a) + K^2} \quad (11.9)$$

These equations describe the dynamics of a dc motor while the field current is constant. This is essentially the requirement of a motor speed control using the Ward–Leonard system.

■ EXAMPLE 11.1

A 240-V, 1-hp, separately excited dc motor is rated at 500 rpm. The armature resistance and inductance are 7.56 Ω and 0.055 H, respectively. The product of machine constant and flux per pole is 3.75, while the moment of inertia of the system is 0.0678 kg-m^2. Determine the variations in the armature current and the motor speed as a function of time under no load when the armature voltage is applied at $t = 0$. Neglect the frictional losses and assume the motor operates in the linear region.

Solution

Since the frictional losses are neglected, $D = 0$. For no load, $T_s = 0$. From Eq. (11.8),

$$\Omega_m(s) = \frac{3.75 \times 240/s}{0.0678 \, s \, (7.56 + 0.055s) + 3.75^2}$$

$$= \frac{241,352^2}{s(s^2 + 137.455s + 3771.12)}$$

which can be reduced by partial fractions as follows:

$$\Omega_m(s) = \frac{A}{s} + \frac{B}{s + 99.59} + \frac{C}{s + 37.87}$$

where

$$A = \frac{241,352}{(s + 99.59)(s + 37.87)}\bigg|_{s=0} = 64$$

$$B = \frac{241,352}{s(s + 37.87)}\bigg|_{s=-99.59} = 39.26$$

and

$$C = \frac{241,352}{s(s + 99.59)}\bigg|_{s=-37.87} = -103.26$$

Therefore, the angular velocity of the motor is

$$\omega_m(t) = 64 + 39.26e^{-99.59t} - 103.26e^{-37.87t} \text{ rad/s}, \qquad t \geq 0$$

From Eq. (11.9), after some simplifications, the Laplace-transformed armature current is

$$I_a(s) = \frac{4363.64}{s^2 + 137.455\,s + 3771.12}$$

$$= \frac{70.7}{s + 37.87} - \frac{70.7}{s + 99.59}$$

Thus, the armature current is

$$i_a(t) = 70.7(e^{-37.87t} - e^{-99.59t}) \text{ A}, \qquad t \geq 0$$

■ EXAMPLE 11.2

The motor given in Example 11.1 is coupled to a load of 14.34 N-m. Determine the speed of the motor as a function of time when the armature is energized at $t = 0$.

Solution

Since the voltage is applied to the armature circuit at $t = 0$, the initial speed and the armature current are both zero. From Eq. (11.8),

$$\Omega_m(s) = \frac{(3.75 \times 240/s) - (7.56 + 0.055s) \times (14.34/s)}{0.0678s(7.56 + 0.055s) + 3.75^2}$$

$$= \frac{212,279.76 - 211.51s}{s(s^2 + 137.455s + 3771.12)}$$

$$= \frac{56.29}{s} + \frac{37.96}{s + 99.59} - \frac{94.25}{s + 37.87}$$

The angular velocity is

$$\omega_m(t) = 56.29 + 37.96e^{-99.59t} - 94.25e^{-37.87t} \text{ rad/s}, \qquad t \geq 0$$

For all practical purposes, the motor will attain its steady operation after five time constants. From the preceding expression, the exponential terms will disappear after 50.2 ms and 132 ms.

■ EXAMPLE 11.3

If the motor discussed in Example 11.1 is loaded with a linear load of $T_s = 0.26\omega_m$ and the armature voltage is impressed at the same time, determine the motor speed as a function of time. What is the torque supplied to the load after the motor attains a steady-state condition?

Solution

In Eq. (11.8), $T_s(s)$ can be replaced by $0.26\Omega_m(s)$. Grouping the speed terms, we obtain

$$\Omega_m(s) = \frac{KV_s(s)}{(D + sJ)(R_a + sL_a) + K^2 + 0.26(R_a + sL_a)}$$

Substituting the parameters of the motor, we get

$$\Omega_m(s) = \frac{241,352}{s(s^2 + 141.29s + 4298.23)}$$

$$= \frac{56.2}{s} + \frac{47.3}{s + 96.96} - \frac{103.5}{s + 44.33}$$

and angular velocity as a function of time is

$$\omega_m(t) = 56.2 + 47.3e^{-96.96t} - 103.5e^{-44.33t} \text{ rad/s}, \qquad t \geq 0$$

When the motor attains the steady-state condition, the angular velocity of the motor is 56.2 rad/s. Therefore, the load torque is

$$T_s = 0.26 \times 56.2 = 14.612 \text{ N-m}$$

11.2 FIELD-CONTROLLED DC MOTORS

Another method of controlling the speed of a dc motor involves an adjustment in the field current while the armature current is held constant. To ensure constant armature current, the armature circuit must be fed by a constant current source, as shown in Fig. 11.2. In the linear operating region, the flux produced by the motor is proportional to the field current (i.e., $\phi_p = k_f i_f$). The torque developed by the motor can be expressed as

$$T_d = K_a \phi_p I_a = K_T i_f \tag{11.10}$$

where $K_T = K_a k_f I_a$ is the constant of proportionality. In the dynamic state, the torque developed by the motor can also be written as

$$K_T i_f(t) = T_s(t) + D\omega_m(t) + J\frac{d\omega_m(t)}{dt} \tag{11.11}$$

The applied field voltage, from Fig. 11.2, is

$$V_f = R_f i_f(t) + L_f \frac{di_f(t)}{dt} \tag{11.12}$$

where R_f and L_f are the effective resistance and inductance of the field circuit. These equations can be expressed in concise matrix form as

$$
\begin{bmatrix} \dfrac{d\omega_m(t)}{dt} \\[2em] \dfrac{di_f(t)}{dt} \end{bmatrix}
=
\begin{bmatrix} -\dfrac{D}{J} & \dfrac{K_T}{J} \\[1.5em] 0 & -\dfrac{R_f}{L_f} \end{bmatrix}
\begin{bmatrix} \omega_m(t) \\[2em] i_f(t) \end{bmatrix}
+
\begin{bmatrix} -\dfrac{1}{J} & 0 \\[1.5em] 0 & \dfrac{1}{L_f} \end{bmatrix}
\begin{bmatrix} T_s(t) \\[2em] V_f \end{bmatrix}
\tag{11.13}
$$

To obtain the dynamic response of the motor, Eqs. (11.11) and (11.12) can be transformed, using the Laplace transform technique, as

$$K_T I_f(s) = (D + sJ)\Omega_m(s) + T_s(s) - J\omega_m(0) \tag{11.14}$$

and
$$V_f(s) = (R_f + sL_f)I_f(s) - L_f i_f(0) \tag{11.15}$$

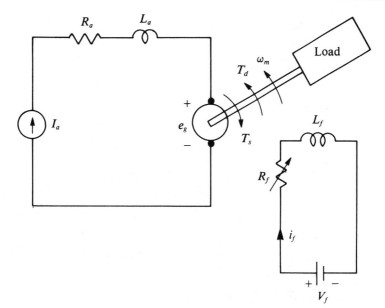

FIGURE 11.2
Separately
Excited dc Motor
with a Constant
Armature
Current

From Eqs. (11.14) and (11.15), the Laplace-transformed angular velocity and the field current are

$$\Omega_m(s) = \frac{K_T(V_f(s) + L_f i_f(0)) + [J\omega_m(0) - T_s(s)](R_f + sL_f)}{(D + sJ)(R_f + sL_f)} \quad (11.16)$$

and

$$I_f(s) = \frac{V_f(s) + L_f i_f(0)}{R_f + sL_f} \quad (11.17)$$

If the motor is at standstill when the voltage is impressed on the field winding with constant current in the armature, Eqs. (11.16) and (11.17) reduce to

$$\Omega_m(s) = \frac{K_T V_f(s) - (R_f + sL_f)T_s(s)}{(D + sJ)(R_f + sL_f)} \quad (11.18)$$

and

$$I_f(s) = \frac{V_f(s)}{R_f + sL_f} \quad (11.19)$$

respectively.

■ EXAMPLE 11.4

A separately excited dc motor has a field winding resistance of 60 Ω and an inductance of 0.3 H. The moment of inertia of all rotating parts is 0.08 kg-m^2 and the coefficient of friction is 0.3 kg-m²/s. In the linear region of operation, the developed torque is 10 times the field current. If the field winding is suddenly subjected to a direct voltage of 240 V, determine the speed as a function of time under no load.

Solution

In the linear operating region, $K_T = 10$ N-m/A. Substituting the parameters in Eq. (11.18), we obtain

$$\Omega_m(s) = \frac{10 \times 240/s}{(60 + 0.3s)(0.3 + 0.08s)} = \frac{100,000}{s(s + 200)(s + 3.75)}$$

$$= \frac{133.33}{s} + \frac{2.55}{s + 200} - \frac{135.88}{s + 3.75}$$

and $\quad \omega_m(t) = 133.33 + 2.55e^{-200t} - 135.88e^{-3.75t}$ rad/s $t \geq 0$

11.3 DYNAMICS OF DC GENERATORS

A separately excited dc generator delivering power to a load is shown in Fig. 11.3. Once again, assume that within the region of operation of the generator the magnetic circuit response is linear. To determine the dynamical behavior of the generator with respect to the changes in the field and armature currents, a considerable simplification in the analysis can be effected by making an assumption that the shaft speed is practically constant.

During the dynamic state, the field voltage is

$$V_f = R_f i_f(t) + L_f \frac{di_f(t)}{dt} \tag{11.20}$$

and the induced voltage is

$$e_g(t) = (R_a + R_L)i_a(t) + (L_a + L_L)\frac{di_a(t)}{dt} \tag{11.21}$$

However, the induced voltage can be expressed as

$$e_g(t) = K_e i_f(t) \tag{11.22}$$

where $K_e = K_a k_f \omega_m$ is the constant of proportionality. Substituting Eq. (11.22) in Eq. (11.21), we obtain

$$K_e i_f(t) = (R_a + R_L)i_a(t) + (L_a + L_L)\frac{di_a(t)}{dt} \tag{11.23}$$

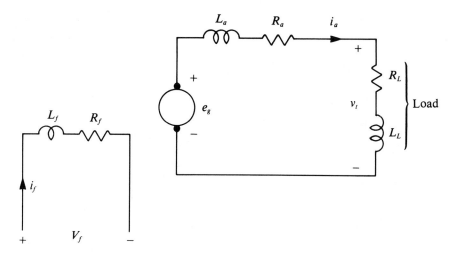

FIGURE 11.3
Separately
Excited dc
Generator

The dynamical representation of the generator–load system can be written in the matrix form as

$$\begin{bmatrix} \dfrac{di_a(t)}{dt} \\[3ex] \dfrac{di_f(t)}{dt} \end{bmatrix} = \begin{bmatrix} -\dfrac{R_a + R_L}{L_a + L_L} & \dfrac{K_e}{L_a + L_L} \\[3ex] 0 & -\dfrac{R_f}{L_f} \end{bmatrix} \begin{bmatrix} i_a(t) \\[3ex] i_f(t) \end{bmatrix} + \begin{bmatrix} 0 \\[3ex] \dfrac{1}{L_f} \end{bmatrix} V_f \quad (11.24)$$

Equations (11.20) and (11.23) can be expressed in the s-domain as

$$V_f(s) = (R_f + sL_f)I_f(s) \tag{11.25}$$

and
$$K_eI_f(s) = (R_a + R_L + sL_a + sL_L)I_a(s) \tag{11.26}$$

with zero initial conditions. From these equations we obtain

$$I_f(s) = \frac{V_f(s)}{R_f + sL_f} \tag{11.27}$$

and
$$I_a(s) = \frac{K_eV_f(s)}{(R_f + sL_f)(R_a + R_L + sL_a + sL_f)} \tag{11.28}$$

■ EXAMPLE 11.5

A separately excited dc generator operating at 1900 rpm has the following parameters: $R_f = 4.5\ \Omega$, $L_f = 0.03$ H, and $K_e = 45$ V/A. If 240-V dc voltage is suddenly applied to the field winding under no load, determine (a) the field current and the induced voltage as a function of time, (b) the approximate time to reach steady-state condition, and (c) the steady-state values of the field current and induced voltage.

Solution

(a) From Eq. (11.27),

$$I_f(s) = \frac{240/s}{4.5 + 0.03s}$$

$$= \frac{8000}{s(s + 150)}$$

$$= \frac{53.33}{s} - \frac{53.33}{s + 150}$$

Therefore, the field current is

$$i_f(t) = 53.33(1 - e^{-150t}) \text{ A}$$

The induced or no-load voltage is

$$e_g(t) = K_e i_f(t) = 2400(1 - e^{-150t}) \text{ V}$$

(b) For all practical purposes, the field current will attain its steady-state value after five time constants. Thus, the time required to reach the steady-state condition is

$$T = \frac{5}{150} = 0.0333 \text{ s} \quad \text{or} \quad 33.3 \text{ ms}$$

(c) The final values of the field current and the no-load voltage are 53.33 A and 2400 V, respectively.

11.4 NUMERICAL ANALYSIS OF DC MACHINE'S DYNAMICS

In the preceding sections the dynamic response of a dc machine was analyzed using the Laplace transform technique. In this section, our aim is to investigate the dynamic behavior of a dc machine by solving the differential equations numerically employing the fourth-order Runge–Kutta algorithm. Prior to its application, a brief discussion of the Runge–Kutta algorithm is given.

Fourth-order Runge–Kutta Algorithm   Consider a set of first-order differential equations given in the matrix form as

$$\frac{d\underline{x}}{dt} = A\underline{x}(t) + B\underline{u}(t) \tag{11.29}$$

where $\underline{x}(t)$ is a state variable vector (column matrix) and $\underline{u}(t)$ is the input variable vector. A and B are constant coefficient matrices of the system. Equation (11.29) can be solved at discrete instants of time as follows:

$$\underline{x}_{n+1} = \underline{x}_n + \frac{\underline{K}_1 + 2\underline{K}_2 + 2\underline{K}_3 + \underline{K}_4}{6} \tag{11.30}$$

where $\underline{x}_n$ and $\underline{x}_{n+1}$ are the values of the state variable vectors at instants of n and $n + 1$, respectively. $\underline{K}_1, \underline{K}_2, \underline{K}_3$, and $\underline{K}_4$ are constant vectors computed at discrete instants of time as outlined next.

Let
$$f(\underline{x}_n) = A\underline{x}_n + B\underline{u}_n \tag{11.31}$$

then
$$\underline{K}_1 = hf(\underline{x}_n) \tag{11.32a}$$

$$\underline{K}_2 = hf(\underline{x}_n + 0.5\underline{K}_1) \tag{11.32b}$$

$$\underline{K}_3 = hf(\underline{x}_n + 0.5\underline{K}_2) \tag{11.32c}$$

$$\underline{K}_4 = hf(\underline{x}_n + \underline{K}_3) \tag{11.32d}$$

where h is the step length, which is defined as the time between two discrete observations. The step length is highly dependent on the time constants of the system. In general, h is chosen less than the smallest time constant of the system in order to maintain the stability of the numerical method. Figure 11.4 shows the flow chart of the algorithm and a sample program is listed in Fig. 11.5. There are, in all, five sets of inputs to the program: A matrix, initial values of the state variables, B matrix, the input variables, and the values for the step length and the total response time.

■ EXAMPLE 11.6

A 240-V, separately excited dc motor having the following rated data operates with a constant field: $R_a = 0.3\ \Omega$, $L_a = 0.002$ H, $K = 0.8$, and $J = 0.0678$ kg-m^2. Determine the motor speed and the armature current as a function of time when the motor is subjected to a torque of 100 N-m after 200 ms of starting at no load. Consider a step length of 0.01 s and observe the response for a period

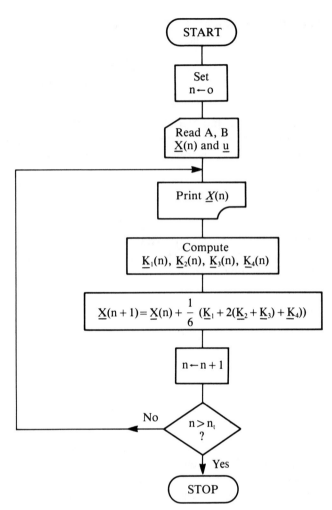

FIGURE 11.4
Flow Chart for
Solving State
Equations

of 0.5 s. Neglect the frictional losses and assume that the motor operates in the linear region.

Solution

From Eq. (11.3), for $t < 200$ ms, we have

$$\underline{x}(t) = \begin{bmatrix} \omega_m(t) \\ i_a(t) \end{bmatrix}, \qquad \underline{u}(t) = \begin{bmatrix} 0 \\ 240 \end{bmatrix},$$

$$A = \begin{bmatrix} 0 & 11.8 \\ -400 & -150 \end{bmatrix}, \quad \text{and} \quad B = \begin{bmatrix} -14.75 & 0 \\ 0 & 500 \end{bmatrix}$$

with initial values of $\omega_m(0) = 0$ and $i_a(0) = 0$.

```
            DIMENSION A(2,2),X(2),B(2,2),AX(2),BU(2),F(2),C1(2),C2(2),C3(2),
           *C4(2),G(2),XG(2),U(2)
            N=2
            READ(5,100) ((A(I,J),J=1,N),I=1,N)
            READ(5,100) (X(I),I=1,N)
            READ(5,100) ((B(I,J),J=1,N),I=1,N)
            READ(5,100) (U(I),I=1,N)
100         FORMAT(2F10.3)
            READ(5,110) T,H
110         FORMAT(2F15.9)
            M=T/H
            WRITE(6,115) M
115         FORMAT(I5)
            DO 5 K=1,M
            TS=H*(K-1)
            CALL PLOT(TS,X(1))
            CALL MVMULT(A,X,AX,N)
            CALL MVMULT(B,U,BU,N)
            CALL VVADDT(AX,BU,F,N)
            DO 10 I=1,N
            C1(I)=H*F(I)
10          G(I)=0.5*C1(I)
            CALL VVADDT(X,G,XG,N)
            CALL MVMULT(A,XG,AX,N)
            CALL VVADDT(AX,BU,F,N)
            DO 11 I=1,N
            C2(I)=H*F(I)
11          G(I)=0.5*C2(I)
            CALL VVADDT(X,G,XG,N)
            CALL MVMULT(A,XG,AX,N)
            CALL VVADDT(AX,BU,F,N)
            DO 12 I=1,N
12          C3(I)=H*F(I)
            CALL VVADDT(X,C3,XG,N)
            CALL MVMULT(A,XG,AX,N)
            CALL VVADDT(AX,BU,F,N)
            DO 13 I = 1,N
13          C4(I) = H*F(I)
            DO 15 I =1,N
15          X(I)=X(I)+(1./6.)*(C1(I)+2.*(C2(I)+C3(I))+C4(I)))
5           CONTINUE
            END
            SUBROUTINE MVMULT(A,B,C,N)
            DIMENSION A(N,N),B(N),C(N)
            DO 5 I=1,N
            C(I) = 0.0
            DO 5 J =1,N
5           C(I)=C(I) + A(I,J)*B(J)
            RETURN
            END
            SUBROUTINE VVADDT(A,B,C,N)
            DIMENSION A(N),B(N),C(N)
            DO 5 I =1,N
5           C(I)=A(I) + B(I)
            RETURN
            END
```

FIGURE 11.5
Computer
Program to
Solve the State
Equations

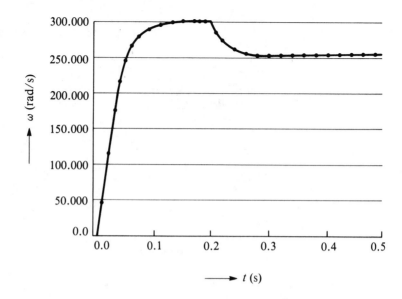

FIGURE 11.6
Variation of the
Speed for
Example 11.6

For $t > 200$ ms,

$$\underline{u}(t) = \begin{bmatrix} 100 \\ 240 \end{bmatrix} \quad \text{and} \quad \underline{x}(0.2) = \begin{bmatrix} 299.93 \\ 0.2s \end{bmatrix}$$

The variations in the speed and the armature current as computed are shown in Figs. 11.6 and 11.7, respectively.

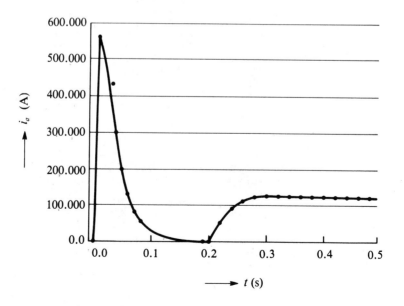

FIGURE 11.7
Variation of the
Armature
Current for
Example 11.6

11.5   SYNCHRONOUS GENERATOR TRANSIENTS

Whenever there is a sudden change in the torque applied to the rotor or in the load current, a finite period of time is needed by a synchronous generator before returning to its steady operation. The operation of the generator during this finite period of time is known as the transient operation, and the transient may be mechanical or electrical in nature. In this section, each type of transient is considered separately.

11.5.1   Electrical Transient

The most severe transient in a synchronous generator is the development of a sudden short circuit between its terminals under no load. To simplify the theoretical development, let us assume that the generator operates in the linear region and the winding resistances are negligible. Just prior to the short circuit, the total flux linkage of the field winding is

$$\lambda_f = N_f \phi_f = L_f I_f \tag{11.33}$$

where I_f is the field-winding current and L_f is its self-inductance. Representing the self-inductance in terms of the leakage inductance, L_{lf}, and the mutual inductance between the field and the armature winding, L_{af}, Eq. (11.33) can be written as

$$\lambda_f = (L_{lf} + L_{af})I_f \tag{11.34}$$

Since the generator is operating under no-load condition, the flux linkage due to the armature winding is

$$\lambda_a = 0 \tag{11.35}$$

When a three-phase short circuit occurs across the armature terminals at $t = 0$, the magnetic axes of the field winding and the armature winding, say phase a, must be orthogonal in accordance with Eq. (11.35). After a short time the rotor will attain a certain angular position α with respect to the magnetic axis of phase a winding, causing current flows of i_a and $I_f + i_f$ in phase a and the field winding, respectively, in order to maintain the same total flux linkages. The flux linkage for phase a can be expressed as

$$\lambda_a = i_a(L_{la} + L_{af}) + (I_f + i_f)L_{af}\sin(90° - \alpha) = 0 \tag{11.36}$$

where L_{la} is the leakage inductance of phase a.

The flux linkages associated with the field winding are

$$\lambda_f = (I_f + i_f)(L_{lf} + L_{af}) + L_{af}i_a\sin(90° - \alpha) = (L_{lf} + L_{af})I_f \tag{11.37}$$

From Eqs. (11.36) and (11.37), it can be shown that

$$i_a = \frac{L_{af}I_f(L_{lf} + L_{af})\sin(90° - \alpha)}{L_{af}^2 \sin^2(90° - \alpha) - (L_{lf} + L_{af})(L_{la} + L_{af})} \tag{11.38}$$

and

$$i_f = -\frac{L_{af}^2 \sin^2(90° - \alpha)I_f}{L_{af}^2 \sin^2(90° - \alpha) - (L_{lf} + L_{af})(L_{la} + L_{af})} \tag{11.39}$$

The most severe transient condition will take place when the currents are maximum. From the preceding equations, the currents are maximum when $\alpha = 0$. Thus, corresponding to the most severe condition,

$$i_a = \frac{L_{af}I_f(L_{lf} + L_{af})}{L_{af}^2 - (L_{lf} + L_{af})(L_{la} + L_{af})} \tag{11.40}$$

and

$$i_f = -\frac{I_f L_{af}^2}{L_{af}^2 - (L_{lf} + L_{af})(L_{la} + L_{af})} \tag{11.41}$$

Equation (11.40) can be used to determine the *transient reactance* of the synchronous generator that governs its behavior during the transient period. Multiplying both the numerator and the denominator of Eq. (11.40) by ω^2, and after some simplifications, we obtain

$$i_a = -\frac{E_\phi(X_{lf} + X_{af})}{X_{lf}X_{la} + X_{lf}X_{af} + X_{af}X_{la}} \tag{11.42}$$

where $E_\phi = I_f\omega L_{af} = I_f X_{af}$ is the generated voltage prior to the fault (short-circuit condition) at no load. The transient reactance can be defined as the ratio of no-load voltage to short-circuit current, or

$$\frac{E_\phi}{-i_a} = X_d' = X_{la} + \frac{X_{af}X_{lf}}{X_{af} + X_{lf}} \tag{11.43}$$

and its equivalent-circuit representation is given in Fig. 11.8.

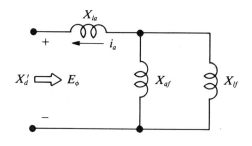

FIGURE 11.8
Equivalent
Circuit for the
Transient
Reactance

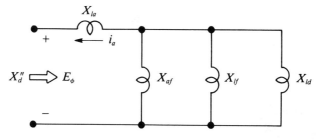

FIGURE 11.9
Equivalent
Circuit for
Subtransient
Reactance

In the preceding development we neglected the damper winding. If we take its effect into consideration, we obtain

$$\frac{E_\phi}{-i_a} = X_d'' = X_{la} + \frac{X_{af}X_{lf}X_{ld}}{X_{lf}X_{ld} + X_{af}X_{ld} + X_{af}X_{lf}} \tag{11.44}$$

which is referred to as *subtransient reactance*. In this expression, X_{ld} is the leakage reactance of the damper winding and can be included in the equivalent circuit as shown in Fig. 11.9. It is evident from Figs. 11.8 and 11.9 that X_d'' is smaller than X_d'. Therefore, during the first few cycles after the short circuit has developed the armature current will be very large. In the preceding development we also neglected the effect of winding resistances. Due to the presence of winding resistance, the current waveform for phase a will follow a damped oscillatory response, as shown in Fig. 11.10. During the first few cycles, the subtransient period

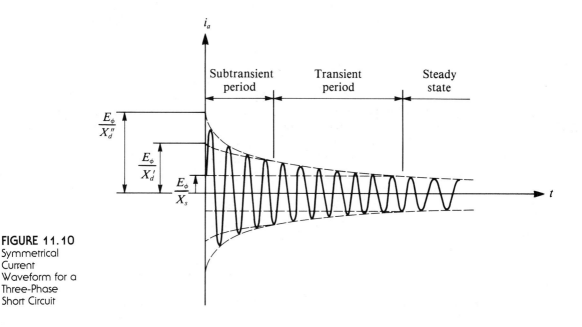

FIGURE 11.10
Symmetrical
Current
Waveform for a
Three-Phase
Short Circuit

caused by the damper winding, the current known as the subtransient current will fall rapidly and can be calculated as

$$I''_a = \frac{E_\phi}{X''_d} \qquad (11.45)$$

After the subtransient period is over, the armature current will continue damped oscillatory response at a slower rate before it reaches the steady state. The time period during which the current decays at a slower rate is known as the transient period, and the transient current, caused by the dc component of current induced in the field winding at the time of the fault, can be obtained as

$$I'_a = \frac{E_\phi}{X'_d} \qquad (11.46)$$

The transient period lasts for 7 to 10 cycles after the subtransient period is over. The transient current is followed by the steady short-circuit current, which is controlled by the synchronous reactance of the generator; its value is

$$I_a = \frac{E_\phi}{X_s} \qquad (11.47)$$

It must be borne in mind that our discussion of electrical transients assumes that all three phases of the synchronous generator developed the short circuit at the same time. If the short circuit occurs only on one or two phases, the discussion of the response is even more complex and is beyond the scope of this book.

11.5.2 Mechanical Transient

When a synchronous generator is made to share the load by connecting its armature terminals to the infinite bus bar, it is possible that prior to the connections being made the rotor may be rotating at a speed somewhat higher or lower than the synchronous speed. Therefore, the frequency of the generator may be higher or lower than that of the infinite bus bar. As soon as the connections for the parallel operation are made, the bus bar frequency will force the generator to slow down or speed up until its frequency coincides with that of the bus bar. The time taken by the generator to adjust its speed is characterized by mechanical transients. This speed adjustment also takes place if a sudden short circuit develops across the generator terminals. In any case, during the transient period,

$$T_m = T_d(t) + D\omega_m(t) + J\frac{d\omega_m(t)}{dt} \qquad (11.48)$$

where T_m is the torque applied to the generator and T_d is the torque developed.

In the absence of the rotational losses, Eq. (11.48) can be written as

$$J \frac{d\omega_m(t)}{dt} = T_m - T_d(t)$$

or
$$J \frac{d^2\theta_m(t)}{dt^2} = T_m - T_d(t) \tag{11.49}$$

where $\omega_m = d\theta_m/dt$ and θ_m is the angular displacement of the rotor with respect to a stationary reference frame. It can be expressed as

$$\theta_m(t) = \omega_s t + \delta_m(t) \tag{11.50}$$

where $\delta_m(t)$ is the power angle at any time t.

From Eq. (11.50), after differentiating it twice, we obtain

$$\frac{d^2\theta_m(t)}{dt^2} = \frac{d^2\delta_m(t)}{dt^2} \tag{11.51}$$

Equation (11.49) can now be expressed as

$$J \frac{d^2\delta_m(t)}{dt^2} = T_m - T_d(t) \tag{11.52}$$

Multiplying both sides of Eq. (11.52) by the rotor speed, $\omega_m(t)$, we get

$$J\omega_m(t) \frac{d^2\delta_m(t)}{dt^2} = \omega_m(t)T_m - \omega_m(t)T_d(t)$$

or
$$J\omega_m(t) \frac{d^2\delta_m(t)}{dt^2} = P_m(t) - P_{dm} \sin \delta_m(t) \tag{11.53}$$

where
$$P_{dm} = \frac{3V_1 E_\phi}{X_s}$$

is the maximum power developed by a round rotor synchronous generator as discussed in Chapter 6. Note that $J\omega_m(t)$ is the angular momentum, $M(t)$. Equation (11.53) can be expressed in terms of per unit quantities with respect to the rated power of the synchronous generator, S_n, as

$$\frac{J\omega_m(t)}{S_n} \frac{d^2\delta_m(t)}{dt^2} = \frac{P_m(t)}{S_n} - \frac{P_{dm}}{S_n} \sin \delta_m(t) \tag{11.54}$$

Let us define the inertia constant for the generator as

$$H = \frac{1}{2}\frac{J\omega_s^2}{S_n} \cong \frac{1}{2}\frac{J\omega_m^2}{S_n} \tag{11.55}$$

In Eq. (11.55), the rotor speed has been replaced by the synchronous speed without introducing any appreciable error. By doing so, we are assuming that the angular momentum is a constant. In terms of the inertia constant of the generator, Eq. (11.54) can be expressed as

$$\frac{2H}{\omega_s}\frac{d^2\delta_m(t)}{dt^2} = p_m - p_{dm}\sin\delta_m(t) \tag{11.56}$$

where p_m and p_{dm} are the per unit powers. Equation (11.56) is a second-order nonlinear differential equation and has been referred to as the swing equation in terms of per unit quantities. It can be solved by using numerical techniques. The eqution can, however, be linearized by substituting δ_m for $\sin \delta_m$ as long as δ_m is a small angle.

11.5.3   Equal-area Criterion

The equal-area criterion is commonly employed to determine the stability of a synchronous generator during the transient period in terms of $\delta_m(t)$. The power developed by a synchronous generator as a function of power angle is shown in Fig. 11.11. Let us assume that the generator operates at a power P_o corresponding

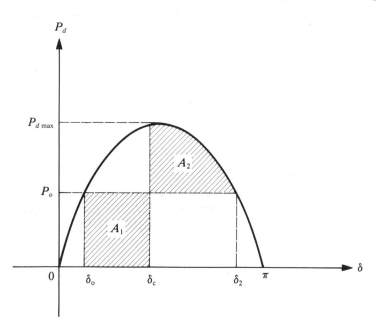

FIGURE 11.11
Equal-Area
Criterion

to power angle δ_o. For a lossless system, the mechanical power supplied must be the same as the power output or power developed under steady state. If a sudden three-phase short circuit occurs, the power developed will become zero. However, the mechanical drive is still supplying power to the generator. The dynamic state of the generator after the onset of the fault can be described as

$$\frac{d\omega_m(t)}{dt} = \frac{\omega_s}{2H} p_m \tag{11.57}$$

yielding
$$\omega_m(t) = \omega_s + \frac{\omega_s}{2H} p_m t \tag{11.58}$$

It is apparent that the shaft speed is higher than the synchronous speed and increases linearly with time. If the fault is not cleared within a short period of time, the rotor may self-destruct. The variation in power angle, from Eq. (11.58), can be given as

$$\delta_m(t) = \delta_o + \omega_s t + \frac{\omega_s}{4H} p_m t^2 \tag{11.59}$$

If the fault is cleared at an instance t_c, the generator will start developing power at a higher level than the prefault state. In other words, the developed power will be larger than the power supplied to the generator. This causes the rotor to slow down.

From Eq. (11.50),

$$\frac{d\delta_m(t)}{dt} = \frac{d\theta_m(t)}{dt} - \omega_s = \omega_m(t) - \omega_s \tag{11.60}$$

and the swing equation can be rearranged as

$$\frac{2H}{\omega_s} \frac{d\omega_m(t)}{dt} = p_m - p_d \tag{11.61}$$

Multiplying Eq. (11.60) by Eq. (11.61), we get

$$\frac{H}{\omega_s} \left(2\omega_m \frac{d\omega_m}{dt} - 2\omega_s \frac{d\omega_m}{dt} \right) = (p_m - p_d) \frac{d\delta_m}{dt} \tag{11.62}$$

Integrating Eq. (11.62), we obtain

$$\frac{H}{\omega_s} \left[\int_{\omega_{m1}}^{\omega_{m2}} 2\omega_m \, d\omega_m - \int_{\omega_{m1}}^{\omega_{m2}} 2\omega_s \, d\omega_m \right] = \int_{\delta_0}^{\delta_2} (p_m - p_d) \, d\delta_m$$

or
$$\frac{H}{\omega_s} [(\omega_{m2}^2 - \omega_{m1}^2) - 2\omega_s(\omega_{m2} - \omega_{m1})] = \int_{\delta_0}^{\delta_2} (p_m - p_d) \, d\delta_m \tag{11.63}$$

where ω_{m1} is the angular velocity prior to the inception of the short circuit and ω_{m2} is the angular velocity corresponding to maximum value of δ_2 after the short circuit has been cleared. Since there is no change in the power angle before the fault occurs, $\omega_{m1} = \omega_s$ from Eq. (11.60). On the other hand, at the maximum value of the power angle δ_2, $d\delta_m/dt = 0$. Thus, from Eq. (11.60), $\omega_{m2} = \omega_s$. Hence, from Eq. (11.63),

$$\int_{\delta_0}^{\delta_2} (p_m - p_d)d\delta_m = 0 \tag{11.64}$$

Equation (11.64) can be rewritten as

$$\int_{\delta_0}^{\delta_c} p_m - p_d\, d\delta_m = \int_{\delta_c}^{\delta_2} (p_d - p_m)d\delta_m \tag{11.65}$$

where δ_c is the power angle at the instant, t_c, the fault is cleared. Equation (11.65) clearly indicates that the area A_1 corresponding to the left side of the equation is equal to the area A_2 equivalent to the right side of the equation.

■ **EXAMPLE 11.7**

A 25-kVA synchronous generator is delivering per unit power of 0.8 with a power angle of 23° at a synchronous speed of 1800 rpm when a fault occurs across its terminals. Determine the per unit power generated by the generator when the fault is cleared 2 ms from its inception. The inertia constant is given as 5 J/VA.

Solution

The angular velocity is

$$\omega_s = \frac{2\pi \times 1800}{60} = 188.5 \text{ rad/s}$$

$$p_{dm} = \frac{0.8}{\sin(23°)} = 2.047 \text{ per unit}$$

The initial rotor angle is $\delta_0 = 23° = 0.401$ rad. The rotor angle at the moment the fault is cleared, from Eq. (11.59), is

$$\delta_m(t) = 0.401 + 188.5 \times 0.002 + \frac{188.5}{4 \times 5} \times 0.8 \times 0.002^2$$

$$= 0.778 \text{ rad} = 44.57°$$

The power generated is

$$p_d = 2.047 \sin(44.57°) = 1.437 \text{ per unit}$$

PROBLEMS

11.1. A 200-V, 10-hp, separately excited dc motor is rated at 1200 rpm. The armature resistance and the inductance are 0.657 Ω and 6.5 mH, respectively. The constant $K = K_a \phi_p$ is 1.36, while the moment of inertia of the rotating system is 0.118 kg-m². Determine the armature current and the motor speed as a function of time when the armature is subjected to the rated voltage under no load. Neglect the effects of saturation and the frictional losses.

11.2. The motor given in Problem 11.1 is operating at steady state under a no-load condition. Suddenly it is subjected to a torque of 5 N-m. Determine the decrease in speed of the motor as a function of time. What is its speed when the steady-state condition is reached?

11.3. A 240-V, 1-hp, 72% efficient separately excited dc motor has the following parameters: $R_a = 19.34$ Ω, $L_a = 147$ mH, $R_f = 500$ Ω, $L_f = 2$ H, $K_a \phi_p = 6.6$, and $J = 0.08814$ kg-m². The characteristic of the mechanical load is given as $T_s = 1.2\omega_m$. Determine the torque developed by the motor as a function of time when the armature winding is suddenly subjected to the rated voltage. Neglect the saturation effect and the frictional losses.

11.4. The motor–load system given in Problem 11.3 is operating under a steady-state condition at its rated values. Determine the variations in the field current and speed as a function of time if the armature current is held constant and the field voltage is instantaneously reduced to 180 V from 240 V. Neglect the frictional losses and consider that $\phi_p = K_f I_f$.

11.5. The separately excited dc motor given in Problem 11.3 is required to operate as a generator. The armature connected to a load resistance of 60 Ω and an inductance of 10 mH was already being driven at a speed of 3000 rpm when the rated voltage is abruptly impressed on the field winding. Determine the variations in the field and armature currents as a function of time. Neglect the frictional losses and the effect of saturation.

11.6. A 240-V, 20-hp, 850-rpm, dc motor is suddenly subjected to its rated voltage and a load torque of 10 N-m under a constant field. Develop the state equations for the state variables i_a and ω_m. The machine parameters are $R_a = 0.76$ Ω, $L_a = 4.6$ mH, $K_a = 2.5$, and $J = 0.41$ kg-m². Ignore the frictional losses and saturation.

11.7. Using the Runge-Kutta algorithm determine the instant at which the motor in Problem 11.6 will achieve its steady-state.

11.8. Repeat Problems 11.6 and 11.7 if the motor is loaded by $T_s = 150$ N-m.

11.9. A 1000-V, 100-kVA round-rotor synchronous generator with subtransient and transient reactances of 150 Ω and 250 Ω, respectively, experiences a three-phase short circuit across its terminals while it is operating at no load. (a) Calculate the per-unit reactances. (b) Calculate the subtransient and transient short-circuit currents in per-unit and in amperes.

11.10. A 750-V, 25-kVA synchronous generator operates a synchronous motor at its rated values of 20-kVA, 700-V and a leading power factor of 0.8. The subtransient reactances of the generator and the motor are given as 0.1 pu and 0.09 pu, respectively. Calculate the fault current in per-unit and in amperes if a three-phase short circuit occurs across the generator terminals.

11.11. Determine the critical fault clearing instant for synchronous generator operating at a mechanical power of P_{emm} with a rotor angle of δ_o by using the equal-area criterion.

11.12. A synchronous generator with $H = 5$ J/VA experiences a short circuit while it was operating at a rotor angle of 20°. (a) Calculate the critical fault clearing rotor angle. (b) Calculate the critical fault clearing time.

CHAPTER 12

Solid-State Control of Electric Machines

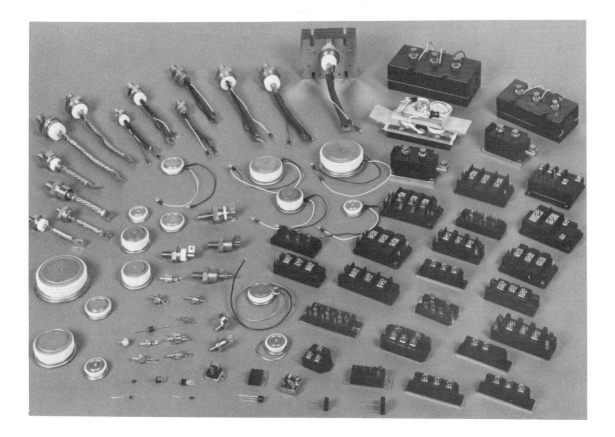

Photograph courtesy of POWEREX, Inc.

The introduction of commercially available, less expensive, more reliable, and more efficient thyristors revolutionized the power industry in the early 1960s. This led to the replacement of conventional motor-control systems using vacuum tubes, relays and contactors, variable resistors, and the like, by more flexible and energy-efficient solid-state controls. In this chapter, we will devote our attention to the fundamental operating principles of some basic circuits widely employed as the control schemes of electric machines.

12.1 DIODES AND THYRISTORS

The basic theory of diodes and thyristors may be found in one of the many text-books on electronics and will not be included in this chapter. Diodes are manu-

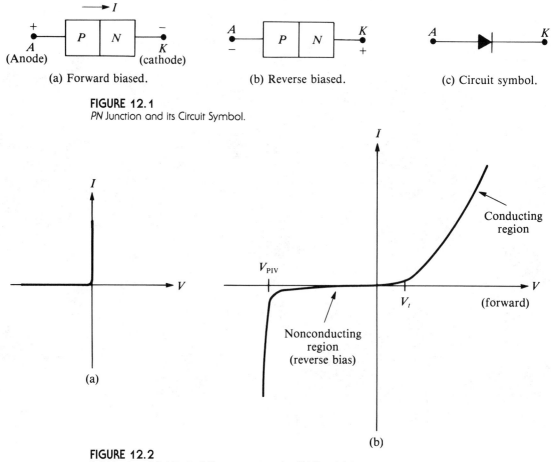

(a) Forward biased. (b) Reverse biased. (c) Circuit symbol.

FIGURE 12.1
PN Junction and its Circuit Symbol.

FIGURE 12.2
(a) Ideal and (b) Typical Characteristics of a *PN* (Diode) Junction

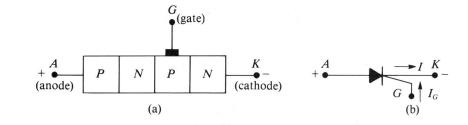

FIGURE 12.3
Thyristor or SCR:
(a) Four-Layer
Construction; (b)
Symbol

factured by joining *P*- and *N*-type semiconducting materials separated by a very thin surface known as a junction. An ideal diode offers no resistance for the current flow when biased in the forward direction and acts as an open circuit in the reverse direction. The *P–N* junction and the circuit symbol of a diode are given in Fig. 12.1. For forward biasing, the anode must be positive with respect to the cathode, as shown in the figure. The ideal and typical diode characteristics are shown in Fig. 12.2. Note from the typical characteristic of the diode that, for it to conduct, the applied voltage in the forward direction must exceed the threshold voltage, which is usually less than 1 V at rated current. On the other hand, when the diode is reverse biased and the applied voltage exceeds the peak-inverse voltage (PIV), the breakdown of the *P–N* junction will take place, thereby resulting in the current flow.

Instead of having a single pair of *P–N* junction, a thyristor has two pairs of *P–N* junctions attached together as shown in Fig. 12.3. An additional electrode

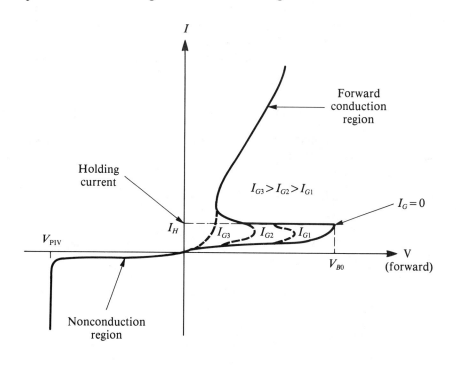

FIGURE 12.4
Typical Thyristor
(SCR)
Characteristics

known as a gate is also connected as shown. Ideally, the combination of two *P–N* junctions offers infinite resistance in both forward and reverse directions. For the conduction to take place in the forward direction, a proper bias must be applied momentarily to the gate terminal. A thyristor will sustain conduction unless the polarity of the applied voltage is reversed or the current is reduced below the holding current. A typical thyristor characteristic is shown in Fig. 12.4. As can be noted, a typical thyristor conducts in the forward direction even in the absence of a gate current when the applied voltage exceeds the breakover voltage, V_{BO}. At this point the thyristor turns on and the voltage drop across it falls sharply. The presence of the gate current helps in switching on the thyristor at voltages lower than the breakover voltage, as shown in the figure. Once conduction takes place, a thyristor acts like a diode and continues conducting unless the current falls below the minimum value known as the hold current, I_H or it is reverse biased.

12.2 RECTIFIER CIRCUITS

To control the speed of a dc motor, proper adjustments are required in the applied voltages to the field and/or armature windings. Such adjustments are facilitated by the rectifier circuits. Prior to probing into the functioning of dc motors employing rectifier circuits, let us first investigate the working of such circuits feeding inductive loads.

Half-wave Diode Rectifier   A half-wave diode rectifier circuit with its voltage and current waveforms is shown in Fig. 12.5 feeding an inductive load. The equation governing the current flow during the conduction period is given by

$$v_s(t) = Ri(t) + L\frac{di(t)}{dt} \tag{12.1}$$

For a sinusoidal voltage input of the form $v_s(t) = V_m \sin \omega t$, the current response is

$$i(t) = \frac{V_m}{\sqrt{R^2 + \omega^2 L^2}} \sin\left(\omega t - \tan^{-1}\frac{\omega L}{R}\right) + \frac{V_m}{R^2 + \omega^2 L^2}\,\omega L\,e^{-(R/L)t} \tag{12.2}$$

For a purely resistive load, the current during the conduction period will simply be

$$i(t) = \frac{V_m}{R}\sin \omega t, \qquad \text{for } 0 \le t \le T/2 \tag{12.3}$$

where T is the time period of the applied voltage.

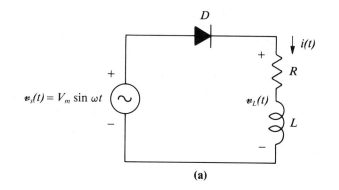

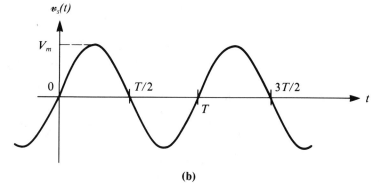

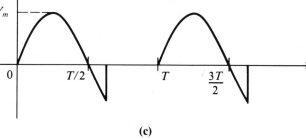

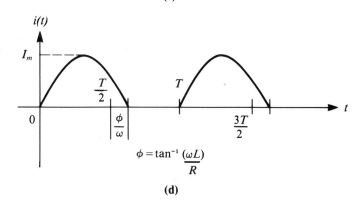

FIGURE 12.5
(a) Half-Wave
Diode Rectifier
Circuit; (b)
Applied
Sinusoidal
Alternating
Voltage; (c)
Voltage Across
the Inductive
Load; (d)
Current Through
the Inductive
Load

465

■ EXAMPLE 12.1

The resistance and inductance of the load in Fig. 12.5a are 2 Ω and 10 mH, respectively. For $v_s(t) = 10 \sin 377t$, (a) sketch the voltage and current waveforms for one period, and (b) calculate the average to peak ratios of the load current and voltage.

Solution

(a) The time period is

$$T = \frac{2\pi}{\omega} = \frac{2\pi}{377} = 16.667 \text{ ms}$$

$$X_L = \omega L = 377 \times 0.01 = 3.77 \ \Omega$$

The load impedance is

$$Z = \sqrt{2^2 + 3.77^2} = 4.27 \ \Omega$$

From Eq. (12.2), the conduction current is

$$i(t) = 2.342 \sin(377t - 62.05°) + 2.068 \ e^{-200t}$$

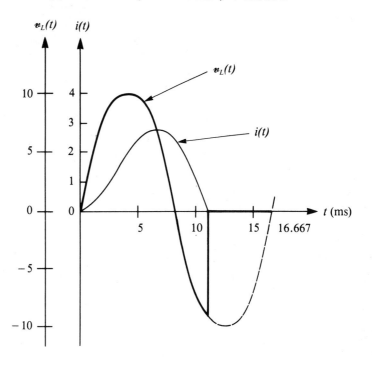

FIGURE 12.6
Voltage and
Current
Waveforms for
Example 12.1

The required voltage and current waveforms are plotted in Fig. 12.6. The maximum values of the current and the load voltage are 2.852 A and 10 V, respectively.

(b) The conduction period is $0 < t < 11.45$ ms. The average load current is

$$I_{av} = \frac{1}{T} \int_0^T i(t)\, dt$$

$$= \frac{10^3}{16.667} \int_0^{0.01145} (2.342 \sin(377t - 62.05°) + 2.068\, e^{-200t})\, dt$$

$$= 1.104 \text{ A}$$

The average load voltage is

$$V_{av} = \frac{10^3}{16.667} \int_0^{0.01145} 10 \sin 377t\, dt$$

$$= 2.2 \text{ V}$$

Thus

$$\frac{I_{av}}{I_m} = \frac{1.104}{2.852} = 0.387$$

and

$$\frac{V_{av}}{V_m} = \frac{2.2}{10} = 0.22$$

The average to maximum ratio gives an indication of the effectiveness of the rectifying circuit. The higher the average to peak ratio, the more effective is the rectifying circuit.

Half-wave Thyristor Rectifier   A diode rectifier starts conduction at the commencement of the positive half-cycle of the source voltage. To control the beginning of conduction and thereby the average values of the load voltage and current, a diode can be replaced by a thyristor. A thyristor when utilized in a rectifier circuit is referred to as a *silicon-controlled rectifier* (SCR). A pulse generator is needed to gate the thyristor at the required instant. For the proper timing of the gating pulses, each pulse must be a time period apart. Figure 12.7 shows the circuit diagram and the waveforms of a half-wave thyristor rectifier.

During the conduction period, which starts at time $t = t_1$, the load current can be obtained from Eq. (12.1) as

$$i(t) = \frac{V_m}{\sqrt{R^2 + \omega^2 L^2}} \sin\left(\omega t - \tan^{-1} \frac{\omega L}{R}\right)$$

$$- \frac{V_m}{\sqrt{R^2 + \omega^2 L^2}} \sin\left(\omega t_1 - \tan^{-1} \frac{\omega L}{R}\right) e^{-(R/L)(t - t_1)} \quad (12.4)$$

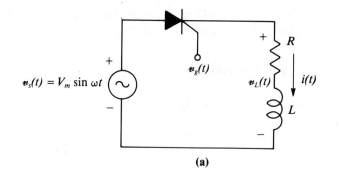

(a)

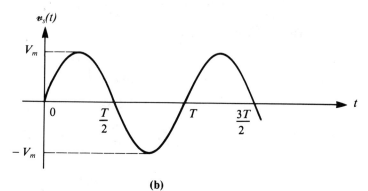

(b)

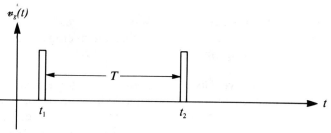

(c)

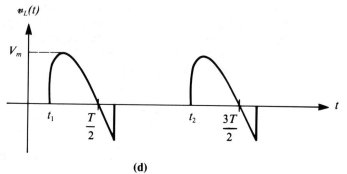

(d)

FIGURE 12.7
(a) Half-Wave
Thyristor Rectifier
Circuit; (b)
Applied Voltage
Waveform; (c)
Gating Pulses;
(d) Load
Voltage

The maximum value of the load current and the time when the current ceases can be determined by plotting $i(t)$ as a function of time.

■ EXAMPLE 12.2

A half-wave thyristor rectifier supplies power to an inductive load as shown in Fig. 12.7a, where $R = 5\ \Omega$, $L = 50$ mH, and $v_s(t) = 120 \sin(377t)$. If the gate firing angle is 30°, (a) sketch the load current and voltage waveforms for one period, and (b) what will be the readings of voltmeter and ammeter connected appropriately on the load side?

Solution

(a) The time period is

$$T = \frac{2\pi}{\omega} = \frac{2\pi}{377} = 16.667 \text{ ms}$$

$$X_L = \omega L = 377 \times 0.05 = 18.85\ \Omega$$

The load impedance is

$$Z = \sqrt{5^2 + 18.85^2} = 19.5\ \Omega$$

The conduction instant is

$$t_1 = \frac{\alpha_1}{\omega} = \frac{\pi/6}{377} = 1.388 \text{ ms}$$

Substituting the values in Eq. (12.4), we obtain

$$i(t) = 6.1535 \sin(377t - 75.14°) + 4.362\ e^{-100(t - 1.388 \times 10^{-3})} \text{ A}$$

during the conduction period, which begins at 1.388 ms and ends at 12.425 ms, as is evident from the waveforms shown in Fig. 12.8.

(b) The reading of meters will correspond to the average values of the load current and voltage. Thus,

$$I_{av} = \frac{10^3}{16.667} \int_{1.388 \times 10^{-3}}^{12.425 \times 10^{-3}} [6.1535 \sin(377t - 75.14°)$$

$$+ 4.362\ e^{-100(t - t_1)}]\ dt = 3.39 \text{ A}$$

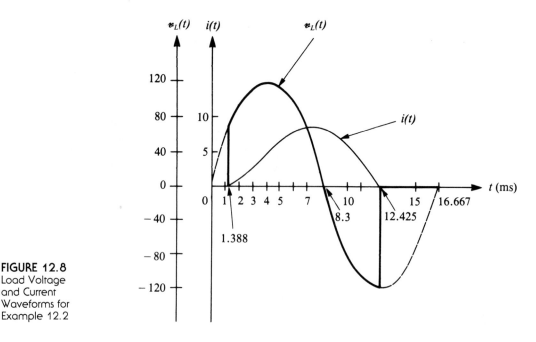

FIGURE 12.8
Load Voltage
and Current
Waveforms for
Example 12.2

and

$$V_{av} = \frac{10^3}{16.667} \int_{1.388 \times 10^{-3}}^{12.425 \times 10^{-3}} 120 \sin 377t \; dt = 17.077 \text{ V}$$

12.2.1 Freewheeling Diode

To improve the average to peak ratio of load voltage and current, a diode, known as a freewheeling diode, is connected across the load as shown in Fig. 12.9 for a half-wave diode rectifier. During the positive half-cycle of the applied voltage, the freewheeling diode acts as an open circuit, but for the negative half-cycle it provides a conducting path for the dissipation of the magnetic energy stored in the inductor. The load voltage can now be expressed as

$$v_L(t) = V_m \sin \omega t, \qquad 0 \le t \le T/2 \qquad (12.5)$$

for one time period. The load current will have two responses: the forced response due to the source voltage for the time interval $0 \le t \le T/2$, and the natural response for the duration $T/2 \le t \le T$. The forced response is given by Eq. (12.2), while the natural response can be written as

$$i(t) = i(T/2) \; e^{-(R/L)(t - T/2)}, \qquad \frac{T}{2} \le t \le T \qquad (12.6)$$

where $i(T/2)$ is the current at $t = T/2$ from Eq. (12.2).

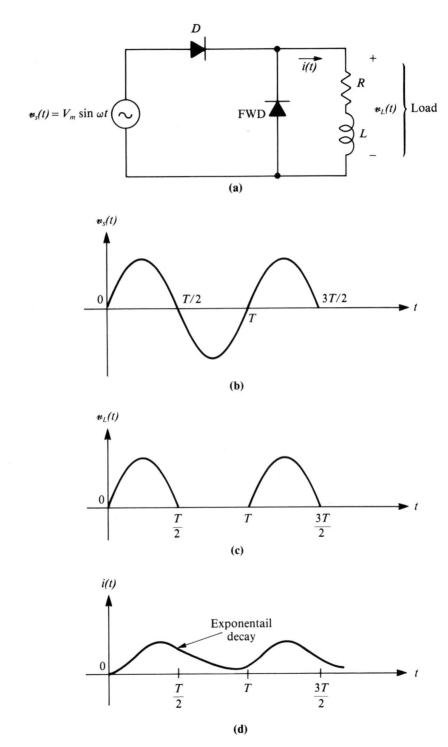

FIGURE 12.9
(a) Half-Wave
Rectifier Circuit
with
Freewheeling
Diode (FWD);
(b) Applied
Voltage
Waveform; (c)
Load Voltage
Waveform; (d)
Load Current
Waveform

The load current at $t = T$ depends on the time constant of the load. The load voltage and current waveforms for a half-wave diode rectifier are given in Fig. 12.9. It is evident that the average values of the current and voltage waveforms are greater than those without the freewheeling diode. Therefore, a freewheeling diode helps to improve the average to peak ratio. The diode in Fig. 12.9 can also be replaced by a thyristor, and corresponding voltage and current waveforms can be obtained during the conduction period.

■ **EXAMPLE 12.3**

Compare the average to peak ratios for the load voltage and current for the half-wave rectifier circuit considered in Example 12.2 with and without the freewheeling diode.

Solution

Without the freewheeling diode, the maximum values of the load voltage and current, from Fig. 12.8, are 120 V and 8.46 A, respectively. The average to peak ratios are

$$\frac{I_{av}}{I_m} = \frac{3.3925}{8.46} = 0.401$$

and

$$\frac{V_{av}}{V_m} = \frac{17.077}{120} = 0.142$$

With the freewheeling diode

$$v_L(t) = 120 \sin 377t, \qquad 1.388 \text{ ms} \leq t \leq 8.33 \text{ ms}$$

The average value is

$$V_{av} = \frac{10^3}{16.667} \int_{1.388 \times 10^{-3}}^{8.33 \times 10^{-3}} 120 \sin 377t \, dt = 35.64 \text{ V}$$

During the conduction period of the thyristor, the load current, from Eq. (12.4), is

$$i(t) = 6.1535 \sin(377t - 75.14°) + 4.362e^{-100(t-t_1)}, \qquad 1.388 \text{ ms} < t < 8.33 \text{ ms}$$

where $t_1 = 1.388$ ms. At $t = T/2 = 8.33$ ms,

$$i(T/2) = 8.126 \text{ A}$$

The current decay through the freewheeling diode is

$$i(t) = 8.126e^{-100(t-t_2)}, \qquad 8.33 \text{ ms} < T < 16.667 \text{ ms}$$

where $t_2 = 8.33$ ms. The average value of the load current is

$$
\begin{aligned}
I_{av} &= \frac{10^3}{16.67}\left[\int_{t_1}^{t_2} (6.1535\sin(377t - 75.14°)\right.\\
&\quad \left.+ 4.362e^{-100(t-t_1)})\, dt + \int_{t_2}^{T} 8.126e^{-100(t-t_2)}\, dt\right]\\
&= 5.008 \text{ A}
\end{aligned}
$$

Thus, $\dfrac{I_{av}}{I_m} = \dfrac{5.008}{8.461} = 0.592$ and $\dfrac{V_{av}}{V_m} = \dfrac{35.64}{120} = 0.297$

12.2.2 Single-phase Controlled Bridge Rectifiers

A single-phase full-wave bridge rectifier circuit having four thyristors is given in Fig. 12.10. During the positive half-cycle of the applied ac voltage, thyristors $T1$ and $T3$ are made to conduct at a predetermined firing angle. During the negative half-cycle, the energy stored in the inductor maintains the current flow through $T1$ and $T3$ until the current falls below I_H or thyristors $T2$ and $T4$ are fired to conduct concurrently. The load voltage and current waveforms are given in Fig. 12.11.

Instead of using four thyristors in a controlled bridge rectifier, two thyristors can be replaced by two diodes as shown in Fig. 12.12. The conduction current is controlled by properly selecting the firing angles for thyristors $T1$ and $T2$. A freewheeling diode is also included in the circuit to provide a path for the decay of magnetic energy in the nonconduction region. Figure 12.13 illustrates the load

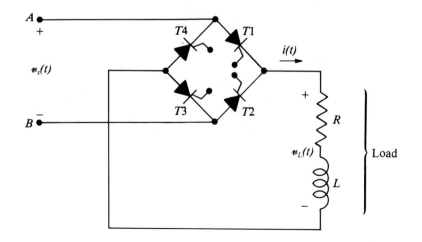

FIGURE 12.10
Single-Phase
Thyristor Bridge
Rectifier with
Four Thyristors

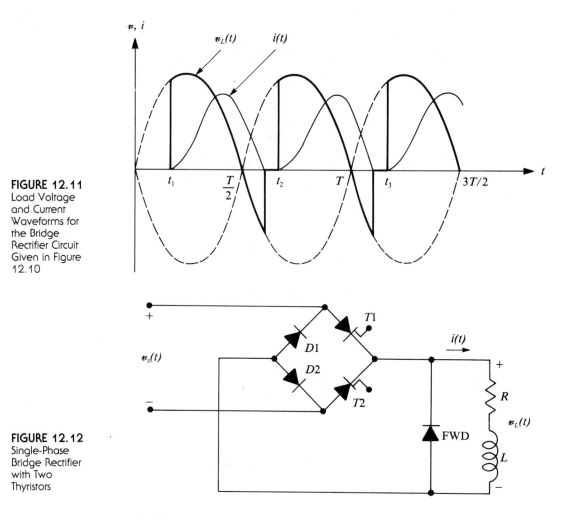

FIGURE 12.11
Load Voltage
and Current
Waveforms for
the Bridge
Rectifier Circuit
Given in Figure
12.10

FIGURE 12.12
Single-Phase
Bridge Rectifier
with Two
Thyristors

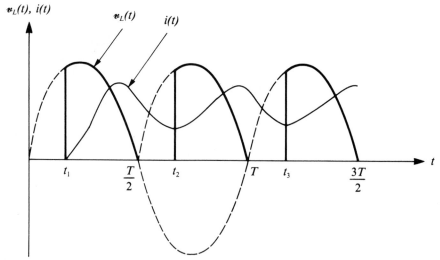

FIGURE 12.13
Load Voltage
and Current
Waveforms for
the Circuit
Shown in Figure
12.12

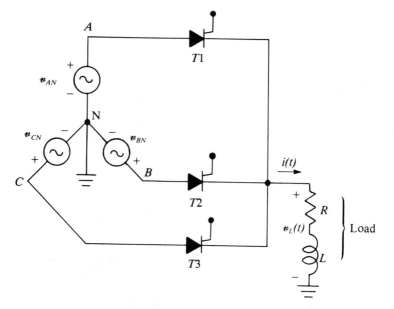

FIGURE 12.14
Three-Phase
Half-Wave
Rectifier Circuit
with *RL* Load

voltage and current waveforms. Many other variations of thyristor–diode combinations with unique features are available in the literature.

12.2.3   Three-phase Half-wave Controlled Rectifiers

A three-phase half-wave controlled rectifier circuit is given in Fig. 12.14. For proper operation, the phase shift between the gating signals must be 120°. The load voltage and current waveforms are shown in Fig. 12.15. For large power machines, in the integral-horsepower range, three-phase full-wave rectifiers can

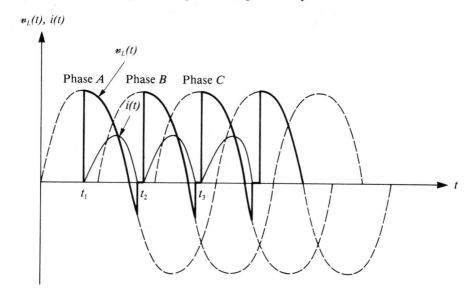

FIGURE 12.15
Load Voltage
and Current
Waveforms for
the Circuit Given
in Figure 12.14

be used. A three-phase rectifier circuit reduces the presence of higher-order harmonics in the load voltage and current as compared to its single-phase counterpart.

12.3 CONVERTERS

A converter is essentially a rectifying circuit designed for the operation of a dc machine from an ac source. To understand the fundamental principle of operation of a converter, consider a separately excited dc motor driven by a half-wave diode converter as shown in Fig. 12.16. During the conduction period,

$$v_s(t) = R_a i_a(t) + L_a \frac{di_a(t)}{dt} + e_g(t) \tag{12.7}$$

and the armature current, from Eq. (12.7), is

$$\begin{aligned}
i_a(t) =\ & \frac{V_m}{\sqrt{R_a^2 + \omega^2 L_a^2}} \sin\left(\omega t - \tan^{-1}\frac{\omega L_a}{R_a}\right) \\
& - \frac{V_m e^{-(R_a/L_a)(t-t_1)}}{\sqrt{R_a^2 + \omega^2 L_a^2}} \sin\left(\omega t_1 - \tan^{-1}\frac{\omega L_a}{R_a}\right) \\
& \qquad\qquad - \frac{e^{-(R_a/L_a)t}}{L_a} \int_{t_1}^{t} e_g(\tau) e^{(R_a/L_a)\tau}\, d\tau \tag{12.8}
\end{aligned}$$

In the beginning of each positive half-cycle the diode conducts as long as $v_s(t) > e_g(t)$. This is shown by time t_1 in Fig. 12.16. After the commencement of conduction, the current continues to flow in the circuit even when the source voltage is less than the induced emf due to the magnetic energy stored by the inductor. Therefore, the time t_2 at which the conduction current ceases depends on the magnitude of the induced voltage and the time constant of the armature circuit.

FIGURE 12.16
(a) Half-Wave Diode Converter Circuit for a Separately-Excited dc Motor; (b) Voltage and Current Waveforms

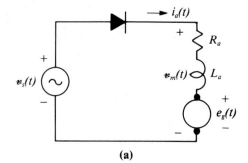

(a)

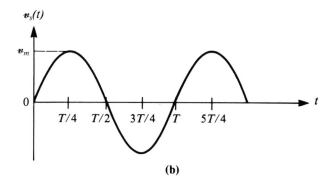

(b)

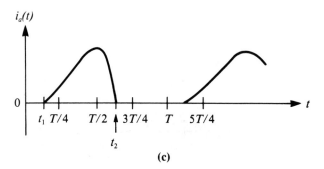

(c)

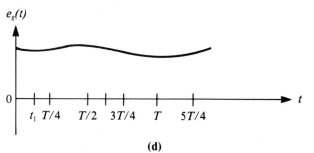

(d)

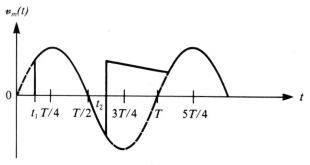

(e)

FIGURE 12.16
(continued)

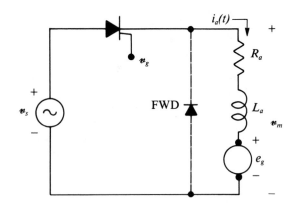

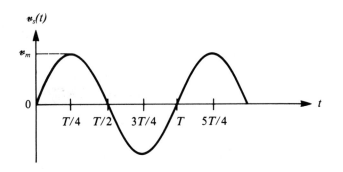

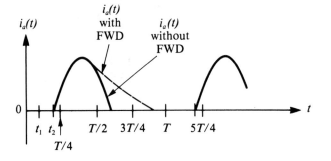

FIGURE 12.17
Half-Wave
Thyristor
Converter Circuit
for a
Separately-
Excited dc Motor
with its Voltage
and Current
Waveforms

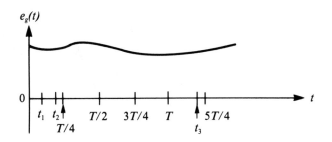

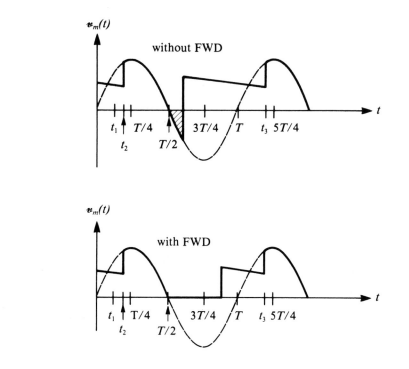

FIGURE 12.17
(continued)

The torque developed by the motor during the conduction interval is

$$T_d(t) = K_a\phi_p i_a(t) = T_s + J\frac{d\omega_m}{dt} \tag{12.9}$$

The developed torque increases linearly with the rise in the conduction current. As long as the developed torque is greater than the load torque, the motor continues to accelerate. However, in the nonconducting period, the load on the motor causes a reduction in its speed, and the motor is said to be coasting.

12.3.1 Thyristor Converters

The principle of operation of a thyristor converter is similar to that of a diode converter except for the fact that the motor voltage can be adjusted by changing the gating time of the thyristor. The circuit diagram and its characteristic waveforms are given in Fig. 12.17. For the thyristor to conduct, the gating pulse must be applied during the time interval when the source voltage is higher than the back emf. To improve the performance of the converter, a freewheeling diode, shown by the dashed lines, can also be included in the circuit. Its effect on the current waveform is shown by the dashed line. The hatched area in the negative half-cycle will also disappear. For further improvements, a full-wave converter (Fig. 12.18) or a three-phase converter (Fig. 12.19) may be used.

■ EXAMPLE 12.4

A separately excited dc motor is fed by a half-wave thyristor converter operating from a source of $v_s(t) = 1200 \sin 377t$. The thyristor gate firing angle is set to 30° for a speed of 1150 rpm. If $R = 1.426 \ \Omega$, $L = 10.4$ mH, and $K_a\phi_p = 1.84$, determine the average armature current and the terminal voltage. Assume the motor losses are negligible and the back emf is constant.

Solution

The angular velocity is

$$\omega_m = \frac{2\pi \times 1150}{60} = 120.42 \text{ rad/s}$$

The back emf is

$$E_g = K_a\phi_p\omega_m = 1.84 \times 120.42 = 221.59 \text{ V}$$

During the conduction period we can use the superposition theorem to determine the armature current. The current due to the applied sinusoidal source can be computed from Eq. (12.4) as

$$i_1(t) = 287.63 \sin(377t - 70°) + 184.88 \ e^{-137.11(t - 1.388 \times 10^{-3})}$$

The contribution of the back emf to the armature current is

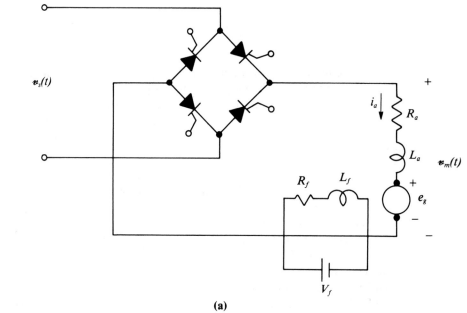

FIGURE 12.18
(a) Full-Wave Converter; (b) Voltage and Current Waveforms for a Separately-Excited dc Motor

(a)

$$i_2(t) = -\frac{E_g}{R_a}[1 - e^{-(R_a/L_a)(t-t_1)}]$$

$$= -155.39[1 - e^{-137.11(t-t_1)}]$$

where $t_1 = 1.388$ ms. The total armature current is

$$i_a(t) = 287.63 \sin(377t - 70°) + 184.88e^{-137.11(t-t_1)} - 155.39(1 - e^{-137.11(t-t_1)})$$

By plotting the preceding equation, it can be shown that the thyristor conduction

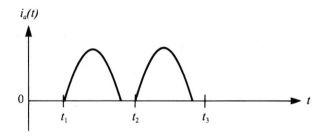

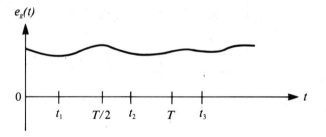

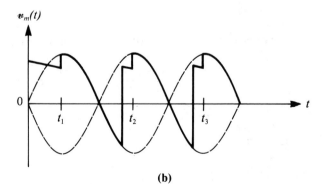

FIGURE 12.18
(*continued*)

(b)

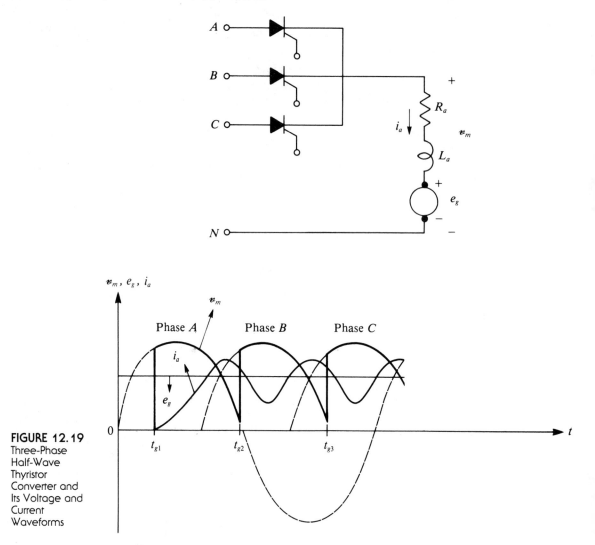

FIGURE 12.19
Three-Phase
Half-Wave
Thyristor
Converter and
Its Voltage and
Current
Waveforms

interval is 1.388 ms $< t <$ 10.976 ms. The average current is

$$I_{av} = \frac{1}{T} \int_0^T i_a(t) \, dt$$

$$= 99 \text{ A}$$

The average voltage is

$$V_{av} = \frac{10^3}{16.667} \left[\int_{1.388 \times 10^{-3}}^{10.976 \times 10^{-3}} 1200 \sin 377t \, dt + \int_{10.976 \times 10^{-3}}^{16.667 \times 10^{-3}} 221.59 \, dt \right] = 344.66 \text{ V}$$

12.4   CHOPPERS

A dc-to-dc converter is known as a chopper. A chopper circuit helps to control the average value of the dc source before applying it to a dc motor. A chopper is basically a static switching circuit that turns the dc voltage on and off across the motor terminals at a certain frequency. Assume that switch S in Fig. 12.20 is initially open at $t = 0$. The closing and opening of the switch results in the application of the source voltage to the armature circuit in a series of pulses, as shown in Fig. 12.21. The armature voltage is on for a period of $t_2 - t_1$ and off during the time $t_3 - t_2$, and the cycle is repeated thereafter. The total time interval from t_1 to t_3 is known as the *time period* of the pulse. The armature current during each on and off time period is also shown in the figure. The average value of the armature terminal voltage is

$$V_{av} = V_s \frac{t_2 - t_1}{t_3 - t_1} = V_s \frac{\text{on-time}}{\text{time period}} \qquad (12.10)$$

It is apparent from Eq. (12.10) that the average value of the applied voltage to the armature terminals of a separately excited dc motor can be controlled by varying either the on-time, the time period, or both. If the time period is held constant and the on-time is increased or decreased, the average value of the applied voltage will also increase or decrease. This scheme of controlling the armature terminal voltage is referred to as the pulse-width-modulation (PWM) technique. On the other hand, by increasing or decreasing the time period while keeping the on-time the same, the average value of the applied voltage can be correspondingly decreased or increased. This technique is known as pulse-frequency control.

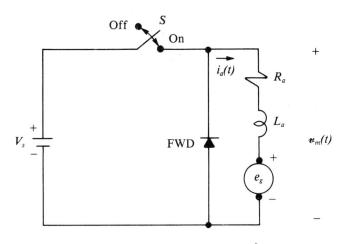

FIGURE 12.20
Elementary
Chopper Circuit

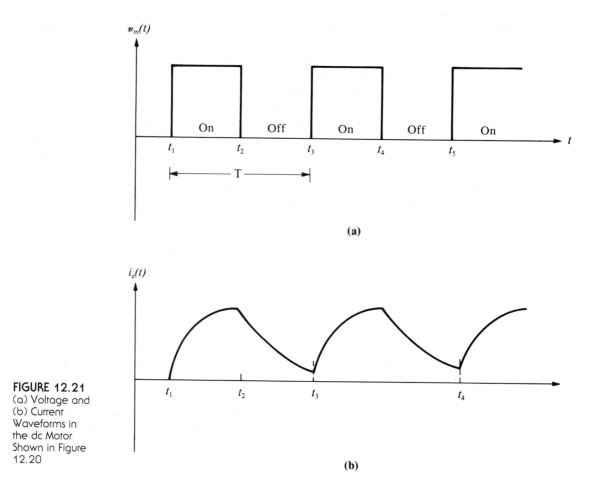

FIGURE 12.21
(a) Voltage and
(b) Current
Waveforms in
the dc Motor
Shown in Figure
12.20

■ EXAMPLE 12.5

The on- and off-durations of a voltage applied from a 24-V dc supply to the armature winding of a separately excited dc motor are 1 ms and 0.2 ms, respectively. Determine the time period, frequency, and average armature terminal voltage.

Solution

Time period: $T = t_{\text{on}} + t_{\text{off}} = 1 + 0.2 = 1.2$ ms

Frequency: $f = 1/T = 1000/1.2 = 833.33$ Hz

Average value: $V_{av} = 24 \times 0.001/0.0012 = 20$ V

■ **EXAMPLE 12.6**

Determine the frequency of a switching operation using pulse-frequency modulation to obtain the average voltage of 20 V across armature terminals of a separately excited dc motor from a 24-V dc source if the on-time is 0.5 ms.

Solution

The time period is $T = t_{on} + t_{off} = 0.5 + t_{off}$ ms. From the average value,

$$20 = \frac{24 \times 0.5}{0.5 + t_{off}}$$

or $t_{off} = 0.1$ ms

Thus, $T = 0.5 + 0.1 = 0.6$ ms

and the frequency is $f = \frac{1000}{0.6} = 1.667$ kHz

Figure 12.22 shows a chopper circuit in its simplest form with one thyristor. A series capacitor–inductor branch is connected in parallel with the thyristor to provide a means for turning off the thyristor. The operation of this circuit can be

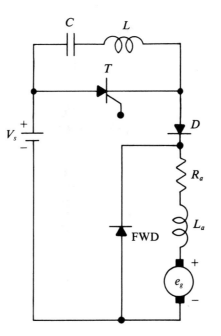

FIGURE 12.22
Chopper Circuit
with One
Thyristor

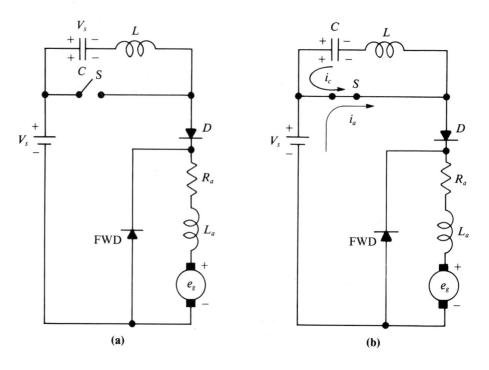

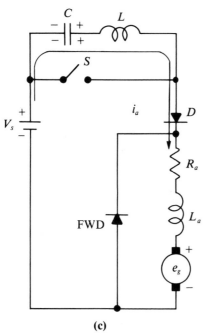

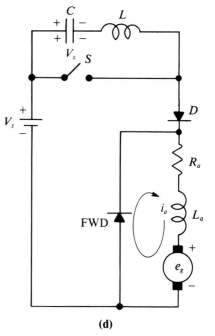

FIGURE 12.23
Operation of
the Chopper
Circuit Given in
Figure 12.22

explained with the help of the switching circuits given in Fig. 12.23. Assume the thyristor is not conducting and the capacitor is fully charged, as shown in Fig. 12.23a. After the thyristor is fired, switch S is closed as shown in Fig. 12.23b; it starts conducting and provides a path for the capacitor energy to flow back and forth between the capacitor and the inductor. The current through the LC branch oscillates at its natural frequency. The net current through the thyristor is equal to the algebraic sum of the oscillating current and the load current. During the oscillations of the capacitor current, the net current through the thyristor reverses and immediately caused the thyristor to turn off. This causes switch S in Fig. 12.23c to open, and the load current follows the new path through the source, capacitor, inductor, diode, and load. During this period the capacitor charges to the source voltage. Thereafter, the stored magnetic energy in the motor decays through the freewheeling diode, as shown in Fig. 12.23d. The off-period of the thyristor can be terminated by refiring it, and the whole cycle repeats itself.

12.5 INVERTERS

Inverters are basically switching circuits used to convert a dc source into an ac source for the operation of ac machines on a dc supply. A rectifier–inverter combination can be used to change the frequency of an ac source. Figure 12.24 illustrates the fundamental operating principle of inverter circuits. Assume that switch S_1 is closed at $t = 0$, while S_2 is still open. The load current can be expressed as

$$i_1(t) = \frac{V_s}{R}(1 - e^{-(R/L)t}) \tag{12.11}$$

The growth of the load current to its final value of V_s/R is shown in Fig. 12.25. If we close switch S_2 and open S_1 at time t_1, the load current can now be written as

$$i_2(t) = \frac{V_s}{R}e^{-(R/L)(t-t_1)} - \frac{V_s}{R}(1 - e^{-(R/L)(t-t_1)}) \tag{12.12}$$

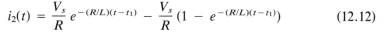

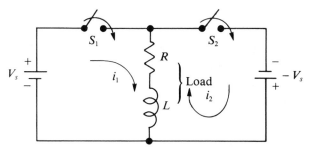

FIGURE 12.24
Basic Inverter
Circuit

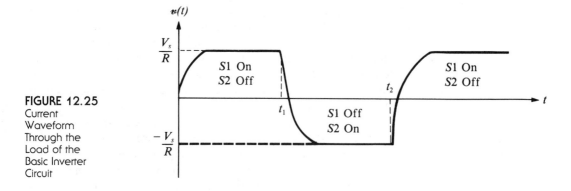

FIGURE 12.25
Current
Waveform
Through the
Load of the
Basic Inverter
Circuit

In Eq. (12.12) it has been assumed that the current $i_1(t)$ has reached its steady-state value prior to closing switch S_2. Depending on the time constant, the current through the load will try to attain its final value of $-V_s/R$. Thus, the current through the load has been reversed. The current can be made to change its direction again by closing switch S_1 and opening S_2 at t_2. By now it should be evident that the load current has a square waveform. The frequency of the square wave can be controlled by varying the frequency of switching. In power circuits, the switching operation may be performed by thyristors.

A simple inverter circuit using thyristors is shown in Fig. 12.26 and its operation can be described as follows: Before the thyristors are activated, the voltage across capacitors C_1 and C_2 are the same since the supply voltages are equal. As soon as T_1 is fired, it provides a path not only for the load current but also for C_1 to discharge. On the other hand, due to capacitive coupling, C_2 charges up to $2V_s$ while C_1 discharges completely. To terminate the present state of the circuit, T_2 can be made to conduct. T_2 provides a path for C_2 to discharge. The voltage

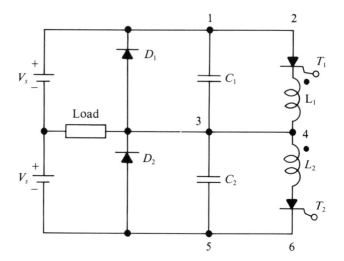

FIGURE 12.26
Inverter Circuit
Operating with
Two Diodes and
Two Thyristors

across C_2 now appears across L_2. Since L_2 is coupled to L_1 and both have the same number of turns, the voltage across C_2 also appears across L_1. This causes T_1 to turn off. The load current now flows through T_2. Also, C_1 starts to charge up while C_2 loses its energy. When C_1 reaches a voltage level of $2V_s$, C_2 becomes uncharged and the circuit is ready to change its state and the polarity of the load voltage by activating T_1. If the load is purely resistive, the current waveform will have the same shape as the voltage waveform. However, for inductive and capacitive loads, the load current will not reverse instantaneously with the voltage. The freewheeling diodes are included to allow the load current to flow after the voltage reversal.

PROBLEMS

12.1. A half-wave diode rectifier supplies power to a resistive load of 10 Ω. If the voltage source connected to the rectifier has an output voltage of 250 sin 314t, calculate the average load current.

12.2. A half-wave diode rectifier feeds an inductive load with an impedance of $100 + j80$ Ω. The input voltage to the rectifier has an amplitude of 100 V and a frequency of 60 Hz. (a) Determine the mathematical expression of the load current for one period. (b) Calculate the first instant at which the rectifier starts to block after the first current conduction. Neglect the transient.

12.3. Calculate the average value of the current for the circuit given in Problem 12.2. Neglect the transient.

12.4. The series combination of a 50-Ω resistor and 100-mH inductor is connected across the output terminals of a half-wave thyristor rectifier. The applied voltage waveform to the thyristor circuit is $v_s(t) = 100 \sin 628t$. Also, it is known that the thyristor gate firing angle is 15°. (a) Determine the mathematical expression of the load current over one period. (b) Plot the waveforms of the load current and voltage. (c) Calculate the average-to-peak ratios for the load current and voltage.

12.5. Repeat Problem 12.4 if a freewheeling diode is connected across the load.

12.6. The inductive load fed by a half-wave thyristor rectifier has an impedance of $4 + j3$ Ω. The alternating voltage connected to the rectifier is given as 50 sin 1000t. (a) Determine the average-to-peak ratios of the load current for gate firing angles of 0°, 30°, 60°, 90°, and 120°. (b) Draw a curve that indicates the variation of the average-to-peak load currents as a function of the firing angles.

12.7. Determine the thyristor gate firing angle of the circuit given in Example 12.4 in order to reduce the average-to-peak ratio of the load voltage to 0.2. Neglect the transient effect to calculate the inception instant of current blocking.

12.8. Determine the mathematical expression of the current through an inductive load for one period if the load is fed by a diode-bridge rectifier. The input to the rectifier is a sinusoidal voltage.

12.9. A single-phase silicon-controlled bridge rectifier shown in Fig. P12.9 supplies power to an inductive load having an impedance of $3 + j1$ Ω. The alternating voltage source has a rms value of 120 V and frequency of 60 Hz. The gate firing angle is 30°. (a) Calculate the gate firing instants for the intervals $0 < t < T/2$ and $T/2 < t < T$. (b) Determine the mathematical expression of load current for the period. (c) Plot the load current and voltage waveforms.

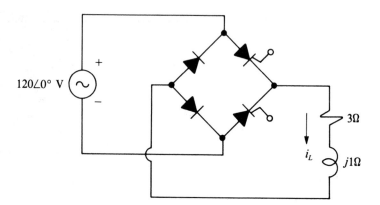

FIGURE P12.9
Circuit for
Problem 12.9

12.10. A half-wave thyristor converter circuit feeds a separately excited dc motor. The alternating voltage applied to the converter has the waveform of $600 \sin 1000t$. The thyristor gate firing angle is 15° while the motor is operating at a speed of 1000 rpm under a constant field. The armature resistance and inductance are given as 10 Ω and 5 mH, respectively, and $K_a\phi_p$ is known to be 1.5. (a) Determine the mathematical expression of the load current for the thyristor conduction period. (b) Plot the motor current and terminal voltage waveforms. (c) Calculate the average armature current and the terminal voltage. Assume that the back emf is constant and the motor losses are negligible.

12.11. Determine the gate firing angle of the thyristor in Problem 12.9 in order to reduce the motor speed to 800 rpm at the same average load current. Assume that the instant at which the conduction stops is the same as in problem 12.10.

12.12. The on- and off-durations of a rectangular waveform are given as 2 ms and 0.5 ms, respectively. (a) Determine the period and the frequency of the waveform. (b) Calculate the average value of the rectangular waveform if the voltage magnitude is 120 V.

12.13. A chopper circuit with a voltage magnitude of 250 V runs a dc series motor. The output voltage frequency of the chopper is 1000 Hz and the on-duration is 0.75 ms. The armature and field resistances of the motor are 1 and 9 Ω, respectively. If the motor operates in the linear region of its magnetization curve at a speed of 230 rpm, calculate the average armature current. Assume that $K_a\phi_p = 4$ for the motor. Neglect transient effects.

12.14. The series dc motor of Problem 12.13 together with the same chopper circuit operates under a load of 1.69 A at a speed of 81 rpm. Calculate the on- and off-durations of the rectangular voltage applied to the motor in order to increase the motor speed to 140 rpm under the same load.

APPENDIX A

Armature Windings

A.1 LAP WINDING

For simplex lap winding, we need as many coils in the armature as there are slots. However, each coil may have N_c turns. Figure A.1 shows two such coils with their starting and ending ends marked as s and f. The two sides of the same coil are a certain distance apart, which is referred to as the *pitch* of the coil. This distance may be expressed in terms of slots or electrical degrees. We may use a full-pitch coil or a fractional-pitch coil. A coil is said to be a full-pitch coil when the distance between its two sides is equal to the distance between two adjacent poles. In other words, a full pitch equals the pole pitch. Any pitch less than the full pitch is known as a fractional pitch. The maximum pitch of the coil may be calculated as follows:

$$Y = \text{integer value of } (S/P) \qquad\qquad (A.1)$$

where

Y = coil pitch, slots
S = total number of armature slots
P = number of main poles.

Once the coil pitch is known, the coil sides can now be placed in armature slots. Assume that side 1 of the first coil is placed in slot 1; then the second side of the same coil must be placed in slot $(1 + Y)$. In simplex lap winding, if the starting end s_1 of the first coil is connected to commutator segment 1, then the ending end f_1 must be connected to commutator segment 2. The starting end s_2 of second coil is now connected to commutator segment 2 and that side of the coil is placed in armature slot 2. The other side of the second coil is now placed in armature slot $(2 + Y)$, and the ending end f_2 of the second coil must be connected to commutator segment 3. It is now obvious that the starting end and the

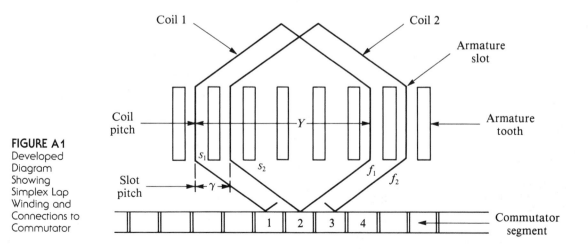

FIGURE A1
Developed Diagram Showing Simplex Lap Winding and Connections to Commutator

ending end of a coil are connected to side-by-side segments of the commutator. This helps define the winding pitch. That is, the winding pitch of the lap winding is equal to one commutator bar (i.e., $y = 1$).

A.2 WAVE WINDING

The wave winding differs from the lap winding only in the way it is connected to the commutator. In the wave winding, the starting and ending ends of a coil are connected almost 2 pole pitches or 360 electrical degrees apart. In fact, for a simplex wave winding, the number of commutator segments must be selected with relation to the pair of poles so that the commutator pitch has a value slightly more or less than 360 electrical degrees. This is done so that, after tracing the winding once around the commutator, the last ending end of a coil is connected to one segment behind or ahead of the starting segment of the commutator. Such an arrangement, as depicted in Fig. A.2 for various pole machines, results in a winding where the starting end of the first coil is connected to the ending end of

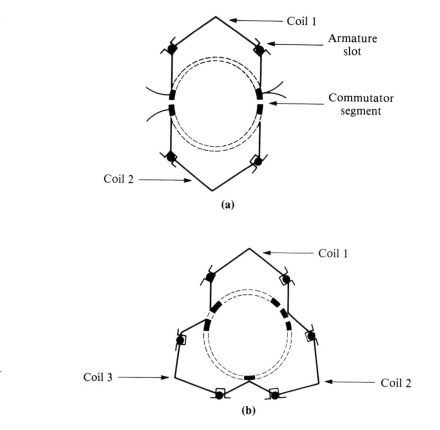

FIGURE A2
Simplex Wave
Winding
Arrangement for
(a) Four-Pole
and (b) Six-Pole
dc Machines

the last coil after going many times around the commutator. Therefore, in a simplex wave winding, the number of commutator segments divided by the pair of poles must not be an integer. Since the commutator pitch must be an integer, then the number of commutator segments must be given as

Commutator segments = (commutator pitch) $\times$ (pair of poles) $\pm$ 1   (A.2)

Production of a Revolving Field

We pointed out the existence of a revolving field in a three-phase wound stator when its phase windings displaced in space by 120 electrical degrees were excited by a balanced three-phase power source with voltages of equal magnitude but 120° apart in time phase. We also assumed that all phase windings are identical and thereby the currents in all windings are equal in magnitude, but displaced in phase by 120°. We have used three-phase windings for the stators in induction machines and synchronous machines. Thus, the following discussion on the production of a revolving field is true for these machines.

Since the mmf and the flux produced by it are in phase under linear conditions, Fig. B.1 shows the instantaneous variations either in the impressed mmf or the flux established by it at any instant. It is assumed that the flux produced by phase 1 leads the flux produced by phase 2 by 120 electrical degrees, which in turn leads the flux produced by phase 3 by another 120 electrical degrees. For the sake of simplicity, let us consider a 2-pole, three-phase wound stator with its six phase groups connected as shown in Fig. B.2.

Let us take the flux produced by phase 1 as the reference; then the expressions for the fluxes produced by the three phases are

$$\phi_1 = \phi_m \sin \omega t \tag{B.1}$$

$$\phi_2 = \phi_m \sin(\omega t - 120°) \tag{B.2}$$

$$\phi_3 = \phi_m \sin(\omega t - 240°) \tag{B.3}$$

where ϕ_m is the amplitude of the flux produced by each of the phases and $\omega = 2\pi f$ is the angular frequency of the flux.

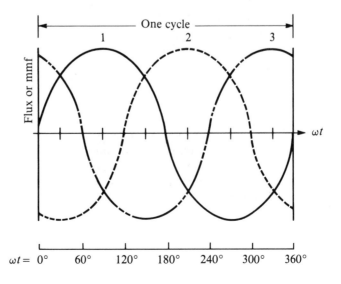

FIGURE B. 1
The mmf and
Flux Waveforms
in a Three-Phase
Stator

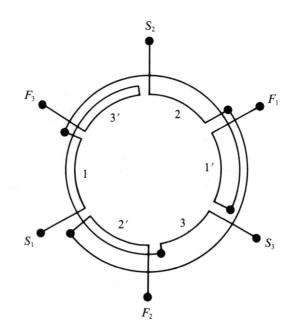

FIGURE B.2
Winding Connections for a 2-Pole, 3-Phase Stator

Just at the outset (i.e., $\omega t = 0$), the flux produced by phase 1 is zero, that by phase 2 is negative, and that by phase 3 is positive. We will follow the notation that when the current in the phase group flows away from the start end and toward the finish end the flux produced by that coil will be positive. In other words, when the coil group is carrying current in the clockwise direction, it will produce a south pole as viewed from the inside of the stator. With this in mind, the directions of the fluxes produced by phases 2 and 3 will be as shown in Fig. B3. The net flux would then be the algebraic sum of these fluxes and will be directed as shown. Mathematically, at $t = 0$

$$\phi_1 = 0 \qquad \text{(B.4)}$$

$$\phi_2 = -\phi_m \sin 120° = -0.866\phi_m \qquad \text{(B.5)}$$

and

$$\phi_3 = -\phi_m \sin 240° = 0.866\phi_m \qquad \text{(B.6)}$$

It should, however, be noted that in the representation of the two fluxes in Fig. B.3 the direction of the flux due to phase 2 has already been taken care of, and the angle between the two flux phasors is 60 electrical degrees. Thus, only the magnitude of the flux phasors is needed to determine the resultant flux.

The resultant flux is

$$\phi_r = \sqrt{\phi_3^2 + \phi_2^2 + 2\phi_3\phi_2 \cos(60°)}$$

or

$$\phi_r = 1.5\phi_m \qquad \text{(B.7)}$$

and it is in phase with the vertical axis drawn through the motor. A simple explanation can be found from the direction of currents in the windings. Phase groups 2 and 3′ act as if they form a single north pole, while phase groups 3 and 2′ act together to form a single south pole. The magnetic axis for each pole then would be along the vertical line.

One-sixth of the cycle later, that is, when $\omega t = 60°$, the current in phase 1 is positive and so is the flux produced by it. That is, phase group 1 will be a south pole, while 1′ will be a north pole. There is no current flow and thereby no flux produced by phase 3. The flux produced by phase 2 is still negative and thereby phase group 2′ will be a south pole and 2 will act as a north pole. Once again, the groups 2′ and 1 form together a south pole, while phase groups 1′ and 2 form a north pole. The center of the combined magnetic poles is then 60 electrical degrees away in the clockwise direction from the vertical axis. Thus, when the time phase has moved ahead by 60°, the field has also advanced spatially by 60 electrical degrees. The strength of the fields are

$$\phi_1 = \phi_m \sin 60° = 0.866\phi_m \tag{B.8}$$

$$\phi_2 = -\phi_m \sin 60° = -0.866\phi_m \tag{B.9}$$

and $$\phi_3 = 0 \tag{B.10}$$

Once again the polarity of the flux phasor ϕ_2 has been taken into consideration

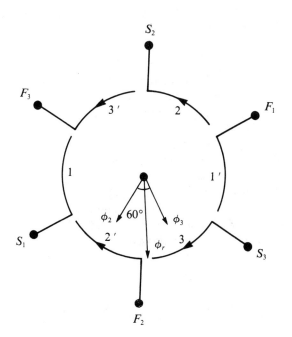

FIGURE B.3
Fluxes at $\omega t = 0°$

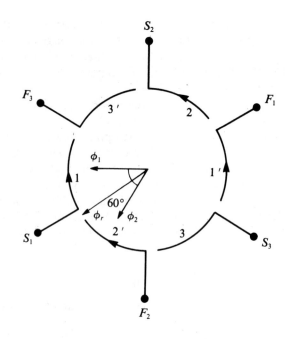

FIGURE B.4
Fluxes at $\omega t =$ 60°

in drawing these phasors in Fig. B.4. The angle between the two flux phasors is 60°. Thus, the resultant is

$$\phi_r = \sqrt{\phi_1^2 + \phi_2^2 + 2\phi_1\phi_2 \cos(60)}$$
$$= 1.5\phi_m \tag{B.11}$$

It must be noted that even when the strength of individual fields has changed with time the resultant field is still 1.5 times the amplitude of the field produced by either phase winding. In other words, the strength of the resultant field has not changed.

One-third of a cycle later, that is, when $\omega t = 120°$, the field produced by phase 1 is positive and thus points from phase group 1′ toward phase group 1, as shown in Fig. B.5. No field is produced by phase 2. The field produced by phase 3 is negative and therefore points from 3 toward 3′. Once again, phase groups 1 and 3′ act together to form a single south pole, while 1′ and 3 form a single north pole. The magnetic axis of the resultant field is now 120° in the clockwise direction from the initial condition of $t = 0$. In fact, as we have progressed in time phase by 120°, the revolving field has also advanced spatially by 120 electrical degrees in the clockwise direction. It can be found that the resultant field strength is still the same.

If this discussion is prolonged enough, one complete cycle later, the magnetic field will also have gone through a complete rotation electrically. Thus 360° of elapsed time phase correspond to 360 electrical degrees of rotation for the magnetic field in a two-pole machine. As can be seen from this discussion, one complete electrical rotation is the same as one complete mechanical rotation for a 2-pole machine. If the machine under discussion has 4 poles, 360 electrical degrees of advancement in the magnetic field actually corresponds to half a revolution (mechanically). Thus the speed of rotation of the magnetic field depends not only on the frequency of the ac power source but on the number of poles as well. The speed with which the flux rotates is known as the *synchronous speed* of the machine. The strength of the magnetic field that revolves around the periphery of the rotor is constant and is 1.5 times the maximum strength of the field produced by one of the phases of the machine.

In our discussion the field windings were tacitly assumed to be arranged in the clockwise direction so as to produce a revolving field that revolves in the clockwise direction. Since the rotor follows the rotating magnetic field, it will also rotate in the clockwise direction. If the phases are wound such that the resulting magnetic field rotates in the counterclockwise direction, the rotation of the rotor will be in the counterclockwise direction as well.

If one pair of source-to-winding leads is interchanged, say phase 2 with phase 3, and the sequence of events is retraced, the resultant field will appear to be revolving in the counterclockwise direction and the rotor, of course, will follow it. Therefore, to change the direction of rotation of a motor, just interchange any two terminal connections. A simple arrangement is shown in Fig. B.6, where the connections for second- and third-phase windings can be reversed with the help of a switch. As shown in the figure, when the switch is in S_1 position, phases 1,

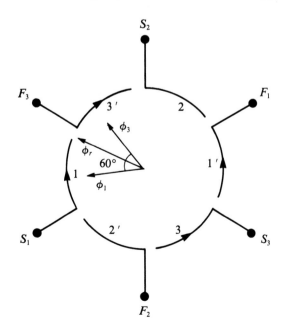

FIGURE B.5
Fluxes at $\omega t = 120°$

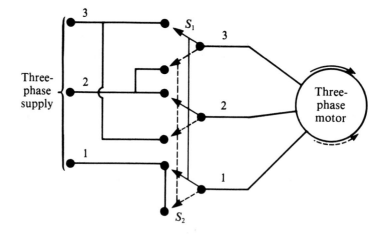

FIGURE B.6
Typical Three-Phase Arrangement for Reversing a Three-Phase Induction Motor

2, and 3 of the motor are connected to the corresponding phases 1, 2, and 3 of the power supply and the motor rotates in the clockwise direction. By throwing the switch into S_2 position, the phase connection for phase 1 remains unchanged, while those of phases 2 and 3 have been reversed. The motor reverses its direction. If this switching technique is used only for stopping the motor, it is called *plugging*. In this case, the supply must be switched off at the time the rotor approaches zero speed. Otherwise, it will start accelerating in the opposite direction.

Remember that the distribution of the flux density around the air gap is sinusoidal with respect to space, and actual density at any point is constant with respect to time as the observation point moves in synchronism with the revolving flux.

Winding Techniques for Polyphase Machines

We already know that it is the primary winding wound into the stator slots that is responsible for the production of electromagnetic flux in induction machines or synchronous motors. Three well-known methods are in common use to place the windings in the stator slots:

1. Two-layer lap winding
2. Single-layer lap winding
3. Concentric winding

C.1 TWO-LAYER LAP WINDING

For a two-layer lap winding, there are as many identical coils as there are slots in the stator. One side of each coil is placed at the bottom half of the slot, while the other side of the same coil fills up the top half of another slot. The bottom half of that slot already contains one side of some other coil. The top of the stator slot is the region that is close to the air gap of the machine. It is therefore evident that these coils have to be prewound on the winding forms and inserted into the slots because each slot contains two different coils. Figure C.1 shows a typical arrangement of a two-layer lap winding in progress. How these windings are internally connected will be explained later.

For winding directly into the slots with automatic winders, the arrangement of the coils has to be changed. The automatic direct winding process requires the winding of both sides of a coil either at the bottom or at the top half of the predesignated slots, instead of placing one side at the bottom and the other at the top. This automatic winding technique may result in somewhat asymmetric winding placement, but for small induction motors the variations in performance are almost negligible.

Since in a two-layer lap winding there are always two different coil sides in a slot, they must be properly insulated.

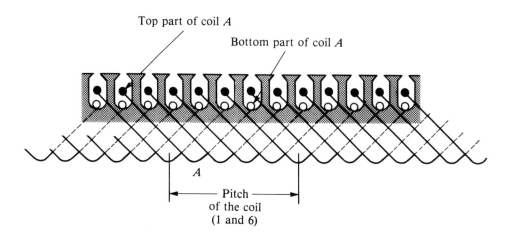

FIGURE C.1
Two-Layer Lap
Winding

C.2   SINGLE-LAYER WINDING

A single-layer lap-winding arrangement requires only half the number of identical coils in comparison with a two-layer lap winding. According to this winding technique, each slot contains only one side of a coil, but each coil has twice as many conductors as in a two-layer lap-winding coil. Since there are only half the number of coils to be wound, inserted, arranged in slots, connected, and taped, a single-layer lap-winding arrangement is both easier and economical. In this case there is no need for in-between slot insulation. However, a single-layer lap winding may not be feasible for all types of three-phase induction motors as it requires an even number of slots to be spanned by each coil, in addition to an even number of stator slots. A schematic development of a single-layer lap winding is shown in Fig. C.2.

FIGURE C.2
Single-Layer Lap
Winding

C.3   CONCENTRIC WINDING

The concentric distribution of coils, although extensively utilized for single-phase induction motors, is not uncommon for fractional- and small-horsepower polyphase induction motors. For such motors this arrangement of windings is rather preferred, particularly by an industry that makes both kinds of motors. The reason is that both single-phase and polyphase induction motors can be wound by making use of the same automatic winding equipment. Depending on the number of slots available per pole per phase (for symmetric winding the number must be an integer), a group of concentric coils may have more than one individual coil. Each individual coil in a coil group spans a different number of slots and therefore exhibits the effect of fractional pitch winding. Whenever there is more than one coil in a coil group, the outermost coil spans the maximum number of slots and the innermost coil spans the minimum number of slots. The coils are said to be *nested*. Most often the innermost coil is wound or inserted first as it spans a minimum number of teeth and has comparatively low end turn buildup. This arrangement is followed by the next bigger coil and so on until the outermost coil has been properly placed in its place. By arranging and placing the coils in this fashion, the coil buildup is kept to a minimum at either end of the stator. Figure C.3 shows a typical concentric winding arrangement with two coils in a phase group. Note that there is only one coil side in each slot. The inner coil spans five

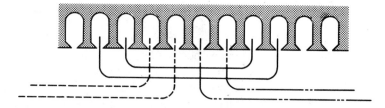

FIGURE C.3
Concentric
Distribution

teeth while the outer coil spans seven teeth. We could have arranged three coils in the phase group. In this case, the innermost coil spans three teeth, the intermediate coil spans five teeth, and seven teeth are spanned by the outermost coil. However, this arrangement results in more than one coil side in each slot. Such an arrangement is not preferred for automatic windings.

C.4 WINDING CONNECTIONS

In all three-phase induction motors, there are three individual phase windings. Each phase winding has as many phase groups as there are poles. Thus, the number of phase groups in a three-phase induction motor is equal to three times its number of poles. In addition, each phase group may consist of one or more individual coils that are spread in a predetermined way in the stator slots. The winding spread depends on the type of winding arrangement, the number of stator slots, and the number of poles. If the total number of coils is known, the number of coils per phase, per phase group, per pole, and so on, can be easily determined.

To calculate the number of coils in each phase, divide the total number of coils in the motor by the number of phases. For a three-phase motor, the number of coils in a phase group is one-third of the total coils in the motor. Dividing the total number of coils in a motor by the number of poles, we obtain the number of coils per pole. To find out the number of coils in a phase group, divide the total number of coils by the number of phase groups. With that many coils in the motor, the following rules must be observed while making internal connections.

1. If there is more than one coil in a phase group, each individual coil must be so connected to give the same polarity. Stated differently, at any instant all

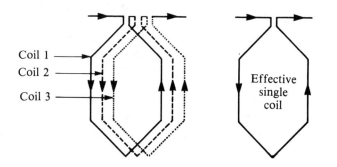

FIGURE C.4
Connections of
Three Coils in a
Phase Group

Coil 1
Coil 2
Coil 3

Effective
single
coil

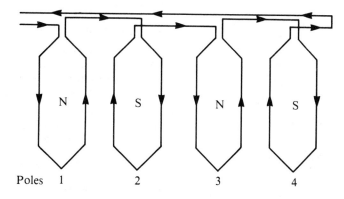

FIGURE C.5
Connections for
One Phase of a
Four-Pole Motor

individual coils in a group must produce magnetic flux in the same direction when properly connected. Figure C.4 shows such a connection for one of the phase groups. In this case, there are three coils in a phase group that are connected to form a single effective coil, as shown.

2. For three-phase motors, the starting ends as well as the finishing ends must be 120 electrical degree apart.

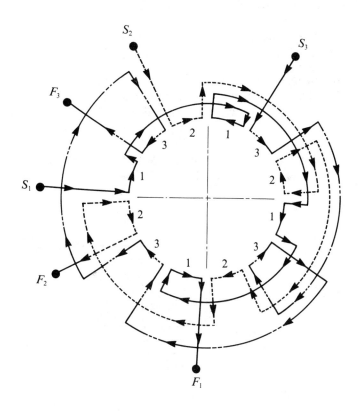

FIGURE C.6
A Four-Pole,
Three-Phase,
Winding
Arrangement

3. Since a phase winding repeats as many times as there are poles, any two adjacent phase windings must be connected to yield the opposite polarity in order to create opposite magnetic poles. In other words, they must carry current in the opposite direction. Figure C.5 illustrates such connections for one of the phases for a 4-pole motor. Note the direction of current in each phase group. At any time, as the phase group winding for one pole, say pole 2, is creating a south pole, the adjacent phase group windings for poles 1 and 3 are producing the fluxes in such a way that they behave like north poles.

4. The direction of current in the adjacent coil groups must be opposite.

Following these rules, the internal winding connections for a 4-pole, three-phase induction motor are shown in Fig. C.6. Since all the individual coils in a group are connected to yield the same polarity, it is convenient to represent each group by a curved line with an arrow pointing out in the direction of current flow. For a three-phase induction motor, the number of phase groups is 12, as shown by the curved lines. For each phase the starting and finishing ends are 120 electrical degrees apart with respect to each other.

C.5 POWER CONNECTIONS

Two types of connections in common use for the three-phase induction motors are (1) delta or mesh connection, and (2) star or wye connection.

Delta Connection   A motor is said to be delta connected if the start end of each phase is connected to the finish end of the adjacent phase. Figure C.7 shows such a connection, where finish end F_1 of the first phase is connected to the start end S_2 of the second phase, the finish end F_2 of the second phase is connected to the start end S_3 of the third phase, and, finally, the finish end F_3 of the third phase is connected to the start end S_1 of the first phase. From every junction so formed, a wire is brought outside the frame of the motor for direct connection to a three-phase power source.

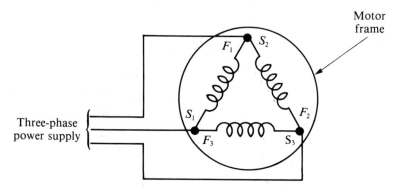

FIGURE C.7
Delta-
Connected
Three-Phase
Induction Motor

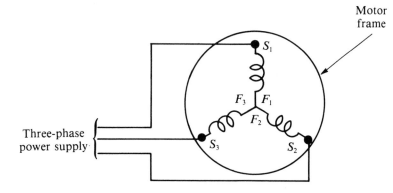

Motor
frame

FIGURE C.8
Star-Connected
Three-Phase
Induction Motor

Star Connection   In a star-connected three-phase winding, shown in Fig. C.8, all the finish ends of the phase windings are connected to form an internal common junction. The start end of each phase winding is brought outside the motor frame for normal three-phase connection. Although it is not a common practice, a neutral wire can also be brought outside from the internal common junction.

When a three-phase induction motor is designed to operate on two supply voltages and/or develops more than 1 hp, all the start and finish ends are brought outside for changing the sequence of connections.

For such a dual-voltage operation, there exists a definite relation, a factor of $\sqrt{3}$, between the high- and low-voltage setting, as pointed out in Chapter 1. For low-voltage operation the motor windings are externally delta connected, while for high-voltage operation the star connection is used. In other words, a given three-phase motor may be used for a 120-V (delta connected) and 208-V (star connected) power source without experiencing any change in performance.

To control the inrush of current at starting, the motor can be started using a star connection and then switched at a predetermined speed to a delta connection with the help of an additional switch, the star–delta starter. This arrangement helps reduce the current intake at starting by a factor of $\sqrt{3}$.

Distribution Factor

It was pointed out in Chapter 6 that a phase group may have more than one coil connected in series. Each coil is displaced by a slot pitch with respect to the other. Therefore, the voltage induced in one of those coils is out of phase by one slot pitch with respect to the next coil. However, all these voltages must be added together to obtain the phase-group voltage. For the sake of general development, assume that a phase group has n coils and the slot pitch is γ. Let the rms value of the induced voltage in coil be E_c. Thus,

$$E_{pg} = \sum_{j=1}^{n} E_c \underline{/-j\gamma}$$ (D.1)

The phasor diagram for Eq. (D.1) is given in Fig. D.1. Since the angle subtended by each voltage phasor is γ, the total angle for the phase group must be $n\gamma$. Therefore, we can write

$$\frac{1}{2} E_{pg} = L \sin \frac{n\gamma}{2}$$ (D.2)

and

$$\frac{1}{2} E_c = L \sin \frac{\gamma}{2}$$ (D.3)

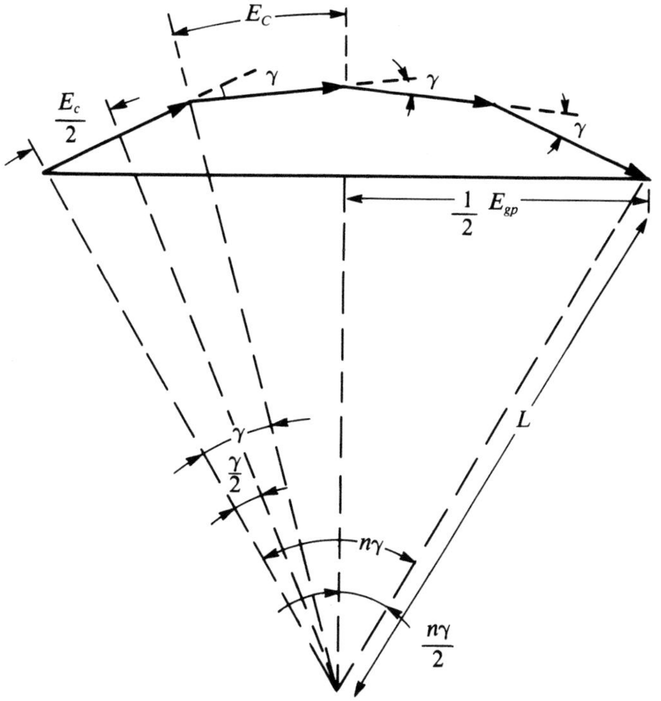

FIGURE D.1
Phasor Diagram
for the Induced
Voltages in a
Coil Group
When $n = 4$

Dividing Eq. (D.2) by Eq. (D.1), we obtain

$$\frac{E_{pg}}{E_c} = \frac{\sin(n\gamma/2)}{\sin(\gamma/2)} = \frac{n\,\sin(n\gamma/2)}{n\,\sin(\gamma/2)} \qquad (D.4)$$

Therefore, the induced voltage per phase group is

$$E_{pg} = nE_c\,\frac{\sin(n\gamma/2)}{n\,\sin(\gamma/2)} = nE_c k_d \qquad (D.5)$$

where
$$k_d = \frac{\sin(n\gamma/2)}{n\,\sin(\gamma/2)} \qquad (D.6)$$

is known as the distribution factor.

APPENDIX E

System of Units

Although the use of the International System of Units (SI) has become more and more prevalent in all areas of electrical engineering, English units are still widely used throughout the profession in dealing with electric machines. For example, it is easy to say that the speed of the motor is 1500 rpm (revolutions per minute) rather than saying 157 rad/s (radians per second). Similarly, a $\frac{1}{4}$-hp (horsepower) motor is very rarely referred to as a 187-W (watt) motor. However, most equations in this book are given in SI system of units, and it sometimes becomes necessary to convert from one unit to another. In the SI system of units, the units of length, mass, and time are meter (m), kilogram (kg), and second (s), respectively. The basic unit of the charge is expressed in coulombs (C). The current is the time rate of change of charge and is expressed in amperes (A). Thus, 1 A = 1 C/s. Among other fundamental units are the unit of temperature, kelvins (K), and the luminous intensity, candelas (cd). In the English system of units, force is expressed in pounds, length in inches or feet, torque in foot-pounds, and time in seconds. The conversion from one unit to the other is given in Table E.1. The units for other quantities that we refer to in this book are given in Table E-2. These are known as the derived units as they can be expressed in terms of the basic units. For example, the fundamental unit of power is the watt (W). It is the rate at which work is done or energy is expended. Thus, the watt is defined as 1 J/s. On the

TABLE E.1
Unit Conversion Factors

From	Multiply by	To obtain
gilbert	0.79577	ampere turns (A-t)
ampere turns/cm	2.54	ampere turns/inch
ampere turns/in.	39.37	ampere turns/meter
oersted	79.577	ampere turns/meter
lines (maxwells)	10^{-8}	webers (Wb)
lines/cm^2 (gauss)	6.4516	lines/in^2
lines/in.^{2}	0.155×10^{-4}	webers/m^2 (T)
webers/m^2 (tesla)	6.4516×10^4	lines/in^2
webers/m^2	10^4	lines/cm^2
inch	2.54	centimeter (cm)
feet	30.48	cm
meter	100	cm
square inch	6.4516	square cm
square inch	1.27324	circular mils
ounce	28.35	gram
pound	0.4536	kilogram
pound-force	4.4482	newton
ounce-force	0.27801	newton
newton	3.597	ounce-force
newton-meter	141.62	ounce-inch
newton-meter	0.73757	pound-feet
revolution/minute	$2\pi/60$	radian/second
horsepower	746	watt
watts/pound	2.205	watts/kilogram

TABLE E.2
Derived Units for
Some of the
Quantities

Symbol	Quantity	Unit	Abbreiation
Y	admittance	siemens	S
ω	angular frequency	radian/second	rad/s
C	capacitance	farad	F
Q	charge	coulomb	C
ρ	charge density	coulomb/meter^3	C/m^3
G	conductance	siemens	S
σ	conductivity	siemens/meter	S/m
J	current density	ampere/meter^2	A/m^2
ϵ_r	dielectric constant	dimensionless	—
E	electric field intensity	volt/meter	V/m
D	electric flux density	coulomb/meter^2	C/m^2
V	electric potential	volt	V
E	electromotive force	volt	V
W	energy	joule	J
F	force	newton	N
f	frequency	hertz	Hz
Z, z	impedance	ohm	Ω
L	inductance	henry	H
ϕ	magnetic flux	weber	Wb
B	magnetic flux density	weber/meter^2	Wb/m^2
		tesla	T
H	magnetic field intensity	ampere-turn/meter	A-t/m
$\mathcal{F}$	magnetomotive force	ampere-turn	A-t
μ	permeability	henry/meter	H/m
ϵ	permittivity	farad/meter	F/m
P	power	watt	W
X, x	reactance	ohm	Ω
$\mathcal{R}$	reluctance	henry^{-1}	H^{-1}
R	resistance	ohm	Ω
B	susceptance	siemens	S
T	torque	newton-meter	N-m
V	velocity	meter/second	m/s
V, v	voltage	volt	V
W	work (energy)	joule	J

other hand, joule (J) is the fundamental unit of work. A joule is the work done
by a constant force of 1 newton (N) applied through a distance of 1 meter (m).
Hence, 1 joule is equivalent to 1 newton-meter (N-m). Above all, 1 newton is the
force required to accelerate a mass of 1 kilogram by 1 meter per second per second.
That is, $1 \text{ N} = 1 \text{ kg-m/s}^2$. Note that the newton is expressed in terms of the basic
units. Therefore, we can now express joules and watts in terms of the basic units.

APPENDIX F

Table of Laplace Transforms

$f(t)$	$F(s)$	$f(t)$	$F(s)$
$\delta(t)$	1	$u(t)$	$\dfrac{1}{s}$
$\delta'(t)$	s	$u(t-a)$	$\dfrac{e^{-as}}{s}$
$\delta^{(n)}(t)$	s^n	$tu(t)$	$\dfrac{1}{s^2}$
$\delta(t-a)$	e^{-as}	$t^n u(t)$	$\dfrac{n!}{s^{n+1}}$
$e^{-at}u(t)$	$\dfrac{1}{s+a}$	$(t-a)u(t-a)$	$\dfrac{e^{-as}}{s^2}$
$te^{-at}u(t)$	$\left(\dfrac{1}{s+a}\right)^2$	$tu(t-a)$	$e^{-as}\left(\dfrac{a}{s}+\dfrac{1}{s^2}\right)$
$e^{-at}t^n u(t)$	$\dfrac{n!}{(s+a)^{n+1}}$	$u(t)\sin\omega t$	$\dfrac{\omega}{s^2+\omega^2}$
$\cosh\beta t$	$\dfrac{s}{s^2-\beta^2}$	$u(t)\cos\omega t$	$\dfrac{s}{s^2+\omega^2}$
$\sinh\beta t$	$\dfrac{\beta}{s^2-\beta^2}$	$e^{-at}\sin\omega t$	$\dfrac{\omega}{(s+a)^2+\omega^2}$
$e^{-at}f(t)$	$F(s+a)$	$e^{-at}\cos\omega t$	$\dfrac{s+a}{(s+a)^2+\omega^2}$
$f(t-a)u(t-a)$	$e^{-as}F(s)$	$tf(t)$	$-\dfrac{dF(s)}{ds}$
$f(at)$	$\dfrac{1}{a}F\left(\dfrac{s}{a}\right)$	$t^n f(t)$	$(-1)^n\dfrac{d^n F(s)}{ds^n}$
$\dfrac{df(t)}{dt}$	$sF(s)-f(0_-)$	$\dfrac{d^2 f(t)}{dt^2}$	$s^2 F(s)-sf(0_-)-\dfrac{df(0_-)}{dt}$
$\displaystyle\int_0^t f(t)\,dt$	$\dfrac{F(s)}{s}$	$\dfrac{f(t)}{t}$	$\displaystyle\int_s^\infty F(s)\,ds$

$$F(s) = \mathscr{L}f(t) = \int_0^\infty f(t)e^{-st}dt$$

Bibliography

Brown, D. & Hamilton, E. P., III. *Electromechanical Energy Conversion*. New York: Macmillan, 1984.

Chapman, S. J. *Electric Machinery*. New York: McGraw-Hill, 1985.

El-Hawary, M. E. *Electric Machines with Power Electronic Applications*. Englewood Cliffs, NJ: Prentice-Hall, Inc., 1986.

Del Toro, V. *Electric Machines and Power Systems*. Englewood Cliffs, NJ: Prentice-Hall, Inc., 1985.

Fitzgerald, A. E., Kingsley, C., & Umans, S. D. *Electric Machinery*. New York: McGraw-Hill, 1983.

Lindsay, J. F. & Rashid, M. H. *Electromechanics and Electric Machinery*. Englewood Cliffs, NJ: Prentice-Hall, Inc., 1986.

Matsch, L. W. & Morgan, J. D. *Electromagnetic and Electromechanical Machines*. New York: Harper & Row, 1986.

McPherson, G. *An Introduction to Electrical Machines and Transformers*. New York: John Wiley & Sons, Inc., 1981.

Nasar, S. A. & Unnewehr, L. E. *Electromechanics and Electric Machines*. New York: John Wiley & Sons, Inc., 1983.

Say, M. G. *Alternating Current Machines*. New York: Pitmann, 1976.

Slemon, G. R. & Straughen, A. *Electric Machines*. Reading, MA: Addison Wesley, 1981.

Index